RESEARCH IN PERSONNEL AND HUMAN RESOURCES MANAGEMENT

RESEARCH IN PERSONNEL AND HUMAN RESOURCES
MANAGEMENT VOLUME 17

RESEARCH IN PERSONNEL AND HUMAN RESOURCES MANAGEMENT

EDITED BY

GERALD R. FERRIS
University of Mississippi, USA

ELSEVIER
JAI

Amsterdam – Boston – Heidelberg – London – New York – Oxford
Paris San Diego – San Francisco – Singapore – Sydney – Tokyo

ELSEVIER B.V.	ELSEVIER Inc.	**ELSEVIER Ltd**	ELSEVIER Ltd
Sara Burgerhartstraat 25	525 B Street, Suite 1900	**The Boulevard, Langford**	84 Theobalds Road
P.O. Box 211	San Diego	**Lane, Kidlington**	London
1000 AE Amsterdam	CA 92101-4495	**Oxford OX5 1GB**	WC1X 8RR
The Netherlands	USA	**UK**	UK

© 1999 Elsevier Ltd. All rights reserved.

This work is protected under copyright by Elsevier Ltd, and the following terms and conditions apply to its use:

Photocopying
Single photocopies of single chapters may be made for personal use as allowed by national copyright laws. Permission of the Publisher and payment of a fee is required for all other photocopying, including multiple or systematic copying, copying for advertising or promotional purposes, resale, and all forms of document delivery. Special rates are available for educational institutions that wish to make photocopies for non-profit educational classroom use.

Permissions may be sought directly from Elsevier's Rights Department in Oxford, UK: phone (+44) 1865 843830, fax (+44) 1865 853333, e-mail: permissions@elsevier.com. Requests may also be completed on-line via the Elsevier homepage (http://www.elsevier.com/locate/permissions).

In the USA, users may clear permissions and make payments through the Copyright Clearance Center, Inc., 222 Rosewood Drive, Danvers, MA 01923, USA; phone: (+1) (978) 7508400, fax: (+1) (978) 7504744, and in the UK through the Copyright Licensing Agency Rapid Clearance Service (CLARCS), 90 Tottenham Court Road, London W1P 0LP, UK; phone: (+44) 20 7631 5555; fax: (+44) 20 7631 5500. Other countries may have a local reprographic rights agency for payments.

Derivative Works
Tables of contents may be reproduced for internal circulation, but permission of the Publisher is required for external resale or distribution of such material. Permission of the Publisher is required for all other derivative works, including compilations and translations.

Electronic Storage or Usage
Permission of the Publisher is required to store or use electronically any material contained in this work, including any chapter or part of a chapter.

Except as outlined above, no part of this work may be reproduced, stored in a retrieval system or transmitted in any form or by any means, electronic, mechanical, photocopying, recording or otherwise, without prior written permission of the Publisher.
Address permissions requests to: Elsevier's Rights Department, at the fax and e-mail addresses noted above.

Notice
No responsibility is assumed by the Publisher for any injury and/or damage to persons or property as a matter of products liability, negligence or otherwise, or from any use or operation of any methods, products, instructions or ideas contained in the material herein. Because of rapid advances in the medical sciences, in particular, independent verification of diagnoses and drug dosages should be made.

First edition 1999
Second impression 2004

ISBN: 0-7623-0489-8
ISSN: 0742-7301 (Series)

∞ The paper used in this publication meets the requirements of ANSI/NISO Z39.48-1992 (Permanence of Paper).
Printed in The Netherlands.

CONTENTS

A NEW BEGINNING: AN OVERVIEW
    Gerald R. Ferris   vii

ORGANIZATIONAL POLITICS: THE STATE OF THE FIELD, LINKS TO RELATED PROCESSES, AND AN AGENDA FOR FUTURE RESEARCH
    K. Michele Kacmar and Robert A. Baron   1

LEGITIMACY IN HUMAN RESOURCES MANAGEMENT
    Maria Carmen Galang, Wolfgang Elsik, and Gail S. Russ   41

CAREER-RELATED CONTINUOUS LEARNING: DEFINING THE CONSTRUCT AND MAPPING THE PROCESS
    Manuel London and James W. Smither   81

TRAINING IN ORGANIZATIONS: MYTHS, MISCONCEPTIONS, AND MISTAKEN ASSUMPTIONS
    Eduardo Salas, Janis A. Cannon-Bowers,
    Lori Rhodenizer, and Clint A. Bowers   123

MANAGERIAL DISCRETION, COMPENSATION STRATEGY, AND FIRM PERFORMANCE: THE CASE FOR OWNERSHIP STRUCTURE
    Henry L. Tosi, Luis Gomez-Mejia, Misty L. Loughry,
    Steve Werner, Kevin Banning, Jeffrey Katz,
    Randall Harris, and Paula Silva   163

PERSON-ENVIRONMENT FIT IN THE SELECTION PROCESS
    James D. Werbel and Stephen W. Gilliland   209

LIFE EXPERIENCES AND PERFORMANCE PREDICTION: TOWARD A THEORY OF BIODATA
    Michelle A. Dean, Craig J. Russell, and Paul M. Muchinsky   245

UNION STRATEGIES FOR REVIVAL: A CONCEPTUAL
FRAMEWORK AND LITERATURE REVIEW
 Marick F. Masters and Robert S. Atkin 283

ABOUT THE CONTRIBUTING AUTHORS 315

OVERVIEW

In writing the Overview for this volume, I am amazed that 15 years have gone by since we began the *Research in Personnel and Human Resources Management* series. The series was designed to fill a need in the field for well-developed conceptualizations, reviews, and important advancements to theory and research in the field of human resources management. As I look back over the 17 published volumes, the range of important topics addressed, and the impressive list of high quality scholars who have contributed papers, I must conclude that the series appears to be accomplishing its objectives.

Volume 17 includes eight papers on key human resources management topics, which continue the mission of the series to both advance theory as well as provide important and productive directions for new research.

The first two papers in this volume, whereas addressing somewhat different subject matter, share in common a perspective that embraces the importance of form (not simply function), images, and impressions in organizations. The volume opens with a paper on organizational politics by Kacmar and Baron. These authors provide an excellent review and analysis of political processes in organizations, and they extend their examination of politics to demonstrate critical linkages and integration with both workplace aggression and retaliatory behavior. The second paper, which also reflects an interest in images and impressions, is on legitimacy in human resources management. Galang, Elsik, and Russ bring to bear legitimacy and structuration theories in an effort to develop a more informed understanding of the operation of human resources policies and practices. Included in their analysis is how organizations acquire legitimacy, maintain it, and even lose it, as well as how organizations respond to their images and reputations being tarnished through efforts at relegitimation.

The next two papers in Volume 17 concern the various processes by which individuals acquire and apply knowledge in organizations, both formally and informally. London and Smither propose a stage model of career-related continuous learning, drawing upon a number of theoretical bases to draw out the foundations of their conceptualization. They then discuss the implications of their model for human resources practices in organizations.

Also reflective of learning processes and knowledge acquision in organizations is the paper by Salas, Cannon-Bowers, Rhodenizer, and Bowers on training. The authors take a hard look at the misconceptions and myths surrounding training, which they argue is a major reason why, despite its importance, training research has not realized a greater impact on both organizational science and practice.

What I believe is quite important to note about these two papers is the implicit message that knowledge and information in organizations is transmitted and aquired in both formal and informal ways, both of which are extremely important. Whether it be the principles conveyed in a formal training program, or the informal rules and understandings regarding how to get ahead, we as organizational scientists desperately need to identify and understand the dynamics of these processes.

There has been increased attention in recent years by organizational scholars directed at developing a better understanding of how human resources systems affect the performance of the firm. The next paper by Tosi, Gomez-Mejia, Loughry, Werner, Banning, Katz, Harris, and Silva examines the nature of managerial discretion and its relationship to both the organization's compensation strategy and its performance. The authors use agency theory and managerial capitalism to articulate the interconnectedness of ownership structure with compensation strategy, and its subsequent effects on firm performance.

The next two papers address issues in personnel selection related to both considerations in the process of personnel selection, as well as specific predictor variables employed. Werbel and Gilliland propose a model of fit in the personnel selection process that incorporates multiple constructs of fit. Whereas prior work has examined notions of fit, such efforts typically focused on a single aspect of either fit to the job, organization, of workgroup. These authors argue that all three componenets need to be incorporated into the selection model with different antecedents and outcomes of the dimensions of fit, and testable propositions are developed to guide future research.

Also related to the area of selection is the next paper by Dean, Russell, and Muchinsky on biodata. Although the concept of biodata has been around for some time, the authors argue that it has been underutilized as a personnel selection device primarily because organizational scientists have regarded it as largely atheoretical. The authors review relevant research on biodata, and they expand upon existing theory to further enrich the conceptual underpinnings of the biodata approach, emphasizing the nature of negative life events. They then discuss important directions for future research to invigorate this important but neglected area of work.

The final paper in Volume 17 acknowledges the historic decline of the union movement in the United States, and proposes strategies for revitalization of unions as key participants and stakeholders in effective organizations. Masters and Atkin review the efforts labor unions have made at startegic revitalization, and they propose their own notion of value-added unionism which positions unions as key partners in efforts to enhance organizational competitiveness. The authors conclude with a discussion of productive avenues for future research in this area.

I believe the papers in Volume 17 represent a fine set of scholarly contributions, which advance our understanding of important human resources management phenomena, and I hope you find it to be a useful addition to your library.

Gerald R. Ferris
Series Editor

ORGANIZATIONAL POLITICS:
THE STATE OF THE FIELD, LINKS TO RELATED PROCESSES, AND AN AGENDA FOR FUTURE RESEARCH

K. Michele Kacmar and Robert A. Baron

ABSTRACT

Virtually all human resources decisions (e.g., promotions, hiring) have the potential to be impacted by political actions and agendas (Ferris & Judge, 1991; Ferris & Mitchell, 1987). When this occurs, the best decisions are not always made. The more that is known about organizational politics, the easier it is to determine why and how politics enter the human resource decision making process and with what consequences. In an effort to reveal the extent of our knowledge of the political process in organizations, this paper begins with a review of past research in this area. Following this review, the principles underlying interpersonal influence are described and related to organizational politics. Using this bridge as a theoretical foundation, links are established between organizational politics and workplace aggression and organizational retaliatory behavior. Finally, suggestions for future research are described which focus on such issues as efforts to identify various motives underlying organi-

Research in Personnel and Human Resources Management, Volume 17, pages 1-39.
Copyright © 1999 by JAI Press Inc.
All rights of reproduction in any form reserved.
ISBN: 0-7623-0489-8

zational politics and the potential links between politics and other organizational processes such as motivation and leadership.

INTRODUCTION

Nearly two decades ago, Porter, Allen, and Angle (1981) wrote a paper on organizational politics. In their introduction, the authors lamented about the lack of published research in the area. They noted that less than a dozen articles out of over 1,700 published by eight of the top journals over the last 16 years were on organizational politics. They further noted that the few articles published on politics were macro (subunit level) versus micro (individual level) in nature. It is clear from our review of the published literature on organizational politics that things have changed. We were able to identify and review over 50 articles published since Porter et al. that examined the political process in organizations. Moreover, the majority of these articles were at the micro level of analysis.

The difference between what Porter et al. (1981) found and what we found may be due to the cyclical nature of research topics. According to Farrell and Petersen (1982), organizational politics was a topic of interest in the early 1900s. However, interest in it waned with the introduction of scientific management. There was a resurgence in interest in the topic in the late 1970s and early 1980s as is evidenced by the paper by Porter et al. being included in the *Research in Organizational Behavior* series. However, once again interest waned. Given the success we had locating empirical research on this topic, it appears that interest in organizational politics once again has increased.

In an effort to capitalize on the rekindled interest in the study of organizational politics and to help it move forward, we offer the present paper. We begin by reviewing the work conducted in this area since Porter et al.'s (1981) review. Specifically, we located all of the articles that referenced Gandz and Murray's (1980) article, one of the first empirical studies on organizational politics. We also examined the reference lists of these articles to locate additional studies on organizational politics. Finally, we requested *in press* work from scholars we knew to be conducting research in this area. We can truly say that we made every effort to be as comprehensive as possible in our literature search. However, despite this fact, we cannot be certain that we have included all relevant sources.

Following our review of the literature, we discuss several issues we feel are important to the further empirical and theoretical development of this field of research with respect to human resource management. First, because organizational politics often involves *influence* (i.e., efforts to change others' attitudes or behavior) we focus on this process, and especially on recent advances in our understanding of its basic nature (e.g., Cialdini, 1994). Second, we examine potential links between organizational politics and other important forms of organizational behavior, especially workplace aggression and related topics. Finally, we offer suggestions for future avenues of research that we feel have the potential for expanding our knowledge of this important topic.

ORGANIZATIONAL POLITICS: A REVIEW OF RECENT LITERATURE

The structure of the first part of this paper is as follows. First, we review various ways in which the concept of organizational politics has been defined and we offer the definition on which this paper is based. The second subsection focuses on psychometric issues. In this section operationalizations of political tactics and organizational politics are reviewed. This is followed by a review of the macro and micro research that has been performed in the area of organizational politics.

Toward A Definition of Organizational Politics

Over the last thirty years, virtually as many definitions of organizational politics have been offered as there are articles on the topic. One of the first was offered by Burns (1961) who indicated that behavior could be considered political if "others are made use of as resources in competitive situations" (p. 257). This definition indicates that in order for politics to occur, competition must be present, and it implies that political behavior is inherently self-centered (i.e., using others). In 1977, Mayes and Allen provided an alternative definition when they suggested that political behavior is enacted to obtain goals that are not sanctioned by the organization or to gain organizationally sanctioned ends through non-sanctioned means. While the idea of competition introduced by Burns disappeared, the self-serving view of political behavior remained. Tushman, also in 1977, offered a definition that incorporated the idea of competition and self-serving behavior. He suggested that organizational politics was the "use of authority and power to effect definitions of goals, directions, and other major parameters of the organization" (p. 207). He felt that politics accentuated differences between individuals and groups and how these differences were resolved. Hence, implicit in his definition was the idea of competition.

These two themes, competition and self-serving, continued to arise in definitions offered by various scholars conducting research in the area (e.g., Drory, 1993; DuBrin, 1988; Farrell & Petersen, 1982; Ferris, Russ, & Fandt, 1989; Gray & Ariss, 1985; Parker, Dipboye, & Jackson, 1995; Ralston, 1985; Vrendenburgh & Maurer, 1984; Zhou & Ferris, 1995). In 1988 and 1990, Drory and Romm delineated ten different components they felt were inherent in any definition of organizational politics. As expected, two of these components were self-serving and conflict or competition. The ten components were divided into three categories: outcomes, means, and situational characteristics. Under outcomes were self-serving behavior, acting against the interests of the organization, securing valuable resources, and attaining power. The four components under the heading of means were influence attempts, power tactics, informal behavior, and concealing one's motive. Finally, the situational characteristics of conflict and uncertainty in the decision-making process rounded out the list.

Interestingly, even though Drory and Romm (1990) wrote an entire article on defining organizational politics, in the end they failed to offer their definition for the term. In addition, they did not specifically address the motivation for engaging in political behavior, an issue we feel is crucial to adequately defining this concept. Drawing on the framework they proposed, plus recent discussions of the motivation behind organizational politics (e.g., Cropanzano & Kacmar, 1995), we offer the following working definition: *Organizational politics involves actions by individuals which are directed toward the goal of furthering their own self-interests without regard for the well-being of others or their organization*. The specific tactics employed in such behavior and the specific targets toward whom it is directed vary greatly, reflecting, in part, the varied motives from which such behavior may stem (e.g., furthering one's own interests; "evening the score" with others for past injustice and other real or imagined wrongs, etc.; e.g., Skarlicki & Folger, 1997). Political behavior often includes activities that are outside the scope of normal job requirements, and so involves actions that are not officially sanctioned by individuals' organization. Moreover, the real reasons for enacting these behaviors are frequently hidden from the target. Finally, political behavior generally occurs when there is competition over limited resources and a lack of clear rules as to how the resources should be allocated.

Psychometric Issues

The diversity found in the definitions of organizational politics is not confined to just that one area. Not surprisingly, it extends to the way in which organizational politics has been operationalized in empirical studies. Similarly, the tactics identified as means of enacting political behavior frequently vary by study. In this section, we examine the many different ways scholars have empirically defined and measured organizational politics, including both perceptions of political environments and the characterization of actual political tactics.

Measuring Organizational Politics

A variety of measures have been introduced into the literature that purport to measure organizational politics. The first was presented in a book entitled *Winning at Office Politics* by DuBrin (1978). The scale consisted of 100 items to which respondents indicated whether or not the item was mostly true or mostly false about themselves. Sample items include "The boss is always right" and "Past promises need not stand in the way of success." Respondents compare their answers to a "key" and award themselves 1 point for every match. Those who score 90 or more are labeled "Machiavellian" and are informed that they are power-hungry, power-grabbing individuals who have an almost uncontrollable tendency to act politically. Respondents who score between 75 and 89 points are labeled "company politicians." These individuals are described as shrewd, successful executives who land on both feet. If an individual scores between 75 and 50 points, he or she is deemed a "survivalist." Survivalists take advantage of good opportunities to act political, but do not seek them out. A score of 49 to 35 places one in the "straight arrow" category. Individuals in this category believe that others are honest and trustworthy and work hard to show they are competent. The final category, "innocent lambs," is reserved for individuals who score less than 35 points. Innocent lambs focus on the task at hand and believe that if they work hard enough they will be rewarded.

Biberman (1985) used DuBrin's (1978) scale to test hypotheses about one's propensity to engage in office politics. He created item-total correlations for each item and selected the top 50 items. The overall alpha for the 50-item scale was .90 based on responses from 59 individuals. Biberman also noted a correlation of .56 between the Mach IV Scale (Christie & Geis, 1970) and the politics scale. This significant correlation was not unexpected given that 11 of the items on the politics scale were paraphrased from the Christie and Geis scale. DuBrin (1988) developed a shorter version of the scale by selecting every other item from the original 100. Using the Kuder-Richardson formula, he reported an acceptable reliability of .78 for the reduced scale.

From 1985 to 1989, Zahra published three different articles on organizational politics. In two of these articles (Zahra, 1985, 1987), the items used were not published. In a third (Zahra, 1989), the items used to measure organizational politics were subjected to a factor analysis and produced 2 usable factors. The first factor had 5 items and dealt with the perceived ethics of politics. Sample items include "work politics is unethical" and "company politicians have no integrity." The second factor had 4 items that focused on the effect politics had on the organization (e.g., "politicking can improve communication" (RS) and "politicking threatens organizational goals"). The internal consistency estimates for the two scales were .84 and .81 respectively.

Organizational politics also has been operationalized via critical incident vignettes. Drory and Romm (1988) developed 15 critical incidents which were all variations of seven of their ten components that constitute the definition of organizational politics. In addition, Kirchmeyer (1990) used 20 vignettes that she developed from Gandz and Murray's (1980) study to measure organizational politics.

In 1991, Kacmar and Ferris developed a measure they called Perceptions of Organizational Politics (POPS). In their two part study they employed factor analysis and classical test theory to reduce 40 items to 12. The final scale consisted of three factors which were named Going Along to Get Ahead, General Political Behavior, and Pay and Promotion Policies. Sample items from each subscale include: Going Along to Get Ahead—"Rewards come only to those who work hard in this organization" (RS); General Political—"There has always been an influential group in this department that no one ever crosses;" and Pay and Promotion Policies—"I can't remember when a person received a pay increase or a promotion that was inconsistent with the published polices" (RS).

Several different groups of scholars have used POPS in their research efforts. First, Nye and Witt (1993) further examined the dimensionality and construct validity of POPS by comparing it to the Survey of Perceived Organizational Support (SPOS) (Eisenberger, Huntington, Hutchison, & Sowa, 1986). Their results indicated that POPS was unidimensional instead of being composed of three distinct factors, and that SPOS and POPS appeared to be strongly and inversely related to one another. In addition, they determined that POPS and SPOS do not differentially correlate with other job attitude measures. In a later study that also incorporated both POPS and SPOS, Cropanzano, Howes, Grandey, and Toth (1997) found POPS and SPOS to predict above and beyond each other indicating that they were indeed different constructs rather than two ends of the same continuum. However, Randall, Cropanzano, Bormann, and Birjulin's (in press) findings were more consistent with Nye and Witt's in that they concluded that POPS and SPOS should be considered mirror images of the same underlying construct.

Ferris and his colleagues (Fedor, Ferris, Harrell-Cook, & Russ, 1998; Ferris et al., 1993; Ferris, Frink, Bhawuk, Zhou, & Gilmore, 1996; Ferris, Frink, Galang, et al., 1996; Ferris, Frink, Gilmore, & Kacmar, 1994; Ferris, Harrell-Cook, & Dulebohn, in press; Ferris & Kacmar, 1992; Gilmore, Ferris, Dulebohn, & Harrell-Cook, 1996; Zhou & Ferris, 1995) used several variations of POPS in their studies while Kacmar, Bozeman, Carlson, and Anthony (in press) used the full 12 item measure. In each of these studies, the internal consistency estimates for POPS were strong. In addition, POPS was consistently and negatively related to satisfaction and positively related to stress, tenure, and intent to turnover.

One additional validation attempt of POPS was offered by Kacmar and Carlson (1997). In their three-part study, the dimensionality, reliability, and validity of POPS was examined via structural equation modeling which was applied to seven different samples with a total of 2559 respondents. Their results suggested that some of the original POPS items were ineffective. These items were removed or replaced and a new version of POPS was offered. The new version retained the three factor structure from the original development of POPS, but new items were added to two of the three factors resulting in a 15 item scale. Further, the overall internal consistency estimate for the full scale indicated that it can be used unidimensionally to measure overall perceptions of organizational politics, or the subscales can be used to examine specific types of politics perceptions.

The political climate in an organization has been measured in at least three other ways. First, Drory (1993) asked respondents how ten different organizational decisions (e.g., promotions, task assignments, performance appraisals) were made in their organizations. The anchors on the graphical rating scale included "mostly influenced by technical professional considerations" on the low end, "equally influenced by both" in the center, and "mostly influenced by political power" on the high end. Using analysis of variance procedures, Drory confirmed that members of each of the five different organizations surveyed responded to the items differently. Hence, he concluded that his scale tapped an organizational rather than a personal attribute.

In 1994, Anderson published two versions of the Dysfunctional Office and Organizational Politics (DOOP) scale: a 21 item version (LFDOOP) and a shorter 10 item version (SFDOOP). The ten item scale was created by selecting the ten top loading items from the longer version. Items (e.g., "A person was lead to believe one thing when the other was clearly true" and "A person's or group's worthwhile efforts or initiatives were intentionally undermined") were rated on a 1 to 7 scale with "never has happened" as the anchor for the low end of the scale and "happens more than once a week" the anchor for the high end. Both scales were deemed to be unidimensional and had strong internal reliability estimates. As with POPS, DOOP was negatively related to satisfaction and positively related to intent to turnover and stress.

Finally, Parker et al. (1995) factor analyzed the items in their survey and found six items that were consistent with Ferris and Kacmar's (1992) and Kacmar and Ferris' (1991) operationalization of perceptions of organizational politics. Sample items include "getting rewarded is political—it's who you know" and "the real world within this organization is one of undercutting and behind the scenes politics." The Cronbach alpha for this measure of politics was .76. Further, the scale was found to correlate significantly and negatively with many of the other variables in the study, including job satisfaction.

Several general conclusions can be drawn from this review of the operationalizations of organizational politics. First, far too many operationalizations of organizational politics have been used for direct comparisons between results from different studies to be made. While several studies have utilized the original POPS (Cropanzano et al., 1997; Kacmar et al., in press; Nye & Witt, 1993; Randall et al., in press), several others have used only portions or various versions of this scale (e.g., Fedor et al., 1998; Ferris et al., 1993; Ferris, Frink, Bhawuk et al., 1996; Ferris, Frink, Galang et al., 1996; Ferris et al., 1994; Ferris et al., in press; Ferris & Kacmar, 1992; Gilmore et al., 1996; Zhou & Ferris, 1995). In addition, others (Anderson, 1994; Drory, 1993; Drory & Romm, 1988; DuBrin, 1988; Kirchmeyer, 1990; Parker et al., 1995; Zahra, 1989) developed a scale to measure politics specifically for their study. Further validation studies or applications of these scale are not available. Hence, one future research effort that can be suggested is an examination of several of the measure of organizational politics in one study. Findings from a study such as this will uncover similarities and differences between the scales. Once the overlap, or lack there of, has been determined, comparisons of findings from past research efforts will be possible. Further, a study that compares available measures of politics may help scholars in this area to converge on or develop one measure that can be used in future studies. Only when we begin to build upon one another's findings in a meaningful way will true progress in this research area be made.

A second conclusion that can be drawn from this summary is that it is still not clear whether organizational politics is a unidimensional or multidimensional construct. While many of the measures used have treated the construct as unidimensional (Cropanzano et al., 1997; Nye & Witt, 1993; Parker et al., 1995; Randall et al., in press), others have shown it to be multidimensional (Drory & Romm, 1988; Ferris & Kacmar, 1992; Kacmar & Carlson, 1997; Kacmar & Ferris, 1991; Zhou & Ferris, 1995). Future research efforts are still needed to determine the most appropriate way to view this construct.

Finally, further construct validation attempts on the various scales appear to be required. For example, the original POPS has been found to be correlated with other variables in the same manner as SPOS, explains no additional variance beyond that explained by SPOS (Nye & Witt, 1993; Randall et al., in press), and it has been found to predict above and beyond SPOS (Cropanzano et al., 1997). Also, not all of the measures of politics used in past studies have correlated in the same way with the same variables across the various studies. This may indicate that different scales are measuring different aspects of the construct of politics. However, for the present, we must leave this speculation to be confirmed or denied by future researchers.

Measuring Political Tactics

Besides measuring organizational politics as an environment or climate construct, some researchers have examined it as an individual level process (Allen, Madison, Porter, Renwick, & Mayes, 1979; Farrell & Petersen, 1982; Ferris & Judge, 1991; Gandz & Murray, 1980; Kacmar & Carlson, 1998; Kipnis & Schmidt, 1982; Kipnis, Schmidt, & Wilkinson, 1980; Ralston, 1985; Schriesheim & Hinkin, 1990; Vredenburgh & Maurer, 1984; Yukl & Falbe, 1990; Zanzi, Arthur, & Shamir, 1991). These scholars have isolated specific tactics that are used by individual politicians. A summary of their findings is offered below.

Allen et al. (1979) generated a list of specific tactics used by organizational politicians by asking chief executive officers, high level staff manages, and supervisors to describe political tactics used by effective politicians. The eight tactics most frequently mentioned include: (1) attacking or blaming others, (2) selective use of information, (3) impression management or image building, (4) generating support for ideas, (5) praising others and ingratiating, (6) building powerful coalitions and strong allies, (7) associating with influential others, and (8) creating obligations and using reciprocity.

In a similar effort to uncover political tactics used in organizations, Kipnis et al. (1980) asked respondents to describe a situation in which they were able to get their way with their boss. Based on the responses, Kipnis et al. developed an influence scale composed of 58 items. Analyses of the responses to these items produced 8 dimensions of influence: assertiveness, ingratiation, sanctions, rationality, exchange, upward appeals, blocking, and coalitions. The 58 original items were later refined and published by Kipnis and Schmidt (1982) as the Profiles of Organizational Influence Strategies (POIS) scale. Several refinements and validations of this scale have been undertaken (Schriesheim & Hinkin, 1990; Yukl & Falbe, 1990).

Using Allen et al. (1979) and Kipnis et al. (1980) as a basis, Vrendenburgh and Maurer (1984) introduced eleven strategies with specific tactics associated with each that organizational politicians can use. These include: (1) accumulating and controlling resources, (2) bargaining aggressively or discreetly, (3) forming coalitions and informal teams, (4) orchestrating events, (5) maintaining flexibility, (6) reducing dependence on others and instilling dependence within others, (7) engaging in conflict, (8) anticipating and preparing for others' actions and reactions, (9) conducting effective interpersonal relations with significant others, (10) exploiting others, and (11) managing one's career. These authors suggested that organizational politicians will determine which strategy and tactics to employ based upon the situation, their experiences, and the likelihood of receiving rewards or being punished for their actions.

Farrell and Peterson (1982) did not offer a laundry list of potential tactics as previous scholars had done. Instead, they offered a typology that could be used to organize the tactics previously mentioned in the literature. The typology was created by combining three dimensions of political behavior: internal-external, vertical-lateral, and legitimate-illegitimate. The internal-external continuum focused on whether the resources used to engage in political activities came from inside (e.g., bypassing the chain of command) or outside (e.g., lawsuits) the organization. The vertical-lateral dimension addressed whether the politician went to a higher organizational level (e.g., complaining to the supervisor) or remained at his or her own level (e.g., forming coalitions). Finally, the legitimate-illegitimate continuum determined whether the political action was a legitimate action taken against the organization (e.g., obstructionism) or an illegitimate activity (e.g., riots). While not all of the tactics offered by previous scholars were incorporated into the typology, they easily could have been.

Ralston (1985) selected one specific category of political behavior, ingratiation, and examined it in detail. Both Allen et al. (1979) and Kipnis and Schmidt (1982) included ingratiation as one of the possible tactics organizational politicians can employ to serve their own self interests. Ralston, relying on past work by Jones (1964), subdivided ingratiation into three, more specific tactics: self-presentation, opinion conformity, and other-enhancement. Self-presentation was defined as enacting behaviors to make oneself more attractive to another or that will be deemed appropriate by another. Opinion conformity, as the name implies, occurs when an individual outwardly agrees with the opinions, judgments, or behaviors of another as a means of illustrating interpersonal compatabilities. Finally, other-enhancements are more direct forms of flattery such as expressing a favorable opinion of another. Kumar and Beyerlein (1991) developed a scale to measure each of these subcomponents of ingratiation as well as favor rendering (i.e., offering to do something not required for another). While it is possible for an individual to use all of these forms of ingratiation effectively, as suggested by Vrendenburgh and Maurer (1984), politicians will select tactics with which they have had past success and that best fit with the situation at hand.

Once again relying on past work (e.g., Kipnis et al., 1980), Zanzi et al. (1991) generated a list of 24 political tactics. The list was presented to their respondents (i.e., business school alumni) who were asked to indicate how frequently they used each tactic. Factor analysis of the items produced a two factor solution with the first factor representing items that dealt with hierarchical structures and dominance relationships. Items from this factor were manipulation, intimidation and innuendoes, rule-oriented tactics, using surrogates, rule evading tactics, ingratiation, and blaming or attacking others. The second factor was described as representing tactics that dealt with lateral social influence (i.e., mentoring, super-ordinate goals, providing resources, using expertise, coalition building, persuasion, and networking). The final factor structure selected incorporated only 14 of original 24 items, and introduced at least three new tactics to the literature: mentor, rule-evasion, and super-ordinate goals.

In 1991, Ferris and Judge published a review article on political influence in human resource management practices in which they provided an overview of the political tactics that had been found in the literature to that date. These authors mentioned several tactics that have not previously been mentioned in this paper. These include exemplification (i.e., offering oneself as an example of appropriate or acceptable behavior), supplication (i.e., acting helpless in an effort to have others support you), entitlements (i.e., claiming responsibility for positive outcomes whether or not you were responsible for them), enhancements (i.e., exaggerating one's accomplishments), apologizing, offering excuses, justifying one's behavior, and using disclaimers (i.e., statements made to prevent negative reactions from others).

Finally, Kacmar and Carlson (1998) asked respondents to describe in their own words a political situation in which they were involved or simply witnessed. The instructions asked the respondents to specifically delineate the action they deemed to be political. The responses were content analyzed and a list of six specific political tactics was uncovered. The tactics included using favoritism (e.g., promoting your best friend), nepotism (i.e., hiring your daughter), gender discrimination (i.e., not promoting or hiring females), use of power-upward (e.g., taking your case to a higher authority), use of power-downward (e.g., a manager forcing a subordinate to wash his or her car), and self-serving (e.g., ingratiation, backstabbing, sabotage).

It seems clear after reviewing the political influence tactics offered in the literature that some agreement exists. There is a great deal of frequency in the appearance of some of the tactics on various lists (e.g., ingratiation or coalition building). This could suggest several things. For example, it could mean that politicians have found the frequently occurring tactics to work effectively for them. In addition, it could mean that these tactics are not as transparent as others and therefore are recognized more frequently by targets and bystanders who have been asked to report political activity they have witnessed. Further, it could mean that these tactics are intuitively pleasing and have been perpetuated by researchers whose tacit knowledge tells them these tactics should be used. To determine which of these explanations, if any, most closely depict the situation, unobtrusive research designs will be needed. That is, future research efforts should attempt to determine which tactics are used first hand rather than asking others to describe political activities. While this type of research is difficult and involves potential pitfalls, it appears that this may be the next logical step in the area of political tactics research.

Macro Studies of Organizational Politics

Precious few organizational politics articles were located that used the organization as the level of analysis. However, several more were uncovered that examined the use of political activities as a means of developing a strong power base for departments or divisions. Further, a variety of articles that explored the politicalness of organizational procedures or policies also were found. Articles from each of these areas are reviewed in this section.

Organizational Level Studies

Two studies, Riley (1983) and Zahra (1987), specifically examined the political process at the organizational level. Riley interviewed 20 people in two different organizations, one routinized and one nonroutinized to determine the symbols used to create political images of these organizations. Her content analysis of the interviews revealed three political structuring mechanisms: signification, legitimation, and dominance. Overall, her structuralist view of the organization revealed that individuals at upper and lower levels in the organization held differing images as did members of one subunit that declared itself as nonpolitical and not like the other subunits. Zahra collected data from 55 different manufacturing firms to investigate how organizational politics influenced the various stages of the strategic process. Even after controlling for company assets and industry type, the intensity of organizational politics was found to be related to various stages of the strategic process and organizational politics was negatively associated with overall company performance. Further, the stage of evolution the company was in moderated the relationship between organizational politics and the strategic process.

Department Level Studies

Two additional studies, Prasad and Rubenstein (1994) and Galang and Ferris (1997), were located that examined how departments or subunits in an organization used political processes to gain influence and power. Prasad and Rubenstein provided a theoretical account of how a division can influence decisions about a new product's characteristics. They suggested that the degree of influence will be determined by the strategy adopted by the company and the political astuteness of the subunit's representatives. The factors they suggested that would influence the level of politicalness in the new product team include uncertainty about the project, the salience of the project, the size and heterogeneity of the team, and the organizational setting in which the team works. Galang and Ferris explored the use of symbols by a department, specifically a department of human resources, to position itself in a better light. Examples of some of the symbols examined include releasing only positive information about the performance of the HR department in official or formal reports, displaying certificates in the office area attesting to the HR staff's training and professional affiliations, and relating stories or anecdotes that portray the importance of HRM to the organization. Their results indicated that symbolic actions by the HR department were stronger predictors of their power than unionization, HR performance, or top management attitudes.

Procedure or Policy Level Studies

At the lowest level of macro analysis are policies and procedures. Specifically, several articles have been written about the political forces at work in human resources policies and procedures. Some of the earliest work was conducted by Madison, Allen, Porter, Renwick, and Mayes (1980) and Gandz and Murray (1980). In both studies, respondents were asked to indicate how frequently they felt politics played a role in different organizational processes. The process respondents deemed as the most frequently influenced by politics was interdepartmental coordination followed by promotions and transfers, and delegation of authority in the Gandz and Murray study and reorganization changes, personnel changes, and budget allocations in the Madison et al. study. The remaining processes included hiring, pay, budget allocations, facility and equipment allocation, personnel policies, disciplinary penalties, work appeals, and grievances and complaints (Gandz & Murray) and dealing with outsiders, setting goals and objectives, and establishing individual standards of performance (Madison et al.). Besides asking their respondents to rate these processes, Gandz and Murray also asked them to provide an example of a political event in their own words. Content analysis of these descriptions resulted in a third of them being categorized as promotion, transfer, demotion, or dismissal decisions.

Kacmar and Carlson (1998) undertook a study similar to Gandz and Murray's (1980) some 15 years later to determine if the human resource practices that were deemed political in 1980 were still considered political. A content analysis of descriptions of 232 political events confirmed some of Gandz and Murray's findings, but uncovered other human resource situations that appear to be deemed political in nature. Similar to Gandz and Murray, promotion, transfer, demotion, and dismissal decisions were still viewed as political. However, new to the list were restructuring decisions, performance appraisals, hiring, work assignments, and compensation issues such as raises and incentives.

Kacmar and Carlson (1998) were not the first to discuss the political nature of many of these human resource processes. For example, Longnecker and his colleagues (Longnecker, 1989; Longnecker, Sims, & Gioia, 1987) noted the political aspects of the performance appraisal process. They found that supervisors readily admit to using the performance appraisal process for reasons other than those generally associated with the process (i.e., accurately assessing performance). Some of the hidden agendas reported included over-rating an individual to avoid confrontation or to give a break to an individual who has improved during the later part of the performance period. Ferris, Fedor, Chachere, and Pondy (1989) reported similar political activities in the examples they offered concerning the design and implementation of a performance appraisal system.

How the performance appraisal process was used in a political fashion also was one of the focuses of Ferris and King (1991) and Ferris and Judge (1991). Another was the selection processes. Specifically, Ferris and King described how many selection decisions come down to the degree of "fit" the candidate has with the job, supervisor, and organization. However, since "fit" is a rather ambiguous term, there is plenty of room for politics to creep into the decision. The idea of fit in selection decisions also was found in Ferris and Judge (1991). These authors describe fit as a mediator in the relationship between political influence and selection decisions. They argue that decision makers are free to define fit any way they choose, and that those who choose to define it in a way that maximizes their own self-interests are using the selection process in a political manner.

Kacmar and Ferris (1993) also reported evidence of political activity in the hiring process in their content analysis of the unprompted anecdotes typed, scrawled, and jotted in the margins of a survey they conducted on politics in the workplace. Besides hiring decisions, respondents provided examples of political decision making in several human resources activities including the distribution of rewards, advancement decisions, and work assignments.

Micro Studies of Organizational Politics

The preceding review of the literature at the macro level of analysis highlighted the fact that all political behavior is initiated by and directed at individuals. Even when the analysis was at the organizational level, individuals were asked to describe and react to the political process. Hence, it is not surprising that the majority of the research in the area of organizational politics has been conducted

Table 1. Predictors of Organizational Politics

Personal Characteristics	Study	Finding
Age	Anderson (1994)	not significant
	Ferris, Frink, Bhawuk, et al. (1996)	not significant
	Ferris, Frink, Galang, et al. (1996)	not significant
	Ferris & Kacmar (1992) Study 1	negative
	Ferris & Kacmar (1992, Study 2)	positive
	Parker et al. (1995)	not significant
	Zahra (1985)	not reported
	Zahra (1989)	not reported
Education	Anderson (1994)	not significant
	Parker et al. (1995)	positive
	Zahra (1985)	not reported
	Zahra (1989)	not reported
Gender	Drory (1993)	not significant
	Fedor et al. (in press)	not significant
	Ferris, Frink, Bhawuk et al. (1996)	negative
	Ferris, Frink, Galang et al. (1996)	negative
	Ferris & Kacmar (1992, Study 1)	not significant
	Kirchmeyer (1990)	not reported
	Parker et al. (1995)	not significant
	Zahra (1985)	not reported
	Zahra (1989)	not reported
Locus of control	Biberman (1985)	not significant
	Kirchmeyer (1990)	positive for males
		not significant for females
Machiavellianism	Biberman (1985)	positive
Need for power	Kirchmeyer (1990)	positive for females
		not significant for males
Race	Ferris, Frink, Bhawuk et al. (1996)	positive
	Parker et al. (1995)	not significant
	Zahra (1989)	not reported
Self esteem	Biberman (1985)	not significant
Self-monitoring	Ferris & Kacmar (1992)	not significant
	Kirchmeyer (1990)	positive
Tenure	Anderson (1994)	not significant
	DuBrin (1988)	negative
	Ferris, Frink, Bhawuk et al. (1996)	positive
	Ferris, Frink, Galang et al. (1996)	positive
	Ferris & Kacmar (1992)	negative
	Fedor et al. (in press)	negative
	Zahra (1985)	not reported
	Zahra (1989)	not reported
Accountability	Ferris et al. (1997)	positive
Ambiguity	Anderson (1994)	positive
	Parker et al. (1995)	negative
Autonomy	Anderson (1994)	negative
	Ferris & Kacmar (1992, Study 1)	negative
	Zanzi et al. (1991)	not significant

(continued)

Table 1. (Continued)

Personal Characteristics	Study	Finding
Career development	Kacmar et al. (in press)	negative
	Parker et al. (1995)	negative
Centralization	Fedor et al. (1998)	positive
	Ferris, Frink, Galang et al. (1996)	positive
	Kacmar et al. (in press)	positive
	Parker et al. (1995)	negative
Cooperation	Kacmar et al. (in press)	negative
	Parker et al. (1995)	negative
Feedback	Ferris & Kacmar (1992, Study 1)	negative
	Kacmar et al. (in press)	negative
	Parker et al. (1995)	negative
Formalization	Fedor et al. (in press)	negative
	Ferris, Frink, Galang et al. (1996)	negative
	Ferris & Kacmar (1992, Study 2)	negative
Hierarchical level	Drory (1993)	positive
	DuBrin (1988)	negative
	Ferris, Frink, Galang et al. (1996)	positive
	Ferris & Kacmar (1992)	positive
	Kirchmeyer (1990)	not reported
	Parker et al. (1995)	not significant
	Zahra (1985)	not reported
Promotional opportunities	Ferris, Frink, Galang et al. (1996)	negative
	Ferris & Kacmar (1992, Study 1)	negative
	Ferris & Kacmar (1992, Study 2)	negative
	Kacmar et al. (in press)	negative
Relationship with supervisor	Ferris & Kacmar (1992, Study 2)	negative
	Kacmar et al. (in press)	negative
Size of organization	Fedor et al. (in press)	negative
Skill variety	Ferris & Kacmar (1992, Study 1)	negative
Span of control	Anderson (1994)	not significant
	Ferris & Kacmar (1992, Study 1)	not significant
	Ferris & Kacmar (1992, Study 2)	not significant
Unionization	Fedor et al. (1998)	not significant

at the micro level. The sections that follow focus on two tasks: (1) identifying personal and organizational characteristics that have been predicted or found to influence individuals to engage in political behavior or view behavior by others as politics and (2) enumerating personal and organizational outcomes that political activities or individuals viewing the behavior of others as political have been found to influence are reviewed. This section ends by outlining the various moderators of the political process that have been offered in the literature.

Antecedents

Predictors of organizational politics can be divided into two categories: personal and organizational characteristics. Common personal and organizational characteristics that have been empirically examined can be found in Table 1. Table 1 also reports the results for each of these antecedents by study. The information presented in the table will be referenced in the discussion that follows along with theoretically supported predictors of organizational politics.

One of the first studies to report personal characteristics of organizational politicians was Allen et al. (1979). In this study respondents were asked to list characteristics that they thought effective politicians possessed. The list, which included only adjectives that at least 10% of the respondents wrote, included: (1) articulate, (2) sensitive, (3) socially adept, (4) competent, (5) popular, (6) extroverted, (7) self-confident, (8) aggressive, (9) ambitious, (10) devious, (11) "organization man," (12) highly intelligent, and (13) logical.

Vrendenburgh and Maurer (1984) added several other characteristics to the list including venturesome, expedient, shrewd, authoritarian, cynical, and locus of control. Only locus of control has been empirically validated. Kirchmeyer (1990) found a positive relationship between locus of control and politics for males, but a nonsignificant relationship for females while Biberman (1985) found no significant relationship between politics and locus of control. Vrendenburgh and Maurer also suggested several individual needs that politicians might possess. These include need for power, autonomy, security, and status. Once again Kirchmeyer provided some empirical support for one of these needs. She found that political women had a strong need for power, but the relationship did not hold for men.

Additional personal characteristics predicted to influence one's perception of organizational politics were offered by Ferris, Russ, and Fandt (1989). These included the demographic characteristics of age and gender and the personality characteristics of Machiavellianism and self-monitoring. All four of these variables have been empirically examined with varying results. In the six studies that reported empirical findings for the relationship between age and politics, four reported nonsignificant results (Anderson, 1994; Ferris, Frink, Bhawuk et al., 1996, Ferris, Frink, Galang et al., 1996; Parker et al., 1995), one reported a positive correlation (Ferris & Kacmar, 1992, Study 2), and one reported a negative correlation (Ferris & Kacmar, 1992, Study 1). For gender, the empirical results are similar. Once again four studies found no relationship between gender and politics (Drory, 1993; Fedor et al., in press; Ferris & Kacmar, 1992, Study 1; Parker et al., 1995), while two reported negative results (Ferris, Frink, Bhawuk et al., 1996; Ferris, Frink, Galang et al., 1996). Machiavellianism was found to have a positive relationship with politics in the one study that included it (Biberman, 1985) and there were mixed results for self-monitoring with one study finding a positive relationship (Kirchmeyer, 1990) and another finding no relationship (Ferris & Kacmar, 1992, Study 2).

Three other demographic characteristics, education, race, and tenure, also have been empirically examined with respect to their association with politics. Two studies have reported a relationship between education and politics. The first study found a positive relationship (Parker et al., 1995) while the second found no significant relationship (Anderson, 1994). Two studies also have reported correlations between race and politics. Ferris, Frink, Bhawuk et al. (1996) found minorities viewed their environments as more political than did white males while Parker et al. (1995) found no significant relationship. One of the most frequently examined demographic characteristics is tenure, with six studies reporting relationships between it and politics. Both Ferris, Frink, Bhawuk et al. (1996) and Ferris, Frink, Galang et al. (1996) reported a positive relationship, DuBrin (1988), Ferris and Kacmar (1992, Study 1), and Fedor et al. (in press) all reported negative relationships, and Anderson (1994) reported a nonsignificant correlation. Finally, Biberman (1985) reported a nonsignificant correlation between self-esteem and politics.

A variety of organizational characteristics also have been offered as predictors of organizational politics. Vrendenburgh and Maurer (1984) suggested that goal and task interdependence, lack of resources, ambiguous goals and roles, organizational climate and history, segmentation of information, and organizational opportunity structure all will influence politics in organizations. Of these, only ambiguity has been empirically tested. Madison et al. (1980) found that 25% of their respondents felt that ambiguity in roles and goals would lead to a more political environment. Further, Anderson (1994) found a positive relationship between politics and ambiguity while Parker et al. (1995) found a negative one.

Ferris, Russ, and Fandt (1989) introduced four additional organizational characteristics and five job/work environment variables that were predicted to influence perceptions of organizational politics. The organizational characteristics include centralization, formalization, hierarchical level, and span of control. The job/work environment variables offered were job autonomy, skill variety, feedback, advancement opportunity, and interactions with co-workers and supervisors. All of these variables have been empirically examined with respect to their relationship with politics. Three of the four studies that examined centralization (Fedor et al., 1998; Ferris, Frink, Galang et al., 1996; Kacmar et al., in press) found a positive relationship while Parker et al. (1995) recorded a negative one. The results for formalization are consistent. All three studies reported a negative relationship between politics and formalization (Fedor et al., in press; Ferris, Frink, Galang et al., 1996; Ferris & Kacmar, 1992, Study 2). Mixed results have been reported for hierarchical level. A positive relationship between politics and hierarchical level were reported in three studies (Drory, 1993; Ferris, Frink, Galang et al., 1996; Ferris & Kacmar, 1992), a negative relationship was found by DuBrin (1988), and a nonsignificant relationship was detected by Parker et al. (1995). Finally, span of control was examined in three studies (Anderson, 1994; Ferris & Kacmar, 1992, both studies) all of which reported a nonsignificant relationship with politics.

A variety of studies also examined the relationships between politics and the job/work environment influences suggested by Ferris, Russ, and Fandt (1989) with most detecting negative correlations. Both Anderson (1994) and Ferris and

Table 2. Outcomes of Organizational Politics

Outcome	Study	Finding
Intent to turnover	Anderson (1994)	positive
	Cropanzano et al. (1996, Study 1)	positive
	Cropanzano et al. (1996, Study 2)	positive
	Kacmar et al. (in press)	positive
	Randall et al. (in press)	positive
Job anxiety	Anderson (1994)	positive
	Cropanzano et al. (1996)	positive
	Ferris, Frink, Bhawuk et al. (1996)	positive
	Ferris, Frink, Galang et al. (1996)	positive
	Kacmar et al. (in press)	positive
Job involvement	Biberman (1985)	not significant
	Cropanzano et al. (1996, Study 1)	negative
	Cropanzano et al. (1996, Study 2)	negative
	Ferris & Kacmar (1992, Study 1)	positive
Job satisfaction	Anderson (1994)	negative
	Biberman (1985)	not significant
	Cropanzano et al. (1996, Study 1)	negative
	Cropanzano et al. (1996, Study 2)	negative
	Ferris, Frink, Bhawuk et al. (1996)	negative
	Ferris, Frink, Galang et al. (1996)	negative
	Ferris & Kacmar (1992, Study 1)	negative
	Ferris & Kacmar (1992, Study 2)	negative
	Kacmar et al. (in press)	negative
	Nye & Witt (1993)	negative
	Parker et al. (1995)	negative
	Randall et al. (in press)	negative
Organization citizenship behavior	Cropanzano et al. (1996, Study 1-compliance)	not significant
	Cropanzano et al. (1996, Study 1-altruism)	not significant
	Randall et al. (in press, individual)	negative
	Randall et al. (in press, organizational)	negative
Organizational commitment	Cropanzano et al. (1996, Study 1)	positive
	Cropanzano et al. (1996, Study 2)	negative
	Drory (1993)	not significant
	Nye & Witt (1993)	negative
	Randall et al. (in press, affective)	negative
	Randall et al. (in press, continuance)	not significant
	Wilson (1995)	negative
Organizational support	Cropanzano et al. (1996, Study 1)	negative
	Cropanzano et al. (1996, Study 2)	negative
	Nye & Witt (1993)	negative
	Randall et al. (in press)	negative
Satisfaction with supervisor	Drory (1993)	negative
	Ferris, Frink, Galang et al. (1996)	negative

Kacmar (1992, in Study 1) reported negative correlations between autonomy and politics while Zanzi et al. (1991) reported a nonsignificant relationship. Skill variety has been examined in only one study (Ferris & Kacmar, 1992, Study 1) with a negative correlation being the end result. Feedback was consistently found to be negatively related to politics by Ferris and Kacmar (1992, Study 1), Kacmar et al. (in press), and Parker et al. (1995). Advancement opportunities also have been found to be negatively correlated with politics according to Ferris, Frink, Galang et al. (1996), Ferris and Kacmar (1992, both studies), and Kacmar et al. (in press). Finally, interactions with co-workers (Kacmar et al., in press; Parker et al., 1995) and supervisors (Ferris & Kacmar, 1992, Study 2; Kacmar et al., in press) were both found to be negatively associated with politics.

Four other organizational characteristics have been empirically related to politics: accountability, career development, organizational size, and unionization. Ferris et al. (1997) reported a positive correlation between accountability and politics while Kacmar et al. (in press) and Parker et al. (1995) reported a negative relationship between politics and career development. Fedor et al. (in press) examined both organizational size and unionization. As predicted by Madison et al. (1980), they found that size correlated positively with politics. However, unionization failed to produce a significant relationship with politics.

The results discussed above and presented in Table 1 reveal some clear and consistent findings. First, politics is negatively associated with the majority of the predictors examined. Also, a negative and significant relationship between politics and feedback, formalization, promotional opportunities, and relationship with supervisor have been found repeatedly. However, some inconsistent findings also have been reported. For example, age, tenure, ambiguity, and hierarchical level have been found to be both positively and negatively related to politics. Hence, future research will be needed to confirm which direction more accurately represents the true relationships between politics and these variables.

Consequences

In addition to exploring the variables that were predicted and found to influence organizational politics, researchers also have been examining what variables organizational politics influences. A list of the most frequently studied outcomes are presented in Table 2 along with the studies that empirically examined each variable and what relationship was found. These outcome variables, and others not so frequently examined in the literature, are summarized in the following paragraphs.

By far the most frequently examined organizational politics outcome variable is job satisfaction. Job satisfaction has been predicted by various scholars to be directly influenced by organizational politics (Ferris, Russ, & Fandt, 1989; Kacmar & Carlson, 1998; Vrendenburgh & Maurer, 1984). While Biberman (1985) found no significant relationship between politics and job satisfaction, a negative relationship between these variables was reported in every other study in which it was examined (Anderson, 1994; Cropanzano et al., 1997, both studies; Ferris, Frink, Bhawuk et al., 1996; Ferris, Frink, Galang et al., 1996; Ferris & Kacmar, 1992, both studies; Kacmar et al., in press; Nye & Witt, 1993; Parker et al., 1995; Randall et al., in press).

A second variable mentioned by Vrendenburgh and Maurer (1980) that has received a great deal of empirical support is organizational commitment or identification with the organization. However, the empirical support for this variable is not as clear as the results for job satisfaction. For example, Cropanzano et al. (1997) found a positive relationship between commitment and politics in their first study and a negative relationship in their second study. Similarly, Randall et al. (in press) reported a negative correlation between politics and affective commitment and a nonsignificant relationship between continuance commitment and politics. Drory (1993) also reported a nonsignificant relationship between politics and commitment. However, Nye and Witt (1993) and Wilson (1995) reported a significant, negative relationship between these two variables.

Ferris, Russ, and Fandt (1989) included three other outcomes in their model of organizational politics: job involvement, job anxiety, and organization withdrawal (i.e., turnover and absenteeism). The relationship between politics and all three of these variables has been empirically examined. The correlation between job involvement and politics was reported to be positive (Ferris & Kacmar, 1992, Study 1), negative (Cropanzano et al., 1997, both studies), and nonsignificant (Biberman, 1985). However, the relationship between politics and job anxiety has been extremely consistent with all reported correlations being positive (Anderson, 1994; Cropanzano et al., 1997, Study 2; Ferris, Frink, Bhawuk et al., 1996; Ferris, Frink, Galang et al., 1996; Kacmar et al., in press).

With respect to organizational withdrawal, intent to turnover has been examined more extensively than absenteeism and with stronger results. All five studies (Anderson, 1994; Cropanzano et al., 1997, both studies; Kacmar et al., in press; Randall et al., in press) reported a positive relationship between intentions to turnover and politics while the one study that examined absenteeism and politics reported a nonsignificant correlation between the variables (Gilmore et al., 1996).

Cropanzano and his colleagues investigated the relationship between politics and organizational citizenship behavior (OCB) in two different studies. In their first study (Cropanzano et al., 1997), they found that politics was not related to OCB when OCB was measured as either compliance or altruism. However, Randall et al. (in press) found a negative relationship between politics and OCB when OCB was measured as outcomes that benefited a specific individual (OCBI) or the entire organization (OCBO).

Another potential outcome for organizational politics that has received some empirical validation is organizational support. A consistently strong, negative relationship between these two variables has been reported in several studies (Cropanzano et al., 1997, both studies; Nye & Witt, 1993; Randall et al., in press). Other outcome variables that have been found to be negatively related to organizational politics include satisfaction with one's supervisor (Drory, 1993; Ferris, Frink, Galang et al., 1996), satisfaction with the organization (Kacmar et al., in press), satisfaction with co-workers (Drory, 1993), satisfaction with promotion and advancement opportunities (Anderson, 1994), perceptions of equal opportunity employment opportunities (Nye & Witt, 1993; Parker et al., 1995), reward accuracy (Anderson, 1994), innovation and creativity (Anderson 1994; Parker et al., 1995), perceptions of supervisor effectiveness (Kacmar et al., in press; Parker et al., 1995), and organizational productivity (Anderson, 1994). One variable, positive psychological withdrawal (Cropanzano et al., 1997, Study 2), was found to be positively related to organizational politics.

As with antecedents to politics, the relationship between politics and several outcome variables has been fairly consistent. For example, job satisfaction and organizational support appear to be negatively related to politics while job anxiety and intent to turnover appear to be positively associated. Some outcome variables in need of further study due to conflicting results include job involvement and organizational commitment.

Moderators

The review of outcome variables for organizational politics indicates that politics appears to be a negative force in organizations. That is, many of the consequences of organizational politics are negative for the individual and the organization (e.g., lower job satisfaction and job involvement, higher job anxiety, and stronger intentions to leave). However, it appears that some of these negative relationships may be moderated by several different variables. An examination of the moderating effects that have been predicted and substantiated follows.

Vrendenburgh and Maurer (1984) suggested that one's sensitivity to organizational politics will moderate the degree and manner in which the antecedent conditions will result in either political behavior or the perceptions of political behavior. They went on to explain that individuals with a heightened sensitivity to organizational politics will be better at acquiring the information they need to determine how to act and react in the various situations that arise. The major components of political sensitivity were thought to be awareness of the importance of norms, one's orientation toward covert processes, and knowledge of significant others.

Ferris, Russ, and Fandt (1989), expanding on Vrendenburgh and Maurer's (1984) ideas, introduced two moderators of the politics-outcomes relationships: understanding and perceived control. They hypothesized that if people recognize that politics exist in their organization, but feel relatively little control over or understanding of these processes, then politics will be a threat to them and negative outcomes will result. However, if employees understand the political process and feel some sense of control of it, they will be more likely to view politics as an opportunity to increase the benefits they receive and more favorable outcomes will result.

The majority of the empirical tests of moderators have been conducted to test Ferris, Russ, and Fandt's (1989) depiction of the moderating process of understanding and perceived control. Further, all but one of the tests for the moderating effect of understanding and perceived control were conducted by Ferris and his colleagues. The first test of the moderating effects proposed by Ferris, Russ, and Fandt (1989) was conducted by Ferris et al. (1994). Using organizational tenure as a proxy for understanding, they found that the relationship between politics and job anxiety, after controlling for age, sex, and supervisory status, was moderated by organizational tenure. Specifically, the relationship between politics and job anxiety at low levels of tenure was positive, nonsignificant at moderate levels, and negative at high levels.

A second study that operationalized understanding as tenure was conducted by Gilmore et al. (1996). However these authors used a more specific type of tenure, tenure with supervisor. They found that the politics - employee attendance relationship was moderated by tenure with supervisor such that those with low levels of tenure with supervisor (i.e., lower understanding) were present at work more when perceptions of politics were low and absent from work more when perceptions of politics were high.

Ferris and his colleagues (Ferris et al., in press) also used organizational tenure as a proxy for organizational socialization or understanding in a third study. In this case organizational tenure was found to moderate the relationship between politics perceptions and self-promotion political behavior. Specifically, the relationship found indicated that under low organizational tenure, there is no significant relationship between politics perceptions and self-promotion political behavior. However, under high organizational tenure increased politics perceptions led to an increased use of self-promotion political behaviors.

Understanding as a moderator of the politics-outcomes relationships also was of interest in Ferris, Frink, Galang, et al. (1996). In this study understanding was measured with an understanding scale by Tetrick and LaRocco (1987). Understanding was tested as a moderator of the relationship between perceptions of politics and three different outcomes: job satisfaction, satisfaction with supervisor, and job anxiety. While findings supported the moderating effect for job anxiety and satisfaction with supervisor, no support was found for the outcome of job satisfaction. The graph of the interactions indicated that as perceptions of politics increased, so did job anxiety, however it increased more for individuals who had a lower understanding of the organization than for those who possessed a higher understanding.

Ferris, Frink, Bhawuk et al. (1996) also examined understanding as a moderator. In this study the authors segregated their sample into white males, white females, and racial/ethical minorities and segmented organizational politics into the subcategories of supervisor political behavior, co-workers political behavior, and political organizations (i.e., policies and procedures). The two outcome variables of interest were job anxiety and general job satisfaction. Co-worker political behavior was not significant in their omnibus test and was not included in the interaction analyses. Understanding was found to moderate the remaining relationships: supervisor political behavior and political organizations with both job anxiety and job satisfaction. Once again, understanding was found to lessen the negative effects associated with politics by reducing job anxiety and increasing job satisfaction. As predicted by the authors, these effects were only found for white males.

One final study that examined the moderating effect of understanding on the perceptions of politics-outcomes relationship was performed by Kacmar et al. (in press). In this study understanding was measured in a manner consistent with Tetrick and LaRocco (1987). Results indicated that understanding moderated the relationship between perceptions of politics and the two outcomes of job satisfaction and self performance evaluation. As previously discussed, higher understanding as compared to lower understanding led to higher satisfaction when perceptions of politics was high. Similarly, individuals who understood the environment rated themselves higher when perceptions of politics were high than those who did not understand the environment as well.

Ferris and his colleagues also have examined perceived control as a moderator of the politics-outcomes relationships. Specifically, Ferris, Frink, Galang et al. (1996) found control, as measured by Tetrick and LaRocco's (1987) scale to moderate the relationships between perceptions of politics and the outcome variables of job satisfaction, satisfaction with supervisor, and job anxiety. In all three cases, perceived control acted as an antidote for potential negative outcomes. That is, under high levels of perceived control, job anxiety did not increase under high perceptions of politics as much as it did under conditions of low control. Further, job satisfaction and satisfaction with supervisor did not decrease as much under high perceptions of politics when control was high than when control was low.

Ferris et al. (1993) undertook two separate studies that examined control as a moderator of the politics-outcomes relationships. In the first study, perceived control was operationalized as supervisor status and was examined as a moderator of the perceptions of politics-job anxiety relationship. Support was found for the hypothesized relationship in that when perceived control was low, the relationship between politics and job anxiety was significant and positive. However, when perceived control was high, the relationship between politics and job anxiety was negative, but not significant. In the second study, perceived control was operationalized as the degree of influence or control employees felt they had over 16 aspects of their work environment. This form of understanding was found to moderate the relationship between perceptions of politics and job satisfaction such that the negative relationship between these variables was less strong when perceived control was high. Understanding also was examined as a moderator of withdrawal behaviors. While it failed to moderate the relationship between politics and intent to turnover, it did moderate two other withdrawal behaviors: the relationships between politics and the negative evaluation of the decision to accept the current job and absenteeism. The predicted relationship for the negative evaluation of the decision to accept employment held. That is, organizational politics perceptions were of greater importance for withdrawal under conditions of low as opposed to high levels of perceived control. However, an opposite result was found for absenteeism. Under conditions of low perceived control, there was no relationship between politics and absences while under high conditions of perceived control, higher levels of politics were associated with greater absences.

One final study (Parker et al., 1995) examined a moderator of the politics-outcomes relationships. The variable selected as the moderator was the level of trust employees had with their co-workers. However, contrary to predictions, this variable failed to moderate any of the politics-outcomes relationships. Parker et al. suggested that various other forms of trust may be more important to these relationships than trust of co-workers. For example, they suggested that trust of those in power may be a more significant moderator and should be considered in future studies.

Two studies examined moderating effects when organizational politics was the dependent variable. Drory (1993) crossed satisfaction with supervisor, satisfaction with co-workers, and commitment with gender and supervisory position to determine the interactive effects these variables might have on politics. All of the interactions, with the exception of satisfaction with co-workers and supervisory position, were significant. However, the interactions were not graphed or fully explained, so the forms of these relationships are unknown.

Fedor at al. (in press) also examined moderating effects using politics as a dependent variable. In this study, organizational politics was segmented into five specific types (i.e., key others, rewards, distortion, self image enhancement, and clarity of pay and promotion policies) which were each used as a dependent variable. Predictors included the four organizational variables of centralization, formalization, organization size, and percent unionized and the two individual variables of gender and organizational tenure. A total of twelve interactions were conducted. Specifically, centralization and formalization were crossed with organizational size, organizational tenure, gender, and unionization, and gender and organizational tenure were crossed with organizational size and unionization. The interaction between formalization and unionization was significant for the political factor of key others. That is, the presence of powerful others was stronger in low unionized firms that are highly formalized and high unionized firms that are low on formalization. The interaction between centralization and tenure was significant for both the reward and image dimensions of politics. For the reward dimension, in highly centralized organizations, newer organizational members are more likely than longer term members to perceive political behavior in the rewards offered. For the political dimension of image management, individuals with low tenure perceived a larger difference in image manipulation between low and high centralized firms than did those with more tenure. Finally, the interaction between gender and organizational size was significant for the political dimension of distortion. This interaction revealed that in larger organizations women perceive more use of distortion than men, whereas in smaller organizations men see more distortion than women.

The preceding review of the literature on political perceptions and behavior provided an in-depth look at the current state of the field. The next step that is needed is to connect additional paths of research inquiry to organizational politics. This step, which is taken in the next section, also provides a means of identifying new ways of examining organizational politics. Several of these potential paths are outlined in the final section of the paper.

INFLUENCE: A THEORETICAL FRAMEWORK FOR UNDERSTANDING ITS ROLE IN ORGANIZATIONAL POLITICS

In order to further their own interests (e.g., whether these involve obtaining a larger share of available rewards, advancing their careers, or exacting vengeance against others) individuals who engage in organizational politics employ a very wide range tactics. As noted earlier in this paper, many of these center around the process of *influence;* that is, somehow inducing other organizational members to change their attitudes or behavior in ways desired by the practitioners of organizational politics (e.g., Cropanzano, Kacmar, & Bozeman, 1995; Ferris & Judge, 1991; Zanzi et al., 1991). This suggests that basic knowledge of the nature of influence might well provide valuable insights into the operation of organizational politics.

In recent organizational research, the study of influence has proceeded along largely empirical lines. Such research has focused on the tasks of identifying the tactics of influence used in organizations, and in establishing their relative frequency. Results obtained in such investigations have been highly informative. For example, a series of carefully-conducted studies by Yukl and his colleagues (e.g., Falbe & Yukl, 1992; Yukl, Falbe, & Youn, 1993; Yukl & Tracey, 1992) found that the most frequently used tactics of influence in organizations are (in decreasing order of use): rational persuasion, inspirational appeals (arousing enthusiasm by appealing to the target person's values and ideals), consultation (asking for input in making a decision or planning a change), ingratiation, and personal appeals (appealing to feelings of loyalty and friendship before making a request). These tactics are used much more frequently than other approaches such as threats, sanctions, blocking, and deceit. As noted earlier, similar actions also have been identified as basic, and frequently used, tactics of organizational politics (e.g., Ambrose & Harland, 1995). The convergence between these two lines of research suggests that influence does indeed play a central role in organizational politics, and that successful practitioners of it are skilled at exerting influence over others (Cropanzano et al., 1995).

While organizational research on influence has yielded valuable information about this topic, it seems fair to say that it has been largely empirical in nature; with a few recent exceptions (e.g., Ambrose & Harland, 1995), it has not focused on the task of establishing a theoretical framework for understanding the nature of the influence process in organizations. This issue, however, *has* been the focus of a large amount of research in other fields. Perhaps most relevant to the present discussion are efforts by social psychologists to identify the basic principles that underlie a wide range influence tactics, and in this way, to elucidate the nature of influence itself (e.g., Shavitt & Brock, 1994). Perhaps the clearest summary of this emerging theoretical framework has been offered by Cialdini (1994).

In order to obtain a better grasp of the basic nature of influence, Cialdini (1988, 1994) adopted a novel and ingenious strategy: he focused his attention on what he described as *influence professionals* (i.e., persons who, in essence, make a career out of influencing others). In order to study such professionals from an "insider's" perspective, Cialdini temporarily concealed his true identify and accepted positions in various settings where exerting influence is a way of life. For example, he worked in advertising, direct sales (e.g., as a used car salesperson), fund-raising, and political campaigns. The focus of his research was on *compliance* (i.e., a basic form of social influence in which individuals attempt to induce others to say "yes" to various requests; e.g., Cialdini, 1988). On the basis of his own work, and that of many other investigators, (e.g., Burger, 1986; Williams, Radefeld, Binning, & Suadk, 1993), Cialdini concluded that many different influence tactics rest on the following set of basic principles:

- **Friendship/Liking**: The more we like other persons or feel friendship for them, the more likely we are to comply with their requests or to accept other forms of influence from them.
- **Commitment/Consistency**: Individuals wish to be consistent in their beliefs and actions. Thus, once they have adopted a position or committed themselves to a course of action, they experience strong pressure to comply with requests that are consistent with these initial commitments; in fact, they may find it virtually impossible to refuse such requests because doing so would force them to reject or disown actions or beliefs they previously adopted.
- **Scarcity**: In general, opportunities, objects, or outcomes are valued in inverse proportion to their scarcity. Thus, requests that emphasize scarcity or the fact that some object, opportunity or outcome will soon no longer be available are difficult to resist.
- **Reciprocity**: Individuals generally experience powerful pressures to reciprocate benefits they have received from others. As a result, requests that

activate this principle are more likely to be accepted than requests that do not.
- **Social Validation**: Human beings have a powerful desire to be correct, and one way of reaching this goal (especially in situations where physical reality provides no hard-and-fast test) is to behave and think like other persons, especially ones they respect or view as similar to themselves. To the extent requests engage the social validation motive, they are often highly effective.
- **Authority**: Individuals are generally more willing to comply with requests from persons possessing authority (real or merely feigned) than from persons lacking in authority.

These basic principles appear to underlie many tactics of influence, including ones identified as strategies of organizational politics. Further, they suggest additional tactics, not yet investigated in the literature on organizational politics, which also may play a role in this process. A brief review of some of these tactics and the principles on which they are based may prove useful at this point.

Ingratiation and impression management, both of which have been identified as tactics of organizational politics (e.g., Allen et al., 1979; Kipnis et al., 1980), are closely related to the principle of liking/friendship. These tactics are often used, and with considerable skill, by practitioners of organizational politics (Ralston, 1985).

The principle of commitment/consistency has been found, in basic research, to play an important role in such successful tactics for gaining compliance as: (1) *the foot-in-the-door*—starting with a small request and, once this is accepted, escalating to a larger one; and (2) the *lowball*—attempting to change a deal or agreement by making it less attractive to the target person after it is negotiated (e.g., Cialdini et al., 1978). Although rational considerations suggest that target persons should reject such changes, they often accept them rather than reverse their initial commitment. Finally, the consistency/commitment principle also plays a role in the *bait-and-switch* tactic, one in which an item, product, or service of lesser value is substituted for another, more desirable item. Again, individuals exposed to this tactic often accept the "poor substitute" rather than begin the process all over again. Although these tactics have not been studied as potential strategies of organizational politics, it seems possible that they are often used by individuals seeking to advance their own interests at the expense of others. For instance, practitioners of organizational politics who bargain aggressively (Allen et al., 1979) and instill dependencies in others (e.g., Vrendenburgh & Maurer, 1984) may be relying, in part on tactics derived from this principle.

Turning to the principle of scarcity, such tactics as *playing hard to get* and *the fast approaching deadline technique*, are widely used in the world of business. Job applicants who mention that they are under consideration for other positions or are very satisfied with their current position (e.g., Williams et al., 1993), are using the "hard-to-get" tactic to manipulate important organizational outcomes in their favor. Similarly, salespersons who suggest that various products are in short supply, or that a price cannot be guaranteed beyond a specific date, are using tactics related to the principle of scarcity, and to the "fast-approaching deadline" approach. To the extent such persons employ these tactics to advance their own interests, they may be viewed as engaging in organizational politics.

The principle of reciprocity, too, is often used for such purposes. Research on organizational politics suggests that persons engaging in such behavior often seek to form coalitions with others or to induce dependencies or obligations in them (e.g., Vrendenburgh & Maurer, 1984; Zanzi et al., 1991). This principle also underlies procedures that have not been explicitly studied in the context of organizational politics, such as the *door-in-the-face* tactic. In this tactic, individuals start with a request that is very large and sure to be rejected. Then they "scale down" their request to a more acceptable one, thus putting the target person under considerable pressure to reciprocate this concession. A related approach is known as the *"that's not all"* strategy. This involves offering a very small concession in order to push target persons into compliance with specific requests. By "sweetening the deal" in a trivial way, practitioners of this tactic invoke the principle of reciprocity, and make it difficult for target persons to refuse. These latter two tactics may be related to political tactics described as manipulation (Zanzi et al., 1991) anticipating and preparing for others' actions and reactions (Vrendenburgh & Maurer, 1984).

The principle of social validation underlies many tactics of influence, including ones based on conformity pressure: "You don't want to stick out like a sore thumb, do you?" "Everyone else agrees; why don't you?" These tactics have been identified quite explicitly as linked to important strategies of organizational politics such as *opinion conformity* (Ralston, 1985; Jones, 1964). Finally, the principle of authority finds expression in such tactics as *obedience* (i.e., simply demanding that others obey direct orders), various forms of *self-presentation* designed to enhance the apparent status or power of users (e.g., Wayne & Liden, 1995), and reliance on one's *expertise* to convince targets to comply (Zanzi et al., 1991).

In sum, many different tactics of influence exist, and a number of these, can be, and probably are, used by persons engaging in organizational politics. Attention to the basic principles that underlie such tactics may provide researchers in this field with a theoretical framework useful for organizing the wealth of data that currently exist; data often obtained, as we noted earlier, with different measuring tools and contrasting procedures (e.g., Drory & Romm 1988; Kacmar & Ferris, 1991). A theoretical framework has been lacking in past research, with the result that at the present time, it is difficult to identify underlying themes or patterns in the many tactics described. Such a framework, especially one focused on basic principles, also may prove useful from the point of view of identifying additional political tactics that have not been investigated to date, and may also suggest factors that will determine the success of such strategies (e.g., Forgas, 1997). Finally, if, as many researchers have suggested (e.g., Ferris, Frink, Beehr, & Gilmore, 1995), influence plays a key role in organizational politics, attention to the basic nature of this process may shed important light on the nature of organizational politics itself. For all these reasons, careful consideration of the principles described above may greatly facilitate the progress of ongoing research in this important area.

POLITICS AND OTHER FORMS OF ORGANIZATIONAL BEHAVIOR: POTENTIAL LINKS TO WORKPLACE AGGRESSION AND ORGANIZATIONAL RETALIATION

Earlier, we noted that participation in organizational politics may stem from several contrasting motives. The one which has received most attention, and which has been central to many definitions, is that of furthering purely personal (selfish) interests, without regard for the welfare of others. However, on close examination, it seems clear that participation in various tactics or strategies labeled as "political" in nature can also stem from other motives as well. For instance, individuals may attempt to influence others in their organizations, and hence important processes and outcomes, in an effort to re-establish justice (e.g., to make up for what they perceive to be unfair treatment at the hands of others). Precisely such links between organizational politics and justice have recently been discussed by several authors who suggest that perceived lack of fairness may be one factor contributing to the occurrence of organizational politics (see, e.g., Cropanzano & Kacmar, 1995; Dulebohn, 1997).

Going further, it seems possible that participation in organizational politics also sometimes may stem from the desire for revenge or retaliation; that is, efforts to "even the score" with others who have inflicted not merely injustice, but other forms of real or imagined harm on the persons who decide to adopt political tactics (e.g., Folger & Baron, 1996; Skarlicki & Folger, 1997). To the extent this is the case, important links may exist between organizational politics on the one hand and workplace aggression on the other. In fact, evidence for such links is provided by the findings of recent studies on workplace aggression.

In a series of closely related studies, Baron and Neuman (e.g., Baron & Neuman, 1996; 1998; Neuman & Baron, 1996, 1998) asked almost five hundred individuals working in several diverse organizations to report on the frequency with which they had witnessed or experienced a wide range of harm-doing behaviors. In this research, workplace aggression was defined in a manner consistent with the large, extant literature on human aggression: as intentional harm-doing. More specifically, workplace aggression encompasses any actions performed by individuals to inflict harm on other members of their organization or on the organization itself. On the basis of past research on human aggression (e.g., Bjorkqvist, Osterman, & Hjelt-Back, 1994), Baron and Neuman (1998) predicted that most aggression

Table 3. Major Forms of Workplace Aggression

Expressions of Hostility

Staring, dirty looks, or other negative eye-contact
Belittling someone's opinions to others
Giving someone "the silent treatment"
Negative or obscene gestures toward the target
Talking behind the target's back/spreading rumors
Interrupting others when they are speaking/working
Intentionally damning with faint praise
Holding target, or this person's work, up to ridicule
Flaunting status or authority/acting in a condescending manner
Sending unfairly negative information to higher levels in the company
Leaving the work area when the target enters
Delivering unfair/negative performance appraisals
Failing to deny false rumors about the target
Verbal sexual harassment
Failing to object to false accusation about the target

Obstructionism

Failure to return phone calls or respond to memos
Failing to transmit information needed by the target
Causing others to delay action on matters important to the target
Failing to warn the target of impending danger
Showing up late for meetings run by target
Failing to defend target's plans to others
Interfering with or blocking the target's work activities
Needlessly consuming resources needed by the target
Direct refusal to provide needed resources or equipment
Intentional work slow downs

Overt Aggression

Attack with weapon
Physical attack/assault (e.g., pushing, shoving, hitting)
Theft/destruction of personal property belonging to target
Threats of physical violence
Failing to protect target's welfare or safety
Damaging/sabotaging company property needed by target
Steals/removes company property needed by target
Destroying mail or messages needed by the target

Note: Forms of harm-doing behavior included under each factor loaded significantly (factor loadings > .40) on that factor.

occurring in workplaces would be *covert* in nature; that is, it would tend to take forms that do, indeed, produce harm, but that are disguised, subtle, and indirect in nature (Bjorkqvist, Lagerspetz, & Kaukianinen, 1992). More direct and open forms of aggression (i.e., actions usually described by the phrase *workplace violence*) would, in contrast, be far less common. Results confirmed these hypotheses: covert forms of aggression were indeed significantly more frequent than more direct and overt forms. Moreover, factor analyses indicated that various forms of workplace aggression fall into three major clusters, two of which appear to be closely linked to organizational politics. The three factors uncovered in these analyses were labeled as shown below; for more complete information on the forms of aggression included in each factor, please refer to Table 3.

Expressions of Hostility: Behaviors which are primarily verbal or symbolic in nature, and which harm the target person or persons in relatively indirect ways (e.g., belittling others' opinions; talking behind the target's back; intentionally damning with faint praise).

Obstructionism: Behaviors that are designed to obstruct or impede the target's performance (e.g., failure to return phone calls or respond to memos, failure to transmit needed information; interfering with activities important to the target; causing others to delay action on matters of importance to the target person). These behaviors are largely passive in nature; that is, they involve harm produced by withholding various actions.

Overt Aggression: Behaviors that have typically been included under the heading "workplace violence" (e.g., physical assault, theft or destruction of property, threats of physical violence; forms of aggression loading significantly on each of these three factors are summarized in Table 3).

As can readily be seen, actions included in the first two factors are closely related to tactics that have often been labeled as "political" in nature. In a sense, this is not surprising, for such behaviors are designed to harm the intended target while simultaneously concealing the identify to the aggressor and, perhaps, even the fact that the harm produced is the result of intentional actions rather than merely bad luck and external circumstances. Organizational politics, when performed proficiently, is designed to achieve similar goals. Further, it should be noted that many forms of workplace aggression, especially those relating to obstructionism, may represent an additional means through which individuals promote their own self-interest. Derailing rivals' careers by interfering with their performance in various ways may yield a double dose of rewards to the persons who successfully pursue such tactics: not only do they gain the satisfaction of harming their victim; but also they may secure increased benefits for themselves as these rivals flounder.

One further point: organizational politics and workplace aggression may not simply resemble one another in terms of the specific forms of behavior they involve; in addition, both may stem from similar motives. For example, in the research described above (Baron & Neuman, 1998; Neuman & Baron, 1998), a strong link was noted between participants' perceptions of the extent to which they had been treated unfairly by their supervisors and the levels of workplace aggression they reported witnessing or experiencing in their organizations. Similarly, in a recent study, Skarlicki and Folger (1997) found that the lower the level of perceived fairness reported by participants, the higher the frequency of organizational retaliatory behavior (i.e., actions by individuals designed to "even the score" for previous unfair treatment). In fact, three basic aspects of perceived justice (i.e., distributive, procedural, and interactional) interacted in producing these effects.

When these findings on workplace aggression and retaliatory behavior are combined with recent suggestions that perceptions of unfairness in an organization may foster increased levels of politics (e.g., Ferris et al., 1995; Dipboye, 1995), it seems reasonable to suggest that important links between these two areas of research exist, and may well be worthy of further study. In essence, we are suggesting that focusing solely on political actions designed to further the politicians' self-interests may result in a somewhat incomplete picture of organizational politics; actions in which such motives are mixed with, or perhaps even largely replaced by, motives for revenge or "evening the score" also should be considered.

We should quickly add that there is no intention here of suggesting that workplace aggression and organizational politics are entirely coincident. On the contrary, it seems clear that they overlap only to a degree. Workplace aggression, like other forms of aggression, stems from a wide range of social, cognitive, and situational factors (cf. Anderson, Anderson, & Deuser, 1996; Baron & Richardson, 1994). While some of these factors may also play a role in organizational politics (e.g., perceived unfairness or mistreatment at the hands of others; unsettling and sudden workplace changes such as downsizing or reorganization; e.g., Baron & Neuman, 1996), other factors probably do not. For instance, workplace aggression, like other forms of aggression, is sometimes relatively *impulsive* in nature; it is evoked in a relatively automatic manner by such factors as high levels of arousal or negative affect. Moreover, this is often the case even when such internal states are generated by conditions or events unrelated to the target persons (e.g., high levels of temperature or irritating noise; Cohn & Rotton, 1997). It seems very unlikely that participation in organizational politics stems from similar sources; rather, such behavior seems more likely to involve various cognitive factors, including a careful analysis of the organization's work environment (e.g., Ferris et al., 1995). However, the overlap between workplace aggression and organizational politics appears to be sufficient in scope to warrant careful examination of potential links between them; doing so may well add to our understanding of both topics.

SUGGESTIONS FOR FURTHER RESEARCH: WHERE DO WE GO FROM HERE?

As noted earlier in this paper, past research on organizational politics has contributed much to our understanding of this process. Such work offers insights into the forms it takes (e.g., Ferris & Judge, 1991), the factors that lead to its occurrence (e.g., Ferris, Russ, & Fandt, 1989; Parker et al., 1995), and the effects it produces (e.g., Cropanzano et al., 1997; Kacmar & Carlson, 1998). However, much remains to be done, and in this final section, we offer some suggestions for future lines of research that may help us to attain a fuller and more comprehensive knowledge of this important process.

First on our "wish list" for further research is systematic efforts to further elucidate the motives behind organizational politics. We, as well as several other researchers (e.g., Cropanzano & Kacmar, 1995), have suggested that advancing one's self-interests is only one such motive, albeit, the most central or important. Additional motives worthy of further attention include several to which we have previously called attention, such as the desire to gain or re-establish justice and the desire for revenge. Still other motives not addressed in detail in this paper also may play a role, including (1) motivation to establish control over one's own outcomes (e.g., Ferris, Russ, & Fandt, 1989), (2) motivation to bolster one's self-esteem or self-efficacy (Gist & Mitchell, 1992), and (3) motivation to protect oneself against real or imagined threats (cf. Baron & Richardson, 1994). Establishing the motives that underlie organizational politics would appear to be an important initial step in the formulation of a comprehensive theoretical framework for understanding such behavior, so such work would be useful from this perspective. In addition, the question of whether different motives lead to the adoption of contrasting political tactics also should be addressed. It seems quite possible that, for instance, the motive for revenge might be reflected in somewhat different tactics than the desire to enhance one's outcomes or benefits. To date, however, no research known to the present authors has been conducted on this and related possibilities. Evidence concerning such links might well provide important insights into the nature of organizational politics and also might offer the added benefit of suggesting specific tactics for reducing or managing such behavior (e.g., Witt, 1995).

A second topic that appears to be deserving of further attention is suggested by our earlier discussion of social influence. Many of the tactics described in that context have not as yet been investigated within the context of organizational politics, despite the fact that several of these appear to be in frequent use in work and business settings. Why have these tactics not been examined in the past? In part, we believe, because of the methodology employed in most studies on organizational influence tactics. Such studies, although excellent in many respects, generally asked participants to rate the frequency with which they used or had observed each of a small set of listed tactics (e.g., Kipnis et al., 1980; Yukl & Tracey, 1992). Although the tactics included in these investigations were ones that are, indeed, in widespread use, the fact that participants could respond only to the tactics named may have restricted the range of information obtained in several ways. More open-ended procedures might well provide evidence for the frequent use of other tactics, including ones based on the principles described above (e.g., commitment/consistency, reciprocity, and scarcity). We believe that such research might add appreciably to our understanding of how organizational politics actually proceeds in many work settings.

A third topic we view as worthy of careful attention is the potential relationships between organizational politics and other forms of work-related behavior. In this paper, we focused primarily on potential links between politics and workplace aggression and organizational retaliatory behavior. However, it seems reasonable to suggest that important links also may exist between organizational politics and two additional processes: *motivation* and *leadership*. With respect to motivation, many previous studies indicate that perceptions of perceived injustice, whether with respect to outcomes, procedures, or personal treatment, can be a virtual "kiss of death" to individual motivation (e.g., Greenberg, 1989). Since such perceptions also appear to play a role in the occurrence of political behavior, studies designed to explore this relationship might prove fruitful.

Turning to leadership, most definitions of this process emphasize the key role played by influence: leaders are, among other things, the most influential members of the groups they head. Thus, successful leaders are, by definition, proficient at exerting social influence; their role requires this, and in fact, most leaders generally report that they prefer to rely on various tactics of influence rather than upon more direct exercise of their authority (in other words, they prefer the "velvet glove" to the "iron fist;" e.g., Huber, 1981; Yukl, 1994). This implies that successful leaders also may be adept in the use of various tactics of organizational politics; indeed, it seems reasonable to suggest that skill in this respect is often one of the factors contributing to the attainment and retention of their leadership positions. Is this actually the case? We believe that this would be an interesting topic for future research. A related issue concerns the relationship between leaders' style and their use of political tactics. One could predict that democratic (i.e., participative) leaders would rely on politics to a greater extent than autocratic (i.e., directive) ones, who would rely, instead, on the authority granted by their positions. Similarly, it seems possible that transformational leaders might rely on political tactics to a greater extent than other leaders and also would be more proficient in their use; after all, they seek to influence their subordinates in very profound ways and these skills may be central to success in doing so.

In sum, there appear to be many potentially valuable lines of investigation open to researchers in the field of organizational politics. Research conducted to date provides a firm foundation for moving ahead on all these intriguing fronts, and we look forward to both reading about and participating in this continuing scientific endeavor. Perhaps it makes sense to close this discussion by paraphrasing former British Prime Minster John Major, who once remarked: "The first requirement of politics is…stamina or patience." Where organizational politics is concerned, we would alter, and expand, his statement to read: "The first requirement of *understanding* organizational politics is perseverance; yes, it is a highly complex topic; but we *will* come to understand it if we continue systematic research."

REFERENCES

Allen, R.W., Madison, D.L., Porter, L.W., Renwick, P.A., & Mayes, B.T. (1979). Organizational politics: Tactics and characteristics of its actors. *California Management Review, 22*, 77-83.

Ambrose, M.L., & Harland, L.K. (1995). Procedural justice and influence tactics: Fairness, frequency, and effectiveness. In. R.S. Cropanzano & K.M. Kacmar (Eds.), *Organizational politics, justice, and support* (pp. 97-130). Westport, CT: Quorum Books.

Anderson, C.A., Anderson, K.B., & Deuser, W.E. (1996). Examining an affective aggression framework: Weapon and temperature effects on aggressive thoughts, affect, and attitudes. *Personality and Social Psychology Bulletin, 22*, 366-376.

Anderson, T.P. (1994). Creating measures of dysfunctional office and organizational politics: The DOOP and short for DOOP scales. *Psychology, 31*, 24-34.

Baron, R.A., & Neuman, J.H. (1996). Workplace violence and workplace aggression: Evidence on their relative frequency and potential causes. *Aggressive Behavior, 22*, 161-173.

Baron, R.A., & Neuman, J.H. (1998). Workplace aggression—The iceberg beneath the tip of workplace violence: Evidence on its forms, frequency, and potential causes. *Public Administration Quarterly, 21*, 446-464.

Baron, R.A., & Richardson, D.R. (1994). *Human aggression* (2nd ed.). New York: Plenum.

Biberman, G. (1985). Personality and characteristic work attitudes of persons with high, moderate, and low political tendencies. *Psychological Reports, 57*, 1303-1310.

Bjorkqvist, K., Lagerspetz, K., & Kaukiainen, A. (1992). Do girls manipulate and boys fight? Developmental trends in regard to direct and indirect aggression. *Aggressive Behavior, 18*, 117-127.

Bjorkqvist, K., Osterman, K., & Hjelt-Back, M. (1994). Aggression among university employees. *Aggressive Behavior, 20*, 173-184.

Burger, J.M. (1986). Increasing compliance by improving the deal: The that's not-all technique. *Journal of Personality and Social Psychology, 51*, 277-283.

Burns, T. (1961). Micropolitics: Mechanisms of institutional change. *Administrative Science Quarterly, 6*, 257-281.

Christie, R., & Geis, F.L. (1970). *Studies in Machiavellianism*. New York: Academic Press.

Cialdini, R., Cacioppo, J., Bassett, R., & Miller, J. (1978). A low-ball procedure for producing compliance: Commitment then cost. *Journal of Personality and Social Psychology, 36*, 463-476.

Cialdini, R.B. (1988). *Influence: Science and practice* (2nd ed.). Glenview, IL: Scott, Foresman.

Cialdini, R.B. (1994). Interpersonal influence. In S. Shavitt & T.C. Brock (Eds.), *Persuasion* (pp. 195-218). Boston: Allyn & Bacon.

Cohn, E.G., & Rotton, J. (1997). Assault as a function of time and temperature: a moderator-variable time-series analysis. *Journal of Personality and Social Psychology, 72*, 1322-1334.

Cropanzano, R.S., Howes, J.C., Grandey, A.A., & Toth, P. (1997). The relationship of organizational politics and support to work behaviors, attitudes, and stress. *Journal of Organizational Behavior, 18*, 159-181.

Cropanzano, R.S., & Kacmar, K.M. (1995). *Organizational politics, justice, and support*. Westport, CT: Quorum Books.

Cropanzano, R.S., Kacmar, K.M., & Bozeman, D.P. (1995). The social setting of work organizations: Politics, justice, and support. In. R.S. Cropanzano & K.M. Kacmar (Eds.), *Organizational politics, justice, and support* (pp. 1-20). Westport, CT: Quorum Books.

Dipboye, R.L. (1995). How politics can destructure human resources management in the interest of empowerment, support, and justice. In R.S. Cropanzano & K.M. Kacmar (Eds.), *Organizational politics, justice, and support* (pp. 56-82). Westport, CT: Quorum Books.

Drory, A. (1993). Perceived political climate and job attitudes. *Organization Studies, 14*, 59-71.

Drory, A., & Romm, T. (1988). Politics in organization and its perception within the organization. *Organization Studies, 9*, 165-179.

Drory, A., & Romm, T. (1990). The definition of organizational politics: A review. *Human Relations, 43*, 1133-1154.

DuBrin, A.J. (1978). *Winning at office politics.* New York: Van Nostrand Reinhold Company.

DuBrin, A.J. (1988). Career maturity, organizational rank, and political behavioral tendencies: A correlational analysis of organizational politics and career experiences. *Psychological Reports, 63*, 531-537.

Dulebohn, J.H. (1997). Social influence in justice evaluations of human resources systems. In G.R. Ferris, (Ed.), *Research in personnel and human resources management* (Vol. 15, pp. 241-292). Greenwich, CT: JAI Press.

Eisenberger, R., Huntington, R., Hutchison, S., & Sowa, D. (1986). Perceived organizational support. *Journal of Applied Psychology, 71*, 500-507.

Falbe, C., & Yukl, G. (1992). Consequences for managers of using single influence tactics and combinations of tactics. *Academy of Management Journal, 35*, 877-878.

Farrell, D., & Petersen, J.C. (1982). Patterns of political behavior in organizations. *Academy of Management Review, 7*, 403-412.

Fedor, D., Ferris, G.R., Harrell-Cook, G., Russ, G.S. (1998). The dimensions of politics perceptions and their organizational and individual predictors. *Journal of Applied Social Psychology, 28*, 1760-1797.

Ferris, G.R., Brand, J.F., Brand, S., Rowland, K.M., Gilmore, D.C., King, T.R., Kacmar, K.M., & Burton, C.A. (1993). Politics and control in organizations. In E.J. Lawler, B.J. O'Brien, & K. Heimer (Eds.), *Advances in group processes* (Vol. 10, pp. 83-111). Greenwich, CT: JAI Press.

Ferris, G.R., Dulebohn, J.H., Frink, D.D., George-Falvy, J., Mitchell, T.R., & Matthews, L.M. (1997). Job and organizational characteristics, accountability, and employee influence. *Journal of Managerial Issues, 9*, 162-175.

Ferris, G.R., Fedor, D.., Chachere, J.G., & Pondy, L.R. (1989). Myths and politics in organizational contexts. *Group & Organization Studies, 14*, 83-103.

Ferris, G.R., Frink, D.D., Beehr, T.A., & Gilmore, D.C. (1995). Political fairness and fair politics: The conceptual integration of divergent constructs. In R.S. Cropanzano & K.M. Kacmar (Eds.), *Organizational politics, justice, and support* (pp. 37-54). Westport, CT: Quorum Books.

Ferris, G.R., Frink, D.D., Bhawuk, D.P.S., Zhou, J., & Gilmore, D.C. (1996). Reactions of diversity groups to politics in the workplace. *Journal of Management, 22*, 23-44.

Ferris, G.R., Frink, D.D., Galang, M.C., Zhou, J., Kacmar, K.M., & Howard, J.L. (1996). Perceptions of organizational politics: Prediction, stress-related implications, and outcomes. *Human Relations, 49*, 233-266.

Ferris, G.R., Frink, D.D., Gilmore, D.C., & Kacmar, K.M. (1994). Understanding as an antidote for the dysfunctional consequences of organizational politics as a stressor. *Journal of Applied Social Psychology, 24*, 1204-1220.

Ferris, G.R., Harrell-Cook, G., & Dulebohn, J.H. (in press). Organizational politics: The nature of the relationship between politics perceptions and political behavior. In S.B. Bacharach & E.J. Lawler (Eds.), *Research in the sociology of organizations.* Stamford, CT: JAI Press.

Ferris, G.R., & Judge, T.A. (1991). Personnel/human resources management: A political influence perspective. *Journal of Management, 17*, 447-488.

Ferris, G.R., & Kacmar, K.M. (1992). Perceptions of organizational politics. *Journal of Management, 18*, 93-116.

Ferris, G.R., & King, T.R. (1991). Politics in human resources decisions: A walk on the dark side. Organizational Dynamics, 20, 59-71.

Ferris, G.R., & Mitchell, T.R. (1987). The components of social influence and their importance for human resources research. In K.M. Rowland & G.R. Ferris, (Eds.), *Research in personnel and human resources management* (Vol. 5, pp. 103-128). Greenwich, CT: JAI Press.

Ferris, G.R., Russ, G.S., & Fandt, P.M. (1989). Politics in organizations. In R.A. Giacalone & P. Rosenfeld (Eds), *Impression management in the organization* (pp. 143-170). Hillsdale, NJ: Lawrence Erlbaum.

Folger, R., & Baron, R.A. (1996). Violence and hostility at work: A model of reactions to perceived injustice. In G.R. VandenBos and E.Q. Bulatao (Eds.), *Violence on the job: Identifying risks and developing solutions* (pp. 51-85). Washington, DC: American Psychological Association.

Forgas, J.P. (1997). Asking nicely? The effects of mood on responding to more or less polite requests. *Personality and Social Psychology Bulletin, 24*, 173-185.

Galang, M.C., & Ferris, G.R. (1997). Human resource department power and influence through symbolic action. *Human Relations, 50*(11), 1403-1426.

Gandz, J., & Murray, V.V (1980). The experience of workplace politics. *Academy of Management Journal, 23*, 237-251.

Gilmore, D.C., Ferris, G.R., Dulebohn, J.H., & Harrell-Cook, G. (1996). Organizational politics and employee attendance. *Group & Organization Management, 21*, 481-494.

Gist, M.E., & Mitchell, T.R. (1992). Self efficacy: A theoretical analysis of its determinants. *Academy of Management Review, 17*, 183-211.

Gray, B., & Ariss, S. (1985). Politics and strategies change across organizational life cycles. *Academy of Management Review, 10*, 707-723.

Greenberg, J. (1989). Cognitive re-evaluation of outcomes in response to underpayment inequity. *Academy of Management Journal, 32*, 174-184.

Huber, V.L. (1981). The sources, uses, and conservation of managerial power. *Personnel, 51*(4), 62-67.

Jones, E.E. (1964). *Ingratiation*. New York: Appleton-Century-Crofts.

Kacmar, K.M., Bozeman, D.P., Carlson, D., & Anthony, W.P. (in press). A partial test of the perceptions of organizational politics model. *Human Relations*.

Kacmar, K.M., & Carlson, D.S. (1997). Further validation of the Perceptions of Politics Scale (POPS): A multi-sample approach. *Journal of Management, 23*(5), 627-658.

Kacmar, K.M., & Carlson, D.S. (1998). A qualitative analysis of the dysfunctional aspects of political behavior in organizations. In R.W. Griffin, A.M. O'Leary-Kelly, & J. Collins (Eds.), *Dysfunctional behavior in organizations* (pp. 195-218). Stamford, CT: JAI Press.

Kacmar, K.M., & Ferris, G.R. (1991). Perceptions of organizational politics scales (POPS): Development and construct validation. *Educational and Psychological Measurement, 51*, 193-205.

Kacmar. K.M., & Ferris, G.R (1993, July-August). Politics at work: Sharpening the focus of political behavior in organizations. *Business Horizons*, pp. 70-74.

Kirchmeyer, C. (1990). A profile of managers active in office politics. *Basic and Applied Social Psychology, 11*, 339-356.

Kipnis, D., Schmidt, S.M., & Wilkinson, I. (1980). Intraorganizational influence tactics: Explorations in getting one's way. *Journal of Applied Psychology, 65*, 440-452.

Kipnis, D., & Schmidt, S.M. (1982). *Profiles of organizational influence strategies*. San Diego, CA: University Associates.

Kumar, K., & Beyerlein, M. (1991). Construction and validation of an instrument for measuring ingratiatory behaviors in organizational settings. *Journal of Applied Psychology, 76*, 619-627.

Longnecker, C.O. (1989). Truth or consequences: Politics and performance appraisals. *Business Horizons, 32*, 76-82.

Longnecker, C.O., Sims, H.P., & Gioia, D.A. (1987). Behind the mask: The politics of employee appraisal. *Academy of Management Executive, 1*, 183-193.

Madison, D.L., Allen, R.W., Porter, L.W., Renwick, P.A., Mayes, B.T. (1980). Organizational politics: An exploration of managers' perceptions. *Human Relations, 33*, 79-100.

Mayes, B.T., & Allen, R.W. (1977). Toward a definition of organizational politics. *Academy of Management Review, 2*, 672-678.

Neuman, J. H., & Baron, R. A. (1997). Aggression in the workplace. In R. Giacalone & J. Greenberg (Eds.), *Antisocial behavior in organizations* (pp. 37-67). Thousand Oaks, CA: Sage.

Neuman, J. H., & Baron, R. A. (1998). Workplace violence and workplace aggression: Evidence concerning specific forms, potential causes, and preferred targets. *Journal of Management, 24*, 391-420.

Nye, L.G., & Witt, L.A. (1993). Dimensionality and construct validity of the perceptions of organizational politics scale (POPS). *Educational and Psychological Measurement, 53*, 821-829.

Parker, C.P., Dipboye, R.L., & Jackson, S.L. (1995). Perceptions of organizational politics: An investigation of antecedents and consequences. *Journal of Management, 21*, 891-912.

Porter, L.W., Allen, R.W., & Angle, H.L. (1981). The politics of upward influence in organizations. In L.L. Cummings & B.M. Staw (Eds.), *Research in organizational behavior* (Vol. 3, pp. 109-149). Greenwich, CT: JAI Press.

Prasad, L., & Rubenstein, A.H. (1994). Power and organizational politics during new product development: A conceptual framework. *Journal of Scientific & Industrial Research, 53*, 397-407.

Ralston, D.A. (1985). Employee ingratiation: The role of management. *Academy of Management Review, 10*, 477-487.

Randall, M.L., Cropanzano, R., Bormann, C.A., & Birjulin, A. (in press). Organizational politics and organizational support as predictors of work attitudes, job performance, and organizational citizenship behavior. *Journal of Organizational Behavior.*

Riley, P. (1983). A structuralist account of political culture. *Administrative Science Quarterly, 28*, 414-437.

Schriesheim, C.A., & Hinkin, T.R. (1990). Influence tactics used by subordinates: A theoretical and empirical analysis and refinement of the Kipnis, Schmidt, and Wilkinson subscales. *Journal of Applied Psychology, 75*, 246-257.

Shavitt, S., & Brock, T.C. (Eds.). (1994). *Persuasion.* Boston: Allyn & Bacon.

Skarlicki, D.P., & Folger, R. (1997). Retaliation in the workplace: The roles of distributive, procedural, and interactional justice. *Journal of Applied Psychology, 82*, 434-443.

Tetrick, L.E., & LaRocco, J.M. (1987). Understanding, prediction, and control as moderators of the relationship between perceived stress, satisfaction, and psychological well being. *Journal of Applied Psychology, 72*, 538-543.

Tushman, M.L. (1977). A political approach to organizations: A review and rationale. *Academy of Management Review, 2*, 206-216.

Vredenburgh, D.J., & Maurer, J.G. (1984). A process framework of organizational politics. *Human Relations, 37*, 47-66.

Wayne, S.J., & Liden, R.C. (1995). Effects of impression management on performance ratings: A longitudinal study. *Academy of Management Journal, 18*, 232-260.

Williams, K.B., Radefeld, P.A., Binning, J.F., & Suadk, J.R. (1993). When job candidates are "hard" versus "easy-to-get": Effects of candidate availability on employment decisions. *Journal of Applied Social Psychology, 23*, 169-198.

Wilson, P.A. (1995). The effects of politics and power on the organizational commitment of federal executives. *Journal of Management, 21*, 101-118.

Witt, L.A. (1995). Influences of supervisor behaviors on the levels and effects of workplace politics. In R.S. Cropanzano & K.M. Kacmar (Eds.), *Organizational politics, justice, and support* (pp. 37-54). Westport, CT: Quorum Books.

Yukl, G. (1994). *Leadership in organizations* (3rd ed.). Englewood Cliffs, NJ: Prentice-Hall.

Yukl, G., & Falbe, C.M. (1990). Influence tactics and objectives in upward, downward, and lateral influence attempts. *Journal of Applied Psychology, 75*, 132-140.

Yukl, G., Falge, C., & Youn, J. (1993). Patterns of influence behavior for managers. *Group and Organizational Management, 18*, 5-28.

Yukl, G., & Tracey, J.B. (1992). Consequences of influence tactics used with subordinates, peers, and the boss. *Journal of Applied Psychology, 77*, 525-535.

Zahra, S.A. (1985). Background and work experience correlates of the ethics and effect of organizational politics. *Journal of Business Ethics, 4*, 419-423.

Zahra, S.A. (1987). Organizational politics and the strategic process. *Journal of Business Ethics, 6*, 579-587.
Zahra, S.A. (1989). Executive values and the ethics of company politics: Some preliminary findings. *Journal of Business Ethics, 8*, 15-29.
Zanzi, A., Arthur, M., & Shamir, B. (1991). The relationships between career concerns and political tactics in organizations. *Journal of Organizational Behavior, 12*, 219-233.
Zhou, J. & Ferris, G.R. (1995). The dimensions and consequences of organizational politics perceptions: A confirmatory analysis. *Journal of Applied Psychology, 25*, 1747-1764.

LEGITIMACY IN HUMAN RESOURCES MANAGEMENT

Maria Carmen Galang, Wolfgang Elsik, and Gail S. Russ

ABSTRACT

Human resource management (HRM) has been the target of recent criticism which, ultimately, could pose a threat to the legitimacy of the HRM department as a separate entity within the organization, to the HRM function itself, and to HRM as a field of study. We propose that a legitimacy framework is appropriate for analyzing and addressing these criticisms. This paper begins by reviewing the relevant organizational legitimacy literature, with an emphasis on how organizational legitimacy can be created, maintained, lost, and repaired. We then propose that structuration theory, from the field of sociology, can be used to increase our understanding of organizational legitimacy. A closer examination of the criticisms of HRM is undertaken next, along with the strategic options available to HR professionals to manage the legitimacy of HRM. Finally, we conclude with directions for future research on the legitimacy of human resource management.

INTRODUCTION

Human resource management: A case of the emperor's new clothes?

—Armstrong (1987)

Listening to the cry that the emperor has no clothes, we dare confront the criticisms that challenge the legitimacy of human resource management (HRM), not so much as to dispute the validity of the observations, but rather to examine why the comment was made at all, and what can be done about it. Admittedly, this can be a risky proposition: paying attention to the boy's cry somehow validates an observation that we would rather ignore. But even at the risk of losing the support of allies,[1] we feel we need to face and listen to these criticisms because of their potential for damaging, and perhaps even destroying HRM's legitimacy. However, our purpose is not to come to the defense of HRM, but to understand why these questions are being raised, thus hopefully helping us address the issues more effectively.

Is Legitimacy a Concern for HRM?

There have been quite a number of publications highly critical of HRM. One need only look at some of the titles to note the serious challenges that face us: in addition to Armstrong's, we have *The Disappearing HR Department* (Brenner, 1996); *HRM: A Case of the Wolf in Sheep's Clothing* (Keenoy, 1990); *Taking on the Last Bureaucracy* (Stewart, 1996b); *In Search of HRM:...or the case of the dog that didn't bark* (Watson, 1995); *Is HR a Dinosaur?* (Wilkerson, 1996). The downsizing of HR departments, the outsourcing of HRM activities or their devolution to line management, the framing of new roles and the corresponding search for new labels to denote the new role also signal that something is amiss.

Nonetheless, given the distinction between department and function (Legge, 1978), one wonders whether it might be that only the department and not the function that is under attack. That is, perhaps the concern is only with regards to the competence and credibility of HR/personnel practitioners,[2] or perhaps critics question the need for a separate unit responsible for managing the organization's human resources. HR practitioners have long lamented that they often become the scapegoat for failures in the organization's management of its human resources (e.g., Peach, 1995; Stewart, 1996a), and yet are not given the credit for successes. As Legge had pointed out, it is difficult to attribute success to the efforts of the HR people alone, and that is an innate condition that HR people need to grapple with. Nevertheless, if it is simply a matter of how that job or function is performed (i.e., by an incompetent job incumbent, or by a separate unit), the problem can easily be remedied by developing competencies, changing the person, or restructuring.

But if it is the job itself that is in question, making changes with respect to the job incumbent or the job context will not settle the matter. As Frost (1989) had advised, what matters is not just doing things right, but also doing the right things, and more importantly, defining what is the right thing. Hence, the attacks on the department may be symptomatic of the "problems" with the function itself.

That the most scathing attacks come mostly from British academics and not from the United States should not come as a surprise. HRM is recognized to have originated from U.S. business schools. That perhaps also explains why much of the debate and discourse in the United States occurs in the area of practice so that it appears that the problem is just a question of practitioners' competence. The only U.S. critiques are by Ichniowski, Delaney and Lewin (1989) and Barkin (1989), which interestingly were both published in a Canadian journal. Ichniowski et al.'s was only mildly critical, questioning the view of HRM as a new and non-union model. Barkin (1989), commenting on the works of Ichniowski et al. (1989), and Kochan and McKersie (1989), echoed much of the concerns of the British scholars.

The criticisms of HRM can be considered to aim at two levels: that of HRM as a field of study, and HRM as a management function. This is not to say that they are not interrelated. An examination of the criticisms against HRM as a field of study gives some indications as to why the difficulties faced by practitioners persist. A closer look at the nature of the debates will show that these criticisms revolve around the matter of definitions and evidence. Further, as a field of study, HRM has been characterized as atheoretical, and hence been dismissed by some academics.

We believe that the nature of the criticisms against HRM can best be examined, understood and addressed within a legitimacy framework. Therefore, we first present an overview of the existing knowledge with regards to legitimacy, then discuss in more detail the criticisms against HRM. As we shall see, taking the legitimacy perspective will enable us to understand why criticisms about HRM have emerged. We can also better assess the effectiveness of current responses to these criticisms, hence providing us a beacon towards which efforts can be directed more fruitfully.

REVIEW OF THE ORGANIZATIONAL LEGITIMACY LITERATURE

Legitimacy has long been recognized to be important for the existence of an organization (Epstein & Votaw, 1978). Legitimacy not only provides the organization with the support that is necessary for its survival, but also deters active challenges to its continued functioning. Support, as do the challenges, comes in various forms as well as degrees of overtness and commitment. While a reading of the literature reveals a full range of possible responses, there is no consensus as to the range that should be included. For example, some, like Della Fave (1986), do not consider resigned acceptance as within the sphere of legitimacy, while others like Stryker (1994) include mere behavioral consent such as passive acquiescence or sullen obedience, and also acknowledge that self-interest and other calculative reasons may play a role. For Della Fave, behavioral acceptance is not enough to indicate legitimacy, but normative approval is the distinguishing element. On the other hand, Stryker recognizes and explicates the differing mechanisms that lead to the various forms of support of cognitive recognition, behavioral consent, or attitudinal approval. It must be noted that while the works of Della Fave and Stryker analyzed the legitimization of broader sociological phenomena (i.e., stratified social order and the use of science in law respectively), their application and extension into organizational legitimacy should prove to be enriching.

Challenges can come in various forms, too. In an experiment, Thomas, Walker, and Zelditch (1986) showed that although the likelihood of collective action was reduced within a legitimate order with a legitimate means for change, illegitimate forms of individual protests, that is, those outside of the legitimate means of change, can still be triggered. In this experiment, the legitimate order was the scientific testing of a communication network, which however resulted in inequitable rewards to participants. Thus, change in the communication network, while available, would destroy the experiment's purpose. The experiment demonstrated that the absence of active overt challenges does not necessarily mean passive support, but that collective action may not have been mobilized because of the fear of sanctions for supporting the challenge (coming from destroying a collective purpose), or because of non-approval of the challenger and/or the form of protest chosen.

Martin (1993) focused on the consequences of organizational illegitimacy, specifically examining the reactions of employees who are disadvantaged by reward allocation. Various responses take place, including acceptance of the legitimacy of inequities. The question of when challenges become effective in causing changes in the organization or of its demise remains however. Institutions have been shown to be quite resilient, as a legitimate order does not succumb to any one incidence of protest (Suchman, 1995). According to Ashforth and Gibbs (1990), "when an organization is accorded legitimacy, it has earned the goodwill of its constituents...management can occasionally deviate from social norms without seriously upsetting the organization's standing" (p.189). No existing framework however has explored the effectiveness of the various forms of challenges.

The concept of legitimacy has been studied in different spheres of social activity and at various levels of analysis; for example, individual business organizations (Allen & Caillouet, 1994; Arnold, Handelman, & Tigert, 1996); industries (Elsbach, 1994; Rao, 1994); voluntary social organizations such as environmental activists (Elsbach & Sutton, 1992), unions (Chaison, Bigelow, & Ottensmeyer, 1993), and non-profit social service organizations (Singh, Tucker, & House, 1986); occupational professions (Richardson, 1985); social institutions such as government and law (Stryker, 1994), and business (Berger, 1981; Boulding,

1978); and more micro events such as organizational change (Brown, 1994; Stjernberg & Philips, 1993), and the exercise of power (Selznick, 1952; Weber, 1978).

Likewise, conceptual frameworks have been developed in the field of organizational studies (Ashforth & Gibbs, 1990; Dowling & Pfeffer, 1975; Neilsen & Rao, 1987; Richardson & Dowling, 1986; Suchman, 1995). The most recent framework by Suchman (1995) integrates much of the previous theoretical and empirical work, and elaborates on the meaning of legitimacy and the various ways to gain, repair, and maintain legitimacy.

Legitimacy Defined

Rather than formulate a new definition, existing definitions of legitimacy will suffice for our purpose. One such definition is by Suchman (1995): "Legitimacy is a *generalized perception or assumption* that the actions of an entity are desirable, proper or appropriate within some *socially constructed* system of norms, values, beliefs, and definitions" (p. 574). We have italicized those elements which Suchman elaborated. It is therefore an evaluation of the acceptability of the entity itself and its actions because it has been perceived to conform to norms and standards held by those making the evaluation.

In operational terms, legitimacy has been defined and measured largely by its consequences. We find the same difficulties as what Pfeffer (1981) and others have with regards to the concept of power. For example, researchers use endorsement and support, either in terms of observed, concrete actions (Elsbach & Sutton, 1992; Singh et al., 1986) or self-reported intentions (Elsbach, 1994). These measures however are more a result of some prior judgment of legitimacy, such as that being captured by Elsbach's (1994) organizational normativity, one of three factors in her 12-item legitimacy scale. Furthermore, as evidence or testimony to legitimacy, they are also imperfect indicators. As Dowling and Pfeffer (1975) pointed out, "economic exchange is not identical with legitimacy...a legitimate purpose will not necessarily ensure resource allocation, nor will resource allocation necessarily ensure legitimacy" (pp. 123-124). Also as earlier indicated, there are various forms and degrees of support, and to limit it to this type ignores the more passive forms.

Clearly, the concept of legitimacy is closely related to that of organizational reputation. In a recent book on the topic of reputation and corporate image, Fombrun (1996) defined corporate reputation as "the overall estimation in which a company is held by its constituents. A corporate reputation represents the 'net' affective or emotional reaction—good or bad, weak or strong—of customers, investors, employees, and the general public to the company's name" (p. 37).

In a manner similar to that of legitimacy, the consequences of corporate reputation are what make it so important. Fombrun (1996) asserted that "reputation is valuable because it informs us about what products to buy, what companies to work for, or what stocks to invest in....The proliferation of such subjective rankings as 'best managed,' 'most innovative,' and 'most admired' attests to the growing popularity of reputation as a tool for assessing companies" (pp. 5-6).

There are differences between organizational legitimacy and corporate reputation, however. For instance, legitimacy stems from conforming to some set of pre-existing norms, as stated above. A superior organizational reputation, on the other hand, is at least in part the result of the firm's uniqueness: "Our best-regarded companies rise above the rest in prestige, status, and fame because they prize, pursue, and achieve *uniqueness*" (Fombrun, 1996, p. 23). It is this quality that makes a good reputation difficult for competitors to imitate, and thus is the basis for competitive advantage (Fombrun, 1996). It can be argued, however, that legitimacy is a necessary, if not sufficient, condition for a superior corporate reputation.

Targets and Types of Legitimacy

Reviewing the literature on types of organizational legitimacy reveals a high degree of convergence although the authors (i.e., Scott, 1995; Stryker, 1994; Suchman, 1995) do not refer to each other. These types of legitimacy imply differences in the responses or support given, in the aspects of the organization being targeted and, particularly in Suchman's framework, also in legitimization strategies. These types of legitimacy could be the basis for constructing a multi-dimensional perceptual measure of legitimacy that could then be related to the consequences of support or opposition, and determinants.

Suchman (1995) differentiates pragmatic, moral, and cognitive legitimacy. Pragmatic legitimacy is conferred when the organization's constituents support the organization because it serves their self-interests. It is calculative because the actors expect more or less concrete benefits from the organization in return for their support, ranging from immediate economic exchange to the belief that the organization will serve their basic interests in general. Moral legitimacy operates on a social in contrast to an instrumental logic. The constituents do not evaluate their potential benefits from the organization and its activities, but instead compare it with moral standards like societal welfare and well-being. Subject to this judgment can be organizational outcomes, techniques and procedures, structures, and members. Although not pointed out by Suchman, these elements are objects of pragmatic and cognitive legitimacy as well. Cognitive legitimacy refers to the comprehensibility and taken-for-grantedness of these aspects of the organization. For Suchman, these three types are interrelated and are somewhat hierarchically ordered. As one moves from pragmatic to cognitive legitimacy, the basis becomes less superficial and less a part of conscious considerations that legitimacy becomes more difficult to attain but also more difficult to lose once established.

Discussing the (de-)legitimizing effects on legal institutions of introducing scientific rationality into law, Stryker (1994) names three approaches to legitimacy that are based on different behavioral processes. Drawing on Sewell's (1992) interpretation of Giddens' (1976, 1979, 1981, 1984) work, these processes are regarded as mechanisms leading to and flowing from legitimacy. The attitudinal approval approach describes legitimacy as anchored in the acceptance of the organization and its outcomes, procedures, and structures and internalized it as the rules of the game. In the behavioral consent approach, legitimacy is not expressed by attitudinal change but by behavioral compliance. Here, constituents follow the rules (e.g., accept decisions or apply procedures) not because they feel that they are right, but mainly because they expect resources and other rewards, or they calculate the costs of opposing higher than the burden of conformance. Legitimacy resulting in behavioral consent, which in turn results in legitimacy, is focused on positive outcomes for the constituents. Compliance is instrumental to rewards. The cognitive orientation approach focuses on the recognition of rules as binding. This does not mean that the actors regard the rules as favorable (instrumental aspect) or have internalized them as the right way of doing (normative aspect), but simply that the actors know that the rules exist and that they are compulsory.

The legitimizing effect of institutions is a central topic of institutional theory. In his comprehensive overview on past and contemporary institutionalism, Scott (1995) depicts three "pillars" of institutions that are emphasized by institutional writers in varying degrees. The three pillars are different bases of legitimacy. The regulative pillar points to the fact that all institutions regulate behavior. Established rules are enforced by monitoring and sanctioning behavior. The basic assumption is that actors, in pursuing their self-interests, try to get rewards and to avoid punishment. The normative pillar embraces normative rules (values and norms) that indicate preferred or desirable goals or objectives and appropriate means to attain them. Acting in accordance with these norms and values is a matter of obligation. The norms are often internalized, but compliance to moral rules is not unreasoned or automatic. As with any kind of rule, social actors always have a choice, even if they act routinely and without much conscious consideration. The cognitive pillar consists of rules for the constitution of meaning (e.g., who is an actor, or what interests does he or she pursue?). The symbols, categories, and frames provide the bricks and mortar for the social construction of reality. Meaning is attributed to events, objects, and activities in interaction. The cognitive rules are rarely recognized because many of them are often taken for granted.

The Determinants of Legitimacy

There are two perspectives with regards to the process of establishing legitimacy. The strategic perspective argues that legitimacy can be acquired through the efforts of key players (e.g., dominant coalitions, or managers of the focal organization), while the institutional perspective downplays their role. The institutional perspective, however, addresses the difficulty in applying legitimacy to contexts where there is an absence of "a single decision-making or motivational locus" (Epstein & Votaw, 1978, p. 73). Suchman (1995) argues for the integration of both perspectives. Thus, while the focal organization can undertake actions intended to establish legitimacy, the concurrence of the audience is necessary. The legitimation process involves the collective construction of meaning (Neilsen & Rao, 1987), and communication of information about the focal organization is central to the process (Suchman, 1995; Terreberry, 1968).

In general, the efforts of the actors have to do with communicating information about its conformity with the standards of society, or that influence which standards to use (Arnold et al., 1996; Dowling & Pfeffer, 1975). The audience, for its part, is not a passive recipient of the information that is being relayed, and may accept and agree or reject and challenge. In addition, there are others whose actions, while not directly targeted at the focal organization and its legitimacy, indirectly influence as it molds the values and norms held by the audience and against which the focal organization's legitimacy is evaluated (Dowling & Pfeffer, 1975). What this implies then is that efforts of the focal organization may or may not be successful in acquiring legitimacy. What this also implies is that legitimacy can be lost, and hence needs to be constantly defended, and at times re-established (Boulding, 1978; Suchman, 1995).

In the same manner, delegitimation takes place as a consequence of what the focal organization does, although unintendedly so, or when the context of social values and norms has changed (Chaison et al., 1993), unless the actor can effectively respond to audience reactions that are triggered by these events. Mostly, "erosion takes place slowly and often almost imperceptibly…Where, however, the issue of legitimacy is related to, or simply contemporaneous with, a social crisis of broad proportions, the survival of the organization may, even in the short run, be very much in doubt" (Epstein & Votaw, 1978, p. 77).

One key to addressing legitimacy issues is by examining the prevailing societal values and norms relevant to the focal organization (Dowling & Pfeffer, 1975), as what Boulding (1978) had done in examining the legitimacy of the institution of business. Boulding argued that the legitimacy of business was rooted in the value placed by society on economic exchange, profit from mere ownership of capital, and private ownership. Herein lies the problem however. As Ashforth and Gibbs (1990) had noted, "social values and expectations are often contradictory, evolving, and difficult to operationalize" and the situation is further complicated by "ambiguities and inconsistencies in their transmission" and "a number of diffuse constituents with frequently conflicting expectations and perceptions" (p. 177). Assuming an accurate reading of who the constituents are and their expectations, this means that different strategies need to be used for different constituents (Allen & Caillouet, 1994). As well, buffering and decoupling (Elsbach & Sutton, 1992; Meyer & Rowan, 1977), and symbolic actions rather than substantive ones (Arnold et al., 1996) help make the use of divergent strategies possible.

The audience of course does not necessarily consist of all of society, only its relevant sectors (Elsbach & Sutton, 1992; Suchman, 1995). The problem is identifying the audience that matters, that is those with the influence and power to affect the focal organization's legitimacy (Allen & Caillouet, 1994). That should include both potential allies as well as those in opposition, especially those who will be effective in mobilizing protest. The theory of stakeholder identification and salience recently proposed by Mitchell, Agle, and Wood (1997) presents a guide to determining the audience that matters. Hybels (1995) also argued that endorsement from those perceived as independent and objective is essential in broadening support.

Human Agents and Non-human Agents[3]

Although there are many human agents involved in the legitimation process, the focus of existing work has largely been on the focal organization. In contrast, not much attention has been given to other players whose importance to legitimation is recognized. Even so, the focal organization has been implicitly viewed as a unified entity, and its efforts directed at external parties. However, potential supporters and opponents also exist within the organization, whose actions may directly or indirectly affect the organization's quest for legitimacy pronouncements from outside audiences.

Events that indirectly influence legitimacy through the shaping of values and norms take place in the larger society, and should be considered in understanding legitimacy issues, though the relevant events are necessarily situation-specific. For example, Chaisson et al. (1993) cited the changing expectations of the workforce as a contributory factor to unions' loss of legitimacy. Berger (1981) attributed the attacks on the legitimacy of business to the rise of a class of knowledge workers, who were not concerned with knowledge related "with the production or distribution of material goods," but "with things...that have to do with the quality of life" (p. 87). Legge's (1995) analysis of the emergence of HRM in the 1980s pointed to changes in the product and labor markets that have led to "the search for competitive advantage" (p. 76).

Strategies of the Focal Organization

Allen and Caillouet (1994), Ashforth and Gibbs (1990), and Suchman (1995) offered typologies of legitimation strategies (see Table 1). Allen and Caillouet derived their listing empirically from an organization undergoing a legitimacy crisis, while Ashforth and Gibbs, and Suchman encompassed the different purposes of acquiring, maintaining, and defending legitimacy. None of these specify the conditions when a strategy will likely be used and in what combination, and when it will be effective. Suchman did recommend that further efforts should be placed in addressing these questions. However, Allen and Caillouet found that the stakeholder being targeted determined the strategy used; on the other hand, Ashforth and Gibbs suggested that the mix and intensity of these strategies depend on the

Table 1. Legitimation Strategies

Allen & Caillouet (1994)	Ashforth & Gibbs (1990)		Suchman (1995)	
		Gain	Maintain	Repair
Excuse	Substantive Management	*General*		
• denial of intention	Role performance	Conform to environment	Perceive change	Normalize
• denial of volition	Coercive isomorphism	Select environment	Protect accomplishments	Restructure
• denial of agency	Altering resource dependencies			
Justification	Altering socially institutionalized practices	Manipulate environment		Don't panic
	Symbolic Management		*Pragmatic Legitimacy*	
• denial of injury	Espousing socially acceptable goals	Conform to demands	Monitor tastes	Deny
• denial of victim	Denial and concealment	Select markets	Protect exchanges	Create monitors
• condemnation of condemner				
• negative events misrepresented	Redefining means and ends		*Moral Legitimacy*	
Ingratiation	Offering accounts	Conform to ideals	Monitor ethics	Excuse/justify
Intimidation	Offering apologies	Select domain	Protect propriety	Disassociate
Apology	Ceremonial conformity		*Cognitive Legitimacy*	
Denouncement		Conform to models	Monitor outlooks	Explain
Factual distortion		Select labels	Protect assumptions	
Institutionalize				

three generic purposes of legitimation, each of which differ in the degree of scrutiny by constituents and as to whether or not legitimacy is problematic.

Ashforth and Gibbs (1990) also pointed to the possibilities for failure, when efforts intended to establish legitimacy can actually backfire, a condition that Suchman (1995) labels as "self-promoter's paradox." An organization with a strong need to be legitimated may be seen to "protest too much," triggering a questioning of its credibility that only serves to further decrease its legitimacy. In addition, being too successful, or what Suchman (1995) termed as the "sector-leader's paradox," also has its risks for the organization as well as the field of similar organizations. Isomorphism provides the base for cognitive and moral legitimacy but also leads to thoughtless, inappropriate or superficial imitation, destroys competitive advantage for one organization, and endangers the entire field's capacity to adapt that only comes from diversity.

A few have empirically tested some of these strategies in the context of repairing damaged legitimacy (Allen & Caillouet, 1994; Elsbach, 1994; Elsbach & Sutton, 1992) and in gaining legitimacy (Brown, 1994; Stjernberg & Philips, 1993). Aside from finding support for some of these strategies, these studies show that: (1) because of the need to challenge prevailing/predominant cultural norms and values, organizations at times may engage in intentionally illegitimate actions, which can then be transformed into support (Elsbach & Sutton, 1992); (2) symbolic actions are effective in securing legitimacy (Arnold et al., 1996; Brown, 1994); (3) acknowledgments may be more effective than denials when the problem is seen to result from deliberate actions by the organization, as it communicates sincerity (Elsbach, 1994); and, (4) the effectiveness of the content of a verbal account depends on the nature of the issue, as well as the nature of the protesting audience (Elsbach, 1994).

Summary

The existing literature in organizational legitimacy, reviewed above, would be helpful in understanding HRM's current state. A few adjustments however are needed to fit the peculiarities of HRM. Such adjustments, too, while focused on the particular case of HRM, would only serve to enrich or refine our understanding of legitimacy in general. For example, the current literature pays limited attention in delineating levels of analysis that might prove to be important distinctions. Wood's (1991) institutional, organizational, and individual levels of framing principles of social responsibility can be applied to HRM. Hence, one can refer to legitimacy issues with respect to the HR manager, with respect to an HR unit in one organization, and with respect to the profession of HRM. In HRM, one can talk about individual versus collective focus; that is, an HR unit in one organization as against the profession of HRM. In addition, the literature treats legitimacy within the context of an entity's relationship with an external audience, because that entity is seen as being homogeneous. Whereas HRM, on the other hand, is embedded in organizations, and such a relationship with the organization in which it is embedded might spell a difference.

Suchman's (1995) typology implies that various aspects of the organization can be targeted, and that the various types are interrelated. In the particular case of HRM, this is indeed an important point to be further explored. The distinction between department and function in HRM and its potential dilemma in justifying existence of one over the other can be seen. It could be that the current state that HR practitioners finds themselves in might have been due to their own doing. Similar to what Suchman (1995) calls as the sector-leader's paradox, the efforts to make HRM important (e.g., as a source of competitive advantage) might have been contributory to line management's reclaiming it as its responsibility. Hence, how the various types or dimensions of legitimacy, and by extension the strategies, interrelate, and how these interrelations affect the outcome need to be fully explicated, as well as tested.

As earlier stated, much of the literature has focused on the focal organization as the major, if not the only, actor involved in the legitimation process. Actions and reactions of constituencies, whether directly countering or supporting the actions of the focal actor, or whether indirectly operating on the cultural frame ultimately used to judge the focal actor's legitimacy have merely been acknowledged as being important. Questions like how effective are the focal actor's strategies in eliciting the various forms and degrees of support or challenges, when legitimate reactions and illegitimate ones are triggered, at what point the (re)actions of the audience would prompt changes in the focal actor, how are the constituencies to be identified, still need to be answered. Jennings (1994) has examined HRM at the macro level, using a sociological perspective that considers both political and institutional processes. In HRM, Tsui (1987) has done much work in empirically identifying its constituencies and their expectations.

Perhaps a general framework is not possible, but one can start with a specific context, much like Stryker's (1994) analysis of the underlying mechanisms of the impact of science on law's legitimacy. Stryker (1994) builds on Giddens' theory of structuration from the field of sociology, and has shown how its application field enriches understanding of the legitimation process of a particular institution. The similarity in conceptions of legitimacy, particularly those of Suchman (1995), Stryker (1994), and Scott (1995), calls for integration, and Giddens' theory offers a convenient way to do so. In fact, Scott refers to Giddens but does not elaborate on his theory. Suchman acknowledges the need to integrate both strategic and institutional perspectives, and Giddens' theory enables the integration of both paradigms within one theory, overcoming the long-standing and pervasive dichotomy in social (as well as management) sciences of structure versus agency (or system versus actor, as the seminal book of Michel Crozier and Erhard Friedberg is titled; or determinism versus voluntarism). Therefore we discuss the basic elements of Giddens' theory of structuration next, and then how it applies to legitimacy.

STRUCTURATIONIST VIEW OF ORGANIZATIONAL LEGITIMACY

Basics of Structuration Theory

Structuration theory was formulated by the British sociologist Anthony Giddens (1984, 1993, 1994). It is a social theory with the universalistic claim to cover the basic aspects of social life and is therefore general and abstract. However, structuration theory has been increasingly applied to organization theory, business administration and even, although just beginning, to HRM (e.g., Boland, 1993; Elsik, 1998; Hanft, 1995; Orlikowski, 1992; Pentland & Rueter, 1994; Riley, 1983; Weaver & Gioia, 1994; Whittington, 1992). For our purposes, it is neither possible nor necessary to present the whole theory but to concentrate on two core aspects which seem to be especially relevant for us: (1) the three dimensions of structure and agency, and (2) the duality of structure.

In organization and management theory, action is usually understood as purposeful, intentional behavior. Giddens argues for the decoupling of purpose/intention and action and defines action not as a series of discrete, intentional acts but as intervention into the flow of events that happens independent of the actor and has no predetermined future. Acting means to make a difference, to change the course of life. Actors use their knowledge of potential consequences to attain certain outcomes. But this knowledge need not be valid in a scientific sense nor conscious from the actor's point of view to constitute action.

The actor needs two abilities to act (see Figure 1). Capability, because of its transformative capacity, is by definition the power to intervene and to make a difference. Knowledgeability refers to the ability of the actors to apply their knowledge to their social action. It is restricted by unacknowledged conditions and unintended consequences of action.

```
                        actor
                       ↙     ↘
            capability         knowledgeability
                ↓              ↙            ↘
             power       practical        discursive
                ↓      consciousness     consciousness
                ↓             ↓                ↓
          intervention in  reflexive monitoring  rationalisation
          the flow of events   of action         of action
```

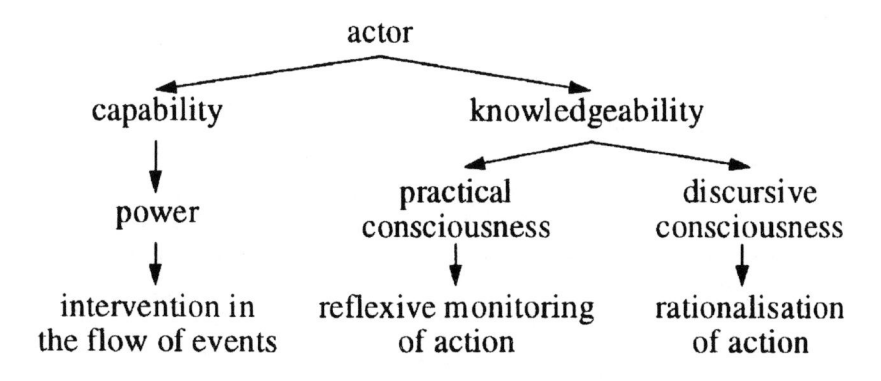

Figure 1. Capability and knowledgeability.

Actors draw on their everyday knowledge to account for their action. This knowledge is the result of one's accumulated experiences and the received wisdom of experts and laypersons. These everyday beliefs provide the actors with ontological security, the basis for the creation of mutual knowledge that enables the actors to constitute their interaction as meaningful. Usually actors are not able to verbalize all the reasons for their action, some of their motives remaining unconscious. Hence, we have three cognitive layers: discursive consciousness, practical consciousness, and unconsciousness.

With reflexive monitoring of action, actors develop an understanding of their own action and its background and "understand what they do while they do it" (Giddens, 1984, p. xxii). Only a little part of that understanding however can be articulated as discursive knowledge, the bigger part is available only as an implicit, tacit, practical knowledge, which enables actors to get along with the routine practices of everyday life. But when actors are asked to explain or justify their routine action, they are in principle able to transform their practical knowledge into a discursive one. This is not true for the motivation of action. It is largely unconscious and only partially, if ever, subject to conscious analysis. Because of the limitations of their consciousness, actors produce unintended consequences and act under partially unacknowledged circumstances.

Social action (interaction) can be analyzed along three interrelated dimensions: communication, power, and sanction. Giddens (1994) maintains that these dimensions are separable only analytically and that all are involved in every moment of social life. Thus, actors communicate meaning, justify or sanction their actions, and use power in every interaction. Successful interaction depends on the mutual understanding of the actors. Understanding does not primarily mean to empathize with the other actor(s), but rather to develop expectations about their future actions in order to align one's own behavior with it. The other actor's actions become meaningful only within a particular context. The meaning of interaction is not some essence inherent in it but depends always on the context within which the interaction takes place. The meaning of interaction has to be actively constructed (negotiated) and maintained by the actors. Most of the time this happens in an unspoken, routinized manner. The actors draw on their mutual knowledge to make their interaction meaningful in order to be able to maintain their communication. But this mutual knowledge is never complete, thus the meaning of some interactions must be negotiated explicitly.

Actors do not accept every action that they understand. Action must not only be comprehensible but also be justified. In every interaction the actors have reciprocal expectations and requirements of proper behavior. To get their expectations met, the actors promise or threaten with sanction. The execution of these sanctions is not like a natural phenomenon, but can be negotiated within certain limits.

In a relational sense, power means the capacity of an actor to realize certain outcomes when this realization also depends on the action of other actors. This does not necessarily mean conflict. The mutual influence in an interaction means that

Legitimacy in Human Resource Management 55

none of the actors is completely autonomous or completely dependent. In the latter case, we could hardly talk of acting but only of behaving in the narrow sense of automatically reacting to external forces. By no means does this imply that the interacting actors have the same share of autonomy and dependence. By mobilizing resources (e.g., skills, knowledge, authority, physical force, etc.) the actors try to control each other's actions in the desired direction. A convenient example is to try to control the production of meaning in the interaction. It points to the fact that the three dimensions operate simultaneously in every interaction. When interacting, actors do not *only* communicate or *only* justify/sanction or *only* influence each other, but rather employ these aspects concurrently.

Structure

The core idea of structuration theory is the duality of structure. The notion of duality of structure refers to the above-mentioned agency-versus-structure dualism in the social sciences. Duality means that structure is at the same time medium and outcome of human action. On one hand, structure (as the consequences of past action) restricts and enables action, and on the other, structure results from and is reproduced by human action. Structure and human action or agency are recursively related, not just cyclically. This is vividly illustrated in many pictures of Escher (e.g. the "Drawing Hands"): We cannot decide which of the two hands draw and which of them is drawn. Giddens (1993) uses the example of language and speaking to illustrate the duality of structure. We need a language (structure) to be able to speak (act). Here the language is the medium for speaking. But any language only exists in the act of speaking it (and in the memories of the actors). This is the "virtuality" of structure. A language that is not spoken (or remembered) by somebody does not exist. It is only brought into existence (or in the words of Giddens, "instantiated") by speaking it. The same holds true for other structures like the law or other rules. They only exist as long as the actors draw upon them in their social practices.

Structures consist of rules and resources (see Figure 2). Rules are generalizable procedures for the (re-)production of social practices. They do not (and cannot) specify all possible situations of their application. To know a rule is not to be able to give a verbal account of it but to apply it to various every day practices. For example, in order to correctly speak a language one need not be aware of its grammar or able to articulate it. Most of the rules are part of our practical consciousness. As part of structure, rules are virtual and exist only in the reproduction of social practices. In any given context, there are always more than one rule set that are partially interconnected and partially overlapping. Rules have to be constantly interpreted by the actors. These interpretations are guided by the particular interests of the actors and thus due to struggle. Here, the actors use resources as a potential sanction tool for carrying through their preferred rule interpretations.

```
                    Elements of Structure
                    /                    \
                Rules                   Resources
                /    \                   /      \
   Constitution   Justification,    Command       Command
   of Meaning    Sanction          over Persons  over Objects
   cognitive     moral and         e.g. hierarchy  e.g. money
   order         legal order
```

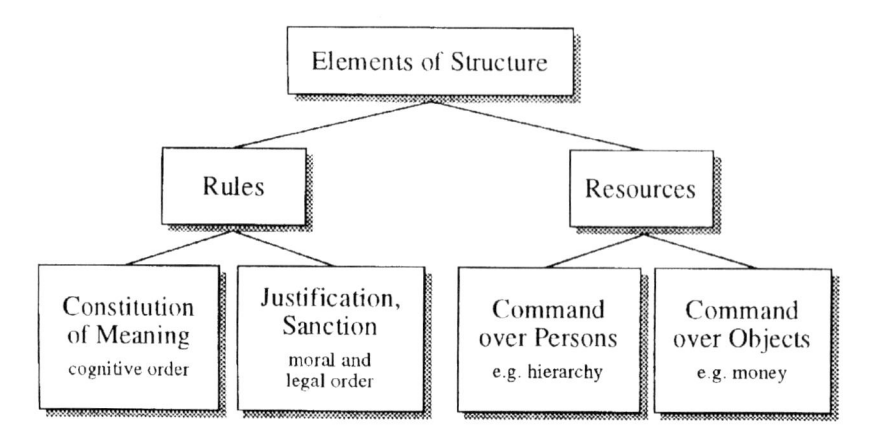

Figure 2. Elements of structure.

Rules pertain to the constitution of meaning as well as to the distribution of rights and obligations and are backed by normative sanctions.

Resources are the second type of structural element. They provide the potential for change. Giddens (1994) distinguishes between allocative and authoritative resources, according to the target of transformation: "By 'authorisation' I refer to capabilities which generate command over *persons*, and by 'allocations' I refer to capabilities which generate command over *objects* or other material phenomena" (p. 100). Like rules, resources are virtual as well. Material and immaterial things become resources only if the actors use them as such in their interaction. Money, knowledge, property, and so forth are not resources *per se*. They are "instantiated" as resources when actors draw on them to exercise authority or allocate other resources. This distinction by Giddens however has been criticized because it is impossible to categorize resources exclusively into one or the other (Becker, 1996; Ortmann et al., 1990). All material and immaterial things can be used to command people as well as things.

Rules and resources can be described and analyzed along three dimensions: signification, legitimation and domination. These dimensions correspond to the action dimensions of communication, sanction and power, respectively. Signification means the semantic rules, codes and "world views" that guide the constitution of meaning and make up the cognitive order of the particular social system (e.g., society, organization). Legitimation are the moral rules of the normative order. Domination depends on the mobilization of allocative and authoritative resources.

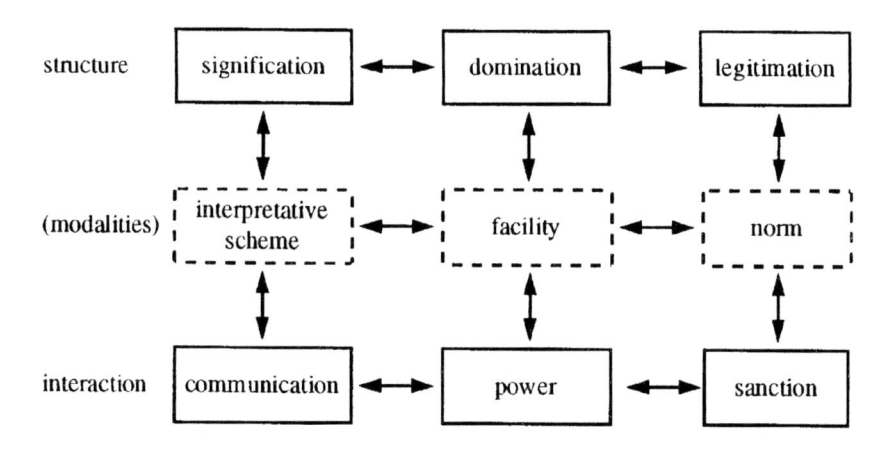

```
structure    [signification] ⇄ [domination] ⇄ [legitimation]
                  ⇅              ⇅                ⇅
(modalities) [interpretative  ⇄ [facility]  ⇄  [norm]
              scheme]
                  ⇅              ⇅                ⇅
interaction  [communication] ⇄ [power]      ⇄  [sanction]
```

Figure 3. Structuration.

Structuration

Action and structure are conceptually interrelated in three ways (see Figure 3). First, they are connected through their three corresponding dimensions (i.e., communication/signification, sanction/legitimation, and power/domination). Second, they are recursively related to each other as depicted in the concept of duality of structure. Third, there is the mediating level of modalities.

There are three modalities of structuration: interpretative schemes, facilities, and norms. By using these modalities, the actors can go on in their interaction and simultaneously reproduce the structural features of the social system. As there is no natural, objective meaning in any aspect of social life, communication calls for the construction of meaning. In order to do so, the actors use interpretative schemes to understand what they are saying and doing. The construction of reality is a social endeavor, not a case of individual discretion (see also Berger & Luckmann, 1966). Interpretative schemes are standardized elements or stocks of knowledge (e.g., social roles) that make up the core of the mutual knowledge. Mostly in a routinized manner they are taken from the cognitive order which is reconstituted by their very use in interaction.

Power is inherent in any interaction. It is applied by the use of allocative and authoritative resources to reach ends that need the cooperation of other actors. Resources are the media by which power is enacted as well as elements of the domination order or the social system which is reproduced by the use of facilities. From the action perspective, facilities are resources employed in the interaction; from the structural perspective, facilities are institutional features of the social system (Giddens, 1994). Albeit distributed asymmetrically, resources are never controlled only by one of the actors. All actors are autonomous *and* dependent, although they need not be equal. Because action is contingent, (i.e., actors always could have acted otherwise), and because action always touches the interests of the actors, it has to be accounted for (and if the accounts are not demanded, the actors must be prepared to do so). To (potentially) account for their action, the actors draw on norms which are related to the legitimation order.

Organization as Structuration

As has been mentioned above, structuration theory deals with society rather than organizations. Giddens says very little about organizations. However, there have been several and increasing attempts to use structuration theory for the benefit of organizational and management issues (Bouchikhi, Kilduff, & Whittington, 1995; Ortmann, Sydow, & Windeler, 1997).

Organizations are social systems. The reproduction of these systems is attained primarily via intentional control and coordination of human actions targeted towards the purposive shaping of formal structures. Thus, relations between positions or organizational units and the related rights and obligations are fixed. Furthermore, these formal structures provide criteria for punctuating the flow of events into discrete actions, and for attributing them to categories like causes/effects or costs/outcomes. Even if actors in organizations behave intentionally, it takes organizational rules and resources (e.g., accounting systems) to punctuate the flow of events and to transform human action into actions that can be attributed to categories and persons. The efficiency and effectiveness in the use of resources legitimize the domination order, and as the dominant interpretative scheme constitutes the cognitive order of the organization. As Ortmann (1995) said, "Organizations are those social systems which have to protect their legitimation via proof or, if necessary, fiction of their rationality" (p. 295).

Organizations differ from other social systems like families or circles of friends in their formality of structure, but it is only the relationships between organizational units and not the social relations between actual actors that are formalized. Organizations do not have goals and strategies. Only individuals are able to act. Collective purposes and strategies usually called "organizational goals" are the result of structuration processes where goals of individual actors become structural properties of the organization, be it intentionally or not, and thus create the context for further action. The actors reproduce these structures by drawing on the rules and resources in their interaction. If the structuration process is intentional, some of the more powerful actors have successfully carried through their goals and strategies, but the implementation of it remains unsure because even the less powerful actors have always the choice to act otherwise.

Goals and strategies can only be implemented by action which is recursively related to structure. From this point of view, organization is structuration because the actors in the organization create the very structure they need as the medium for

Legitimate order consists of	Sets of rules		Combination of resources	
Dimensions of structure	Signification	Legitimation	Domination	
Types of rules and resources	Rules for the constitution of meaning	Rules for the sanctioning of social action	Authoritative resources	Allocative resources

Modalities	Interpretative schemes	Norms	Authoritative facilities	Allocative facilities
Examples of modalities	• perception patterns • organizational vocabulary • models	• legal norms • decision premises • standard operating procedures	• formal authority • work design • bureaucratic procedures • planning tools	• money • raw materials • information-/production technology

Dimensions of legitimizing action	Communicating	Sanctioning, Justifying	Authoritative Use of Power	Allocative Use of Power

Legitimizing

Figure 4. Legitimation as structuration.

further action. Contrary to "everyday structuration," structures in organizations are created intentionally and reflexively, although this does not rule out the possibility of unintentional consequences of action in organizations. Especially repeated and routinized practices in organizations (e.g., administration of compensation systems, reporting, budgeting, etc.) require and result in structures which restrict action and thus enable it. For example, work time schedules and the norm of punctuality are restrictions of action necessary to enable coordinated behavior. In organizations, structuration is not just the natural by-product of human agency but is monitored reflexively: "Organization is structuration which has lost its naivety, its naturalness-reflexive structuration" (Ortmann et al., 1997, p. 315). Like any structure, organizational structure is virtual and dual, too. It only exists in the (inter-)action of actors in the organization or stored in human and organizational memories. Organizational structure is thus not the static counterpoint to organizational processes but structured processes.

Legitimacy from the Structurationist Perspective

Applying structuration theory to legitimation results in two modifications of prior writings on organizational legitimacy: the types and the process of legitimation. In two of the three types of legitimacy, convergence meets the eye. Suchman (1995), Stryker (1994), and Scott (1995) describe the normative and cognitive aspects of organizational legitimacy. To appear legitimate, organizational activities, structures, procedures, and actors must be understood and accepted. The third type could also be linked without too much force. Suchman (1995) and Stryker (1994) stress the instrumental aspect of actors calculating the costs and benefits of rule obedience or violation, and then acting in a way that meets their self-interest. Scott (1995) picks up that logic of instrumentality but adds the coercive mechanism to it. This is a very Giddensian way of reasoning: (cognitive and normative) rules can be resources for pursuing one's interests, but at the same time resources are needed to enforce the rules.

Legitimacy does not only have to be incorporated in structure but also has to be realized in action. According to structuration theory, the distinction of structure as something fixed and action as something dynamic does not make sense. Structure only comes into life when instantiated in social practices, thus the term structuration. And to reiterate: action means intervention in the flow of social life, and that means power. Legitimizing action can thus be described with the three interrelated dimensions of social interaction: communicating, sanctioning/justifying, and the use of power (both authoritative and allocative) (see Figure 4). As these dimensions are intertwined, legitimacy (as any aspect of structure) is not only a matter of moral rules, as legitimizing (as any kind of action) is not only a matter of justifying and sanctioning (Becker, 1996).

To legitimize (i.e., to justify and sanction action), the actors within and outside the organization use material and immaterial resources according to constitutive and moral rules. By drawing on these rules and resources in their recurrent practices, the actors reproduce the moral order. Legitimation efforts are not restricted to the use of norms, just as the reproduction of the domination order does not just involve the use of authoritative and allocative facilities to exercise power, or just as the interpretative schemes are not solely used to reproduce the cognitive order. Legitimizing encompasses all three dimensions, the communication of meaning by using interpretative schemes like organizational vocabulary or models, the reference to legal and organizational norms like decision premises, and the use of facilities like hierarchy, work design, planning techniques, budgets, or technology. Norms have to make sense and must be enforced. A given legitimate order justifies the domination order (i.e., the structural inequalities in the distribution of resources) as well as the cognitive order ("the way we see things around here"). Because the three dimensions of agency and structure are intertwined, the lifting out of legitimation in Figure 4 is only to indicate it as the focus of our analysis while taking into account that legitimizing action is not restricted to sanctioning and justifying but also involves the use of power and the communication of meaning (Ortmann et al., 1990).

HRM AND LEGITIMACY

Human Resource Management in a Structurationist Perspective

The categories of structure and agency can easily be found in HRM. HRM encompasses two basic tasks: systems design and behavior control (Berthel, 1995) that mirror the organizational dilemma of innovation versus routine. On the structural level, HRM systems for recruiting, appraising, compensating, and developing employees have to be created and refined. On the interactional level, these systems have to be implemented and handled (i.e., the tools, manuals, and procedures have to be used when making decisions about recruiting, appraising, compensating, and developing employees). A second aspect of HRM at the interactional level is leadership. As HRM is not only what the folks from the HR department do (their share is getting smaller and smaller following the trend of decentralization and putting HR tasks and responsibilities back to the line managers), leading subordinates on an everyday, face-to-face basis is a genuine HRM task of supervisors.

As we have learned from structuration theory, action always takes place within a structural context to which it is recursively related. Actions labeled as HRM practices, too, require and reproduce organizational resources handled according to the organizational rules of signification and legitimation. Viewed from that theoretical perspective, HRM systems are modalities of structuration. They are those formalized combinations of organizational rules and resources that have been intentionally developed and implemented to enable and restrict HRM practices in

```
                    Organization Structure
              Signification - Domination - Legitimation

                         HRM - Systems
              Interpretative Schemes - Facilities - Norms

         Communicating - Use of power - Justifying/Sanctioning
                    Managing Human Resources
```

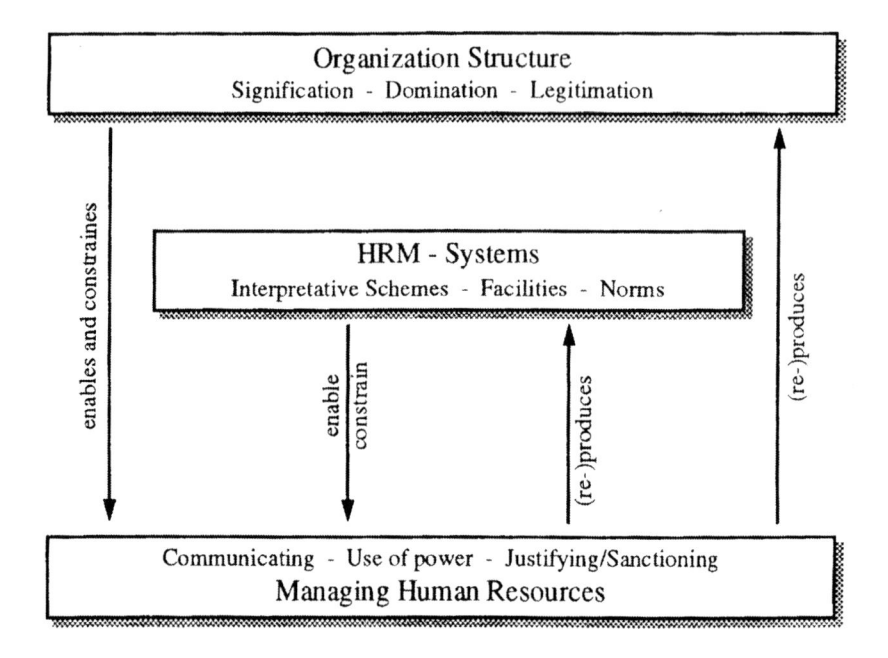

Figure 5. HRM systems as modalities of structuration.

organizations. Organizational actors like HR managers, executives, and supervisors can draw upon them in their interaction as interpretative schemes, as allocative and authoritative means of power (i.e., facilities), and as norms (see Figure 5).

The use of HRM systems as interpretative schemes can be observed in various aspects of managing human resources. That HRM serves functions other than what it purports was proposed by Trice, Belasco, and Alutto (1969), who argued that the primary role of HRM was ceremonial, underscoring the organization's adherence to desirable social values such as social justice, employee welfare, and industrial democracy. HR planning and programming covers all areas of HRM, from recruitment and selection to training and appraisal (e.g., Schuler & Huber, 1993). Personnel planning tools for HR analysis do not depict an objective reality but are means for defining and interpreting the objects of analysis. This is also true pertaining to quantitative instruments of HR planning which seem to be especially objective and impartial. The choice of a particular planning instrument is at the same time the choice of a specific definition of the problem and of the relations between variables considered relevant while ignoring others. By reducing complexity, this procedure also reduces uncertainty for the planners. Strategic HR planning especially serves to orient operative planning and implementing activities by making sense of numerous possible and perhaps confusing alternatives, directions, and focal points.

Selection devices are aimed at gathering information on the knowledge, skills, and abilities of an applicant, and to compare it with the requirements of the job. Neither the job requirements nor the qualifications however are objective realities that just have to be detected and measured properly (however technically difficult it may be), but the results of precisely that very same "measuring" device. Whether some knowledge, skill, or ability is regarded as qualification does not lie in the nature of these characteristics. For the time being, they are just some knowledge, skill, or ability. They only become a qualification if perceived through the lenses of job analysis and selection devices, or in other words, by employing these HRM tools as interpretative schemes.

The same holds true for appraisal systems. Like qualification, performance is no objective entity but the result of applying particular performance evaluation techniques. They focus the attention of the evaluator to predefined aspects like work outcomes, behaviors, and/or traits of the ratee, and help to recognize good and bad performance. For example, the mere observation that an employee is talking to and laughing with a colleague from another department during working hours does not in itself bear the information whether this is hanging about or fostering interdepartmental working relationships in order to improve cooperation in future projects. Appraisal techniques help to make sense of that ambiguous observation.

Training and development does not only upgrade the knowledge, skills or abilities that the employees need now or in the future, but also provides them with cues about "how we see the things around here." Sending an employee to some off-the-job training can have very different meaning. It might be regarded as a reward for outstanding performance or long loyal service, or as a punishment for having made mistakes, signaling that one might be in danger of being on the scrap heap soon. Statements of training policies, the contents of the training, and the trainers themselves (e.g., are they high ranked managers who initiate the trainees into the secrets of the company, or some MBA greenhorns recruited to execute some standard training design?) convey the meaning of a particular training program.

HRM systems are normative as well. They contain legal and organizational norms that actors can utilize to manage human resources. One important norm incorporated in all HRM systems is rationality. It is a widely accepted and expected standard of for-profit and, with some reservations, non-profit organizations. HRM practices that operate rational HRM systems are very likely to be regarded as legitimate. Rational HRM systems and the decisions and actions flowing from them are equipped with procedural justice. In many cases it is not rationality per se but rationality *myths* (Meyer & Rowan, 1977) that is sufficient for conferring legitimacy. For example, selection interviews are used in virtually every hiring process in spite of the fact that the resulting decisions do not substantially exceed random selection (Finzer & Mungenast, 1992). But a hiring decision without an interview would hardly be regarded as legitimate.

There are other norms than rationality that are ingrained in HRM systems. Recruiting policies state whether to recruit from the internal or the external labor market. For example, the norm of "promotion from within" supersedes the norm of filling the vacancy with the most qualified person available, because it eliminates external job applicants (one or more of whom may be more qualified than any internal applicant) from consideration. It is of course justified with rational criteria as it runs the risk of being accused of nepotism. Hence, it is argued that only internal labor markets provide the organization with the qualified and committed workforce it needs, or with opportunities for growth and that enhances morale of the workforce. Appraisal systems contain the norm of fairness. The mere existence of an appraisal system signals that HR decisions are achieved objectively, impartially, systematically, and not arbitrarily. But appraisal systems also convey norms of how to arrange the relationship between the supervisor (in most of the cases, the evaluator) and the subordinate (in most cases, the ratee). However they might conduct their relationship, they are reminded on a regular basis that it is the right and obligation of the supervisor to judge and the duty of the subordinate to accept being judged by the boss. Townley (1993) elaborates that appraisal systems are devices of discipline, albeit from a Foucaultian perspective. Career systems have explicit and (more often) implicit rules about what it takes to climb up the hierarchical ladder in the organization. Scott-Morgan (1994) describes some of these tacit career rules. These norms can be used to justify one's career-related behavior (e.g., "No offense, old pal, but you know it's every (wo)man for him/herself").

The notion that HRM are tools to exercise power and domination should come as no surprise (e.g., Legge, 1995). HRM is the very function to manage, that is, influence and command people in the organizations. However, this view is rarely presented in HRM textbooks. HRM systems are (among other things) devices to allocate resources in organizations, like money, time, information, and people. HRM systems that are sophisticated and difficult to handle call for experts to manage them. These experts control a "zone of uncertainty" (Crozier & Friedberg, 1980) and thus gain power. HR forecasting tools and the like can be employed in budgeting to control financial resources because "HR demand" is not an objective quantity but a political outcome. As mentioned before, appraisal systems are not only means to measure performance but also to exercise power and discipline, both on the evaluator and the ratee. Systematic performance appraisal (i.e., as far as it exists in the organization) is the basis of a number of resource allocation decisions, like pay increases, promotions, status, developmental opportunities, challenging assignments, and so forth. And to end with the most obvious, compensation systems, too, are means to allocate resources and thus to exercise power.

Interpreting HRM systems as modalities of structuration in organizations focuses on two kinds of recursive relations. First, systems and systems operation mutually constitute each other. HRM systems enable the management of human

Legitimacy in Human Resource Management 65

Table 2. HRM's Questioned Legitimacy

Main Issues Raised According to Suchman's (1995) Types of Legitimacy

Pragmatic Legitimacy	
Exchange legitimacy	• Too managerial: Evidence has shown that workers rarely reap the benefits of improved organizational performance (e.g., Barkin, 1989)
Influence legitimacy	
Dispositional legitimacy	• Advocates of HRM (academics, line managers, personnel managers) are driven by self-interests (e.g., Legge, 1995, the hyping of HRM)—research grant, publications, university department.
Moral Legitimacy	
Consequential legitimacy	• Little or conflicting evidence on impact on organizational performance (e.g.,Guest & Hoque, 1994) • In reality, HRM is mainly a cover-up for the intensification of the labor process, regardless of its rhetoric (e.g., empowerment, participation but more intensified, more subtle control from the top) • Bleakhouse: claims that HRM substitutes for unions in protecting employee interests are not true (Sisson's bleakhouse, 1993; Guest's black hole, 1995 in Legge, 1995) • No evidence of wide use, either across firms, or across different employee groups (Godard, 1991)
Procedural legitimacy	• Strategic HRM is too rational/unitary—strategy formulation is often an emergent process, involving various political interests (Legge, 1995—matching strategy approach is problematic; Boxall, 1996) • Too prescriptive and a one best way approach, applied universally • Downsizing has created anorexic organizations; eroded commitment of employees that is crucial in today's competitive world
Structural legitimacy	• Much of what HR departments can do can be done by outsiders, and even more efficiently (Adams, 1991) • HRM is the responsibility of line management, and not just some staff people (Keenoy, 1990—not the exclusive responsibility of the personnel function)
Personal legitimacy	• HR specialists are too bureaucratic, not innovative enough; they do not have the necessary skills, competence, outlook, attitudes, that is needed by organizations (Guest, 1991) • Lack senior rank (Hendry, Pettigrew, & Sparrow, 1988)
Cognitive Legitimacy	
Comprehensibility	• There is confusion in definitions—just exactly what is HRM? (Boxall, 1993) by logical extension, what does effectiveness of HRM mean? • HRM is atheoretical (Wright & McMahan, 1992) too focused on individual practices (Boxall, 1993) • Fundamental employer-employee conflict limits effectiveness of HRM (Godard, 1991)
Taken-for-grantedness	• It's just a trendy new name for personnel management—nothing new, old wine in new bottles, glitzy packaging (Armstrong, 1987) • It's just a fad—eventually it too will fade away as nothing being significant enough

resources by providing interpretative schemes, facilities, and norms. In turn, by operating HRM systems, actors reproduce them because otherwise these systems would remain virtual and without impact on action. But as the rules inherent in HRM systems allow and call for negotiated interpretation when drawn upon in interaction, this reproduction does not lead to identical reproduction of the systems. Second, because managing human resources by using HRM systems does not take place within a vacuum, implementing and operating these systems is enabled and restricted by organizational rules and resources and in turn reproduced by them. To manage human resources, actors utilize other organizational modalities like business goals and strategies, legal requirements, symbols, decision premises, budgets, work design, decision and operating procedures, too. Thus, it is more likely that the cognitive, moral, and domination order of the organization will be reinforced and maintained by the use of HRM systems and less likely that it will be altered. But we can think of intentionally introducing new, incompatible HRM systems (or changing existing ones) in order to trigger change of organizational rules and the distribution of resources, respectively.

But there are those who subscribe to other interpretative schemes and norms, and it is to them, or rather, to the issues they have raised, that we turn to now.

The Nature of HRM's Questioned Legitimacy

The writings relevant to the question of HRM's legitimacy can be categorized into (1) those that criticize HRM, including those that propose an analysis as to why such criticisms developed, and (2) those that address the issues by providing arguments or evidence to the contrary, or suggesting ways to deal with the criticisms. We first examine the issues that have been raised against HRM, because by doing so, we would be in a better position to address the questions. We also discuss and evaluate the efforts that have so far been made to resolve the ongoing debate.

Using Suchman's (1995) types of legitimacy, Table 2 outlines the major issues that bring into question HRM's legitimacy. Enumerating these points in this manner is not intended to imply the independence of these issues. In fact, interconnections can be made, as we shall show later. Although not all of the ideal types can be properly illustrated (i.e., influence legitimacy), Suchman's (1995) typology however is still useful as it enables us to cover the full range of issues that have been raised so far. We also do not assume these issues are of equal weight in their capacity to affect legitimacy. Some might indeed be of paramount importance that priority need to be given to it. Nonetheless we note Epstein and Votaw's (1978) observation that "erosion takes place slowly and often almost imperceptibly" (p. 77) such that what seems to be trivial at the moment might in fact be the one that deals the crucial blow. Obviously too, some issues may appear to be more of a concern primarily only to particular groups or stakeholders.

A closer examination reveals that these issues stem from the question of definition and the corresponding evidence that supports the claims made in accordance with the definition. Different conceptions of what HRM is (i.e., its nature) and what it is meant to do (its domain) abound, and this gives rise to the debates that the field of HRM is undergoing. One's views about HRM in essence reflect one's values and expectations, and this becomes the foundation or starting point of one's subsequent actions and opinions. In other words, different interpretive schemes and norms are being used to make evaluations in regard to legitimacy.

Lees (1997) had argued that there are currently two contrasting positions governing the views with respect to HRM. Under what he calls the product market position, internal control and economic efficiency are the chief concerns of businesses because of the imperative to remain competitive, with the firm's human resources being offered as the only advantage that firms may have left over its competitors (e.g., Pfeffer, 1994; Wright, McMahan, & McWilliams, 1994). It is within this position that we locate the prevailing definition of HRM that has lately triggered much of the controversy we are seeing. The labor market position on the other hand puts stress on non-economic considerations such as social justice. Lees proposed a third position which he argues as being both appropriate and realistic for HRM to focus on: the legitimacy market, which acknowledges the existence of different expectations and standards that need to be met. Although Lees depicted the legitimacy market as different from the first two but overlapping (i.e., capturing the notion that these perspectives are not inherently in conflict), the concept of legitimacy, as discussed in the literature, does encompass the value systems reflected in both the product market and labor market positions. Be that as it may, we can use Lees' notion of contrasting market positions to group the issues in Table 2, and make sense of the interconnections among these seemingly disparate issues.

As Lees (1997) had observed, "...internal efficiency and product market considerations are all part of the everyday rhetoric of management—a common and almost universally shared language.." (p. 232), or by Hendry, Pettigrew, and Sparrow (1988): "...firms that have made developments in their HRM have done so under the pressure of competitive forces....much of the detail of the what, why and how of transformation in HRM is dictated by considerations inner to the firm" (p. 41). Similarly, Legge (1995) traces the evolution of this view of HRM to the competitive business environment as well as concurrent innovations in management techniques such as TQM, team-based work systems, and decentralization. This is the predominant environmental and organizational context that shapes HRM actions. It is within this product market position that the critical issues primarily relate to personal legitimacy and structural legitimacy, because these are seen as the constraints to achieving the goal of building human resources as the firm's competitive advantage.

There are those who find the prevailing definition of HRM problematic either because they find little or contradictory empirical support for its claims, or because they hold a different view as to what it should be, the latter giving rise to the observations of confusion of or inconsistent definitions of HRM (e.g., Boxall,

1993; Lees, 1997). Not only is there a lack of strong evidence as to its wide usage and promised benefits, but that there might in fact be adverse consequences, which need not necessarily have been intended. In effect, critics say, the way it is conceived remains "good" only on paper, but not in reality. Awareness of negative outcomes of course reflects the observer's values and expectations used as one's interpretive scheme. For example, what one would present as "empowerment," another would interpret as "making someone else take the risk and responsibility," or "core and periphery" as "reducing the organization's commitment" (Sisson, 1994). So from this perspective, the critical issues lie in the areas of pragmatic legitimacy, consequential legitimacy, procedural legitimacy, and cognitive legitimacy, and the personal and structural legitimacy issues arising from the product market position would either disappear, either because they are no longer relevant or need to be redirected.

Countering the Criticisms

Some of the charges outlined above are slowly being addressed. For example, empirical evidence with respect to HRM's impact on organizations have begun to be available (e.g., Becker & Huselid, 1998), although it is limited to the predominant definition of HRM, and does not effectively address those who hold a different view of HRM. One of the problems is also conflicting evidence that is difficult to resolve because of the lack of theory. However, some have begun to address this issue as well. For instance, Godard (1991) argues that the observed "limited adoption and effectiveness" of the so-called progressive HRM paradigm is not simply a sign of irrationality on the part of organizations, underscoring the fact that under certain conditions, it is only reasonable that its claims will not be realized. The same message is relayed by Stone and smith's (1996) contingency theory of human resource management devolution. Several scholars (see Legge, 1995) have also started to identify different typologies of HRM systems, recognizing again that there are different ways of managing human resources, possibly because that is what is called for by the situation.

Within the predominant view of HRM, the area of what competencies are required and how to develop such competencies have also been addressed (e.g., Ulrich, 1997).

Where to, HRM?

Legitimizing HRM in Organizations

Legitimacy is a resource that is provided to HRM by its constituents. The constituents judge and sanction the goals, activities, systems, procedures, and sometimes (relevant) actors of HRM. Legitimation strategies are actions of HR managers who attempt to justify the various aspects of HRM by the application of

Table 3. Legitimation Strategies in HRM

Modality	Instrumental Legitimacy	Normative Legitimacy	Cognitive Legitimacy
Interpretative Schemes	• Explain to constituents their benefits from HRM • Use "utilitarian" language	• Explain the appropriateness of HRM evaluated by accepted norms • Describe and explain change of values and norms	• Select labels • Touch well-known models (contagion) • Advertise HR innovations
Norms	• Justify utility as a proper organizational norm in evaluating work relationships	• Justify HRM by pointing to legal and organizational norms and practices • Select favorable norms	• Justify HRM innovations or failures with accepted norms • Relate to legal norms
Facilities	• Conform HRM to constituents' demands • Coopt constituents • Safeguard mechanisms like ombudspersons, grievance procedures	• Conform HRM to constituents' norms • Change constituent' norms by advertising, persuasion, rewards, threats of sanctions • Allocate budgets	• Conform HRM to constituents' models • Formalize HR operations • Institutionalize through persistent practices • Allocate budgets to promote new practices

cognitive schemes, norms, and facilities. These legitimation endeavors can be aimed at different outcomes (i.e., to gain, maintain, or repair legitimacy) and focused at different dimensions of legitimacy (e.g., cognitive, normative, and instrumental). The constituents likewise sanction by using norms, facilities, and interpretative schemes. They draw on norms that are meaningful to them and reward or punish HR managers by providing or withdrawing support. The legitimacy they confer rests on three bases: meaning, morality, and interests.

Talking about legitimation strategies relies on the premise that cognitive and moral rule sets are contingent, multiple, and overlapping, and therefore allow and require their selection and interpretation. Otherwise, legitimation strategies would be restricted to conforming to the one and only rule set, or being refused any legitimacy at all, which could hardly be labeled a strategy.

Table 3 summarizes the strategic options for HR managers to gain, maintain, or repair the legitimacy of HRM. The primary modality for each dimension of legitimacy is indicated by a bold frame. The potential for these strategies to be effective is demonstrated by an empirical study by Galang and Ferris (1997) on the HR department's use of symbolic actions to gain power within the organization. Although the thrust of that study was in relation to gaining power within the organization, the notion of symbolic actions to influence perceptions and construct the "reality" on which other organizational actors base their decisions and actions is relevant to our discussion. Galang and Ferris (1997) found that regardless of the existence of objective bases of power, HR departments were still able to gain power by engaging in such symbolic actions.

The thrust of cognitive legitimation is to make HRM goals, systems, practices, and so forth more meaningful, appear comprehensible, and taken-for-granted to the constituents. The primary way is to use appropriate interpretative schemes. One option is to select labels for HRM the constituents know well and probably take for granted. It is a matter of vocabulary. If the organization is sales-driven, then HR activities should be expressed in sales terms like customerization, meeting client demands, and so forth. In a finance-driven organization, HR accounting might be the right word. In times of TQM, seeking certification of HRM processes leads to meaningful labels for describing proper and modern managerial activities. It takes advantage of the legitimation process of contagion where "highly institutionalized aspects of organizations are simultaneously legitimate and a source of contagion" (Zucker, 1988, p. 38). Other institutionalized elements of the organization the HR manager may wish to influence are organizational mission and strategy (and then call their activities strategic HRM). To maintain the legitimacy once gained, HR should not change their labels too often. Although innovation and change are often praised in management circles, not changing things for some time helps to attain that taken-for-grantedness that will provide legitimacy for quite a long time. To advertise new or existing HRM systems and practices means to offer interpretative schemes to the constituents. This might be done in articles in the corporate newspaper as well as in everyday small talk, preferably with opinion leaders. Using simple accounts and avoiding an indigestible HR jargon helps the constituents to understand, remember and use these labels when talking about HRM. Further, HR practices that are presented as "progressive" may influence shareholders (current and prospective) to infer a better run company, who then might invest more in the organization, thus leading to higher stock prices.

In the case of repairing legitimacy, explaining the problem or failure by laying down the facts and reasons for the problem, or uncovering the contradictions HR managers have to deal with, or telling about the complex and non-linear cause-effect relationships in HRM might improve the constituents' ability to understand the complexity of the situation, but might well backfire because of the lack of simplicity.

Interpretative schemes need not only be comprehensible, they also have to be acceptable. HR managers can attempt to justify the (new) way to describe and explain their work by relating it to legal and organizational norms. For example, if "quality" in HRM does not necessarily mean sophistication or comprehensiveness but "meeting the customers' needs," then it needs the concept of TQM to lend its legitimacy to this change of meaning. By the same token, selecting labels is a matter of positive connotations. Instead of just treating all employees equal HR managers might install the position of an employment equity officer, outplacement programs instead of just helping the persons fired, or career management programs instead of advising employees for their future. If the organization sees itself as a pioneer (maybe as part of its strategic orientation), then changing HRM systems and practices and adopting latest fashions nearly need no more extra justification. In addition, new HRM practices, tools, or policies can be justified by referring to legal requirements.

It takes action as well as talk to gain cognitive legitimacy. Instead of selecting or changing the accounts, HRM structures and practices themselves can be altered to conform to the constituents' models and schemes. This can be done by benchmarking other (non-)HR departments and imitating their standards. Formalizing HR operations, too, can contribute to cognitive legitimation because it is a visible form of standardization, institutionalization, and hence legitimation. The potential downside of this strategy is that something unwelcome becomes taken for granted, viz. the image of HR managers as bureaucrats. Institutionalizing HRM practices by persistence can lead to taken-for-grantedness but this strategy is better suited at the interorganizational level (e.g., for professional associations). The promotion of new labels for describing (and inevitably simultaneously evaluating) HRM can be fostered by allocating budgets to advertise, distribute, and popularize these interpretative schemes.

Reference to widely accepted norms and standards is the main road to normative legitimacy. Decisions, procedures, policies, and practices in HRM will be regarded as desirable, right, and just if they can be related to legal and organizational norms and practices. Within certain limits, HR managers are free to select the norms to which they want to conform. But too much freedom of choice might lead to a very fragile legitimacy. HRM is constantly confronted with a lot of different constituencies with different, sometimes contradicting ideas of what should be. Managers, unions, and white-collar and blue-collar workers are very likely to evaluate HRM activities and outcomes from different perspectives. Even if they agree on the label they might have different ideas about the actual meaning of quality, innovation, cost reduction, or social responsibility. From a legitimation point of view, HR managers must behave opportunistically and conform to the norms of the group that controls the most relevant resources, but they can hardly neglect the claims of the other because they need their support and collaboration as well to fulfill their mission.

The adherence to norms must be recognized and realized. The contribution of HRM to organizational norms is seldom self-evident but has to be shown and explained to the audience. It is not enough that the HR managers know how they contribute to organizational goals and strategies, how much they take into account social responsibility and equal treatment, or how often they simplify HRM tools and procedures. To gain normative legitimacy, the constituents have to be aware of these valued accomplishments. Sometimes the relevant norms and values change without being noticed by powerful groups in the organization. Here HR managers have to find the balance between conforming to the (partially outmoded) norms represented by these groups and the yet-to-be-noticed norms and values the HR managers regard as relevant for their success. Consulting outside experts might add the required credibility, even if their report does not differ from what the HR managers would say.

There are various ways to realize conformity to organizational and societal norms. HR managers can produce highly valued outcomes like cost reduction, quality improvement, quick delivery of services, or social balance. They can promote the creation of ethics programs or "ethics officers," as is being done in many organizations today. They can reengineer their processes and restructure the HR department. They can link their HRM procedures and tools to existing institutions like accounting, strategic planning, or legal rules, and thus infect it with their respective legitimacy. They can replace some of their colleagues and revise some of their practices. They might disassociate from persons by replacing the guilty person or the scapegoat, from practices and procedures (e.g., omit sexist questions in selection interview, remove promotion criteria that clearly discriminate against minorities, or simplify appraisal procedures). The second basic option is to change the norms they are compared with by advertising, persuasion, offering rewards, and credible threatening of sanctions. This requires that the HR managers control sufficient material and immaterial resources which is a function of the "zones of uncertainty" (Crozier & Friedberg, 1980) they control.

Striving for instrumental legitimacy is to make HRM seem to be useful and proper for the realization of the constituents' self-interests. The direct way is to conform to the constituents' demands by responding to their needs, that is, developing a service and marketing orientation to HRM by providing personnel in the required quantity, quality, time and place, doing the dirty jobs for the line managers, simplifying and reducing bureaucratic procedures, and so forth. Russ, Galang, and Ferris (in press) have developed an extensive model of HRM's boundary spanning function to explain the rise of HRM's strategic importance to the overall organization, and the role of the federal government in contributing to the increase in HRM's power has previously been noted (e.g., Jennings, 1994).

Another way to achieve instrumental legitimacy is to coopt constituents, such as by involving unions more than legally required, creating project groups with representatives of all relevant parties when implementing/reforming HR systems, staying in touch with recent developments like managerial fads (e.g., lean management, business reengineering), attending conferences, or reading non-HRM popular business journals. Constituents, like norms, are not naturally given. HR managers can attempt to select (some of their) constituents by identifying the well meaning actors, coopting these friendly constituents and intensifying relationships with them (e.g., installing an HR advisory committee).

To maintain legitimacy, it is helpful to protect exchanges by building up trust through demonstrating the HR department's reliability by delivering on time, or keeping (side) arrangements. To make up ground, instrumental legitimacy can be enhanced by creating safeguard mechanisms like installing the position of an ombudsperson or grievance procedures to minimize the danger of repeated damaging events, and to convey the message that future exchanges will not be disturbed again.

Instrumental legitimacy needs normative and cognitive backup. Exchange relationships where the participants attempt to realize their interests needs to be related to norms that justify utility, instrumentality and usefulness as acceptable criteria for evaluating HRM. This is more likely within the framework of 'hard HRM' (e.g., quantitative, economic, rational), as opposed to its "soft" version (e.g., human-relationship oriented aspects, as discussed by Legge, 1995). If the HR department has been restructured as a profit center, it is much easier for the HR manager to focus on (economic) exchange relationships. As with the normative dimension, instrumental legitimacy, too, needs interpretative schemes to be communicated and understood. HR managers have to explain their usefulness for the constituents. For example, they have to argue that the new appraisal system will lead to better HR decisions without putting too much bureaucratic burden on the shoulders of the supervisors, or that the existing compensation system provides fair pay and enhances the firm's competitiveness at the external labor market. It is advertising for products, services innovations, accomplishments of the HR department and doing impression management to create a favorable image. To build up their reputation, HR managers might refer to prior successes, to services delivered, to well accepted and effective HR systems, or to offers for participation and involvement.

DIRECTIONS FOR FUTURE RESEARCH

Throughout this paper, we have noted areas that suggest the need for future research. The following, then, are additional thoughts for a future research agenda. As we have demonstrated in the course of our discussion, considerable attention has been paid to theoretical constructs of organizational legitimacy. Legitimacy of subunits of the organization, on the other hand, has received less attention, both theoretically and empirically. Are the processes that result in the creation, maintenance, and repair of a subunit's legitimacy simply the same as at the organizational level of analysis, or are the dynamics different? Are there additional factors that are critical within the organization? Are strategies that are successful at the organizational level equally appropriate at the subunit level? Considering the dilemma of a HR manager who wants to improve the perceived legitimacy of the HR department, for instance, is the successful strategy one that focuses on managing impressions of key internal or key external constituencies? Or, are both equally important? Are legitimizing strategies equally effective on both sets of constituencies? If those most in need of engaging in legitimizing strategies are likely to suffer from the "self-promoter's paradox" (Suchman, 1995), is establishing or repairing legitimacy in such cases simply impossible, or are more subtle strategies possible? Clearly, both the conceptual and empirical aspects of subunit legitimacy are far from being resolved.

We also need to understand the relationship between organizational (and subunit) legitimacy and organizational reputation. As noted earlier, it is possible that legitimacy is a necessary, but not sufficient, condition for a superior reputation. It seems likely that a paradox exists in the legitimacy/reputation relationship: Legitimacy depends on conforming to accepted norms, whereas an excellent reputation may be a result of some aspect of uniqueness. If this is true, it can be argued that reputation can be a source of competitive advantage, whereas legitimacy can, at best, only be a source of competitive parity (see Barney, 1986). If Fombrun (1996) is correct and reputation's importance as an assessment tool continues to increase, then understanding the legitimacy/reputation relationship (at both the organizational and subunit levels) becomes more critical.

With regard to HRM specifically, monitoring of downsizing and outsourcing trends is warranted, particularly as some firms begin to quietly re-establish previously downsized HR departments. (It is interesting to note that downsizing itself may have been, at least in part, a legitimizing strategy at the organizational level, as firms imitated industry leaders in an attempt to signal to external stakeholders their determination to get "lean and mean.")

Further, we suggest a broader definition of HRM be employed, and that research in this area not be restricted to a definition that prescribes only one way of managing people in organizations, which has excluded the interest of others. In effect, we are subscribing to and reiterating a contingency approach. A contingency theory then can be developed that takes into consideration the circumstances under which one model of HRM is more effective. Hence, we avoid the risk of failing to deliver not because the people running it are incompetent, or HRM itself is branded as invalid. Some have started to work along this line (e.g., Stone & smith, 1996). The most important contingency for HRM would be the different stakeholders who hold different values, standards, expectations such that what the HR person needs to develop is skill in recognizing the stakeholders that matter, what their values are, and flexibility to try and balance the different interests. It may be of course that it makes no sense to talk about HRM systems (i.e., there will never be a coherence or internal consistency to HRM) simply because of the fact that there are different stakeholders whose expectations may sometimes be in conflict. But using the words of Giddens (1984), our role in the academic sector is to uncover more of the unacknowledged conditions of actions, and the unintended consequences of action, thereby providing a more informed basis for decision-making by expanding the actor's knowledgeability.

CONCLUSION

The primary purpose of this paper has been to discuss how a legitimacy framework potentially can help us to understand, and perhaps even to counter, recent attacks that have been made on HRM as a function, organizational department, and academic field of study. We have discussed the basic elements of Giddens' structuration theory, from the field of sociology, and suggested that our general understanding of organizational legitimacy can be enhanced by integrating these two conceptual areas. Further, we believe that structuration theory can stimulate the development of a more in-depth analysis of the legitimation process for HRM, specifically. With the controversies that have been raised with respect to HRM, it is not only a particularly appropriate undertaking, but also a timely one.

NOTES

1. Legge (1995) referred to the "three major groups, or human actors, that have a vested interest in hyping HRM: academics, line managers and, more ambiguously, personnel managers themselves" (p. 318). Our use of the term "allies" refers to them as well.

2. The terms "human resources management" and "personnel management" will be used interchangeably in this article, although some scholars and practitioners have distinguished them.

3. In analyzing HRM as a postmodernist discourse, Legge (1995) pointed to the factors responsible for creating the rhetoric of HRM, including those which she labeled "non-human actors:" operational management techniques, organizational design initiatives, information technology, labor legislation. Inasmuch as the term "actor" is normally applied to humans, we use "agents" instead as a more apt term to include both human and such non-human actors in the legitimation process.

REFERENCES

Adams, K. (1991). Externalisation vs. specialisation: What is happening to personnel? *Human Resource Management Journal, 1*, 40-54.
Allen, M.W., & Caillouet, R.H. (1994). Legitimation endeavors: Impression management strategies used by an organization in crisis. *Communication Monographs, 61*, 44-62.
Armstrong, M. (1987). Human resource management: A case of the emperor's new clothes? *Personnel Management, 19*, 31-35.
Arnold, S.J., Handelman, J., & Tigert, D.J. (1996). Organizational legitimacy and retail store patronage. *Journal of Business Research, 35*, 229-239.
Ashforth, B.E., & Gibbs, B.W. (1990). The double-edge of organizational legitimation. *Organization Science, 1*, 177-194.
Barkin, S. (1989). Human resources management examines itself and its limitations. *Relations Industrielles, 44*, 691-700.
Barney, J.B. (1986). Organizational culture: Can it be a source of sustained competitive advantage? *Academy of Management Review, 11*, 656-665.
Becker, A. (1996). *Rationalitat strategischer entscheidungsprozesse (Rationality in strategic decision processes)*. Wiesbaden: Deutscher Universitatsverlag.
Becker, B.E., & Huselid, M.A. (1998). High performance work systems and firm performance: A synthesis of research and managerial implications. In G.R. Ferris (Ed.), *Research in personnel and human resources management* (Vol. 16, pp. 53-101). Greenwich, CT: JAI Press.
Berger, P.L. (1981). New attack on the legitimacy of business. *Harvard Business Review, 59*(5), 82-89.
Berger, P.L., & Luckmann, T. (1966). *The social construction of reality*. New York: Doubleday.
Berthel, J. (1995). *Personal-management* (4th edition). Stuttgart: Schaeffer-Poeschel.
Boland, R. J. (1993). Accounting and the interpretive act. *Accounting, Organizations and Society, 18*(2/3), 125-146.
Bouchikhi, H., Kilduff, M., & Whittington, R. (Eds.) (1995). *Action, structure, and organizations*. Coventry: Warwick Business School Research Bureau.
Boulding, K.E. (1978). The legitimacy of the business institution. In E.M. Epstein & D. Votaw (Eds.), *Rationality, legitimacy, responsibility*. Santa Monica, CA: Goodyear Publishing Company, Inc.
Boxall, P.F. (1993). The significance of human resource management: A reconsideration of the evidence. *The International Journal of Human Resource Management, 4*, 645-664.
Boxall, P. (1996). The strategic HRM debate and the resource-based view of the firm. *Human Resource Management Journal, 6*, 59-75.
Brenner, L. (1996, March). The disappearing HR department. *CFO*, 61-64.
Brown, A.D. (1994). Politics, symbolic action and myth making in pursuit of legitimacy. *Organization Studies, 15*, 861-878.
Chaison, G.N., Bigelow, B., & Ottensmeyer, E. (1993). Unions and legitimacy: A conceptual refinement. In R. Seeber & D. Walsh (Eds.), *Research in the sociology of organizations* (Vol. 12, pp. 139-166). Greenwich, CT: JAI Press.
Crozier, M., & Friedberg, E. (1980). *Actors and systems: The politics of collective action*. Chicago: University of Chicago Press.
Della Fave, L.R. (1986). Toward an explication of the legitimation process. *Social Forces, 65*, 476-500.
Dowling, J., & Pfeffer, J. (1975). Organizational legitimacy: Social values and organizational behavior. *Pacific Sociological Review, 18*, 122-136.
Elsbach, K.D. (1994). Managing organizational legitimacy in the California cattle industry: The construction and effectiveness of verbal accounts. *Administrative Science Quarterly, 39*, 57-88.
Elsbach, K.D., & Sutton, R.I. (1992). Acquiring organizational legitimacy through illegitimate actions: A marriage of institutional and impression management theories. *Academy of Management Journal, 35*, 699-738.

Legitimacy in Human Resource Management

Elsik, W. (1998). *Personalmanagement als spiel (Personnel management as game)*. Stuttgart: Schaffer-Poeschel.

Epstein, E.M., & Votaw, D. (1978). *Rationality, legitimacy, responsibility*. Santa Monica, CA: Goodyear Publishing Company, Inc.

Finzer, P., & Mungenast, M. (1992). Personalauswahl (Personnel selection). In E. Gaugler & W. Weber (Eds.), *Handworterbuch des personalwesens* (Vol. 2, pp. 1583-1596). Stuttgart: Schaffer-Poeschel.

Fombrun, C.J. (1996). *Reputation: Realizing value from the corporate image*. Boston: Harvard Business School Press.

Frost, P.J. (1989). The role of organizational power and politics in human resource management. In A.N.B. Nedd, G.R. Ferris, & K.M. Rowland (Eds.), *International human resource management, Supplement 1, Research in personnel and human resources management* (pp. 1-21). Greenwich, CT: JAI Press.

Galang, M.C., & Ferris, G.R. (1997) Human resource department power and influence through symbolic action. *Human Relations, 50*, 1403-1426.

Giddens, A. (1976). *New rules of sociological method: A positive critique of interpretive sociologies*. London: Hutchinson.

Giddens, A. (1979). *Central problems in social theory: Action, structure and contradiction in social analysis*. Berkeley and Los Angeles: University of California Press.

Giddens, A. (1981). *A contemporary critique of historical materialism*. London: Macmillan.

Giddens, A. (1984). *The constitution of society: Outline of the theory of structuration*. Berkeley and Los Angeles: University of California Press.

Giddens, A. (1993). *New rules of sociological method* (2nd edition). Stanford, CA: Stanford University Press.

Giddens, A. (1994). *Central problems in social theory*. Berkeley: University of California Press.

Godard, J. (1991). The progressive HRM paradigm: A theoretical and empirical re-examination. *Relations Industrielles, 46*, 378-398.

Guest, D.E. (1991). Personnel management: The end of orthodoxy? *British Journal of Industrial Relations, 29*, 149-175.

Guest, D.E. (1995). Human resource management, trade unions and industrial relations. In J. Storey (Ed.), *Human resource management: A critical text* (pp. 110-141). London: Routledge.

Guest, D.E., & Hoque, K. (1994). Yes, personnel does make a difference. *Personnel Management, 26*, 40-44.

Hanft, A. (1995). *Personalentwicklung zwischen Weiterbildung und "organisationalem Lernen" (Personnel development between training and "organizational learning")*. Munich and Mering: Hampp.

Hendry, C., Pettigrew, A., & Sparrow, P. (1988). Changing patterns of human resource management. *Personnel Management, 20*, 37-41.

Hybels, R.C. (1995). On legitimacy, legitimation, and organizations: A critical review and integrative theoretical model. *Academy of Management Best Papers Proceedings*.

Ichniowski, C., Delaney, J.T., & Lewin, D. (1989). The new human resource management in US workplaces: Is it really new and is it only nonunion? *Relations Industrielles, 44*, 97-119.

Jennings, P.D. (1994). Viewing macro HRM from without: Political and institutional perspectives. In G.R. Ferris (Ed.), *Research in Personnel and Human Resources Management* (Vol. 12, pp. 1-40). Greenwich, CT: JAI Press.

Keenoy, T. (1990). HRM: A case of the wolf in sheep's clothing? *Personnel Review, 19*, 3-9.

Kochan, T.A., & McKersie, R.B. (1989). Future directions for American labor and human resource policy. *Relations Industrielles, 44*, 224-248.

Lees, S. (1997). HRM and the legitimacy market. *The International Journal of Human Resource Management, 8*, 226-243.

Legge, K. (1978). *Power, innovation, and problem-solving in personnel management*. London: McGraw-Hill.
Legge, K. (1995). *Human resource management: Rhetorics and realities*. London: Macmillan Business.
Martin, J. (1993). Inequality, distributive justice, and organizational illegitimacy. In J.K. Murnighan (Ed.), *Social psychology in organizations* (pp. 296-321). Englewood Cliffs, NJ: Prentice-Hall.
Meyer, J.W., & Rowan, B. (1977). Institutionalized organizations: Formal structure as myth and ceremony. *American Journal of Sociology, 83*, 340-363.
Mitchell, R.K., Agle, B.R., & Wood, D.J. (1997). Toward a theory of stakeholder identification and salience: Defining the principle of who and what really counts. *Academy of Management Review, 22*, 853-886.
Neilsen, E.H., & Rao, M.V.H. (1987). The strategy-legitimacy nexus: A thick description. *Academy of Management Review, 12*, 523-533.
Orlikowski, W.J. (1992). The duality of technology: Rethinking the concept of technology in organizations. *Organization Science, 3*, 398-427.
Ortmann, G. (1995). *Formen der produktion (Forms of production)*. Opladen: Westdeutscher Verlag.
Ortmann, G., Sydow, J., & Windeler, A. (1997). Organisation als reflexive strukturation (Organization as reflexive structuration). In G. Ortmann, J. Sydow, & K. Turk (Eds.) *Theorien der organisation* (pp. 315-354). Opladen: Westdeutscher Verlag.
Ortmann, G., Windeler, A., Becker, A., & Schulz, H. (1990). *Computer und macht in organisationen: Mikropolitische analysen*. Opladen: Westdeutscher Verlag.
Peach, L. (1995, June 29). Don't blame HR for the company's failure. *People Management*, p. 21.
Pentland, B.T., & Rueter, H.H. (1994). Organizational routines as grammars of action. *Administrative Science Quarterly, 28*, 484-510.
Pfeffer, J. (1981). *Power in organizations*. Marshfield, MA: Pitman.
Pfeffer, J. (1994). *Competitive advantage through people: Unleashing the power of the work force*. Boston: Harvard Business School Press.
Rao, H. (1994). The social construction of reputation: Certification contests, legitimation, and the survival of organizations in the American automobile industry: 1895-1912. *Strategic Management Journal, 15*, 29-44.
Richardson, A.J. (1985). Symbolic and substantive legitimation in professional practice. *Canadian Journal of Sociology, 10*, 139-152.
Richardson, A.J., & Dowling, J.B. (1986). An integrative theory of organizational legitimation. *Scandinavian Journal of Management Studies, 3*, 91-110.
Riley, P. (1983). A structurationist account of political culture. *Administrative Science Quarterly, 28*, 414-437.
Russ, G.S., Galang, M.C., & Ferris, G.R. (in press). Power and influence of the human resources function through boundary spanning and information management. *Human Resource Management Review*.
Schuler, R.S. & Huber, V.L. (1993). *Personnel and human resource management* (5th ed.). St. Paul, MN: West.
Scott, W.R. (1995). *Institutions and organizations*. Thousand Oaks, CA: Sage.
Scott-Morgan, P. (1994). *The unwritten rules of the game*. New York: McGraw-Hill.
Selznick, P. (1952). *Leadership in administration*. Berkeley: University of California Press.
Sewell, W.H. (1992). A theory of structure: duality, agency, and transformation. *American Journal of Sociology, 98*, 1-29.
Singh, J.V., Tucker, D.J., & House, R.J. (1986). Organizational legitimacy and the liability of newness. *Administrative Science Quarterly, 31*, 171-193.
Sisson, K. (1993). In search of HRM? *British Journal of Industrial Relations, 31*(2): 201-210.
Sisson, K. (1994). Personnel management: Paradigms, practice and prospects. In K. Sisson (Ed.), *Personnel management* (pp. 3-50). Oxford: Blackwell.

Stewart, T.A. (1996a, May 13). Human resources bites back. *Fortune*, p. 175.
Stewart, T.A. (1996b, January 15). Taking on the last bureaucracy. *Fortune*, p. 105.
Stjernberg, T., & Philips, A. (1993). Organizational innovations in a long-term perspective: Legitimacy and souls-of-fire as critical factors of change and viability. *Human Relations, 46,* 1193-1219.
Stone, T.H., & smith, f.l. (1996). A contingency theory of human resource management devolution. *Canadian Journal of Administrative Sciences, 13,* 1-12.
Stryker, R. (1994). Rules, resources, and legitimacy processes: Some implications for social conflict, order, and change. *American Journal of Sociology, 99,* 847-910.
Suchman, M.C. (1995). Managing legitimacy: Strategic and institutional approaches. *Academy of Management Review, 20,* 571-610.
Terreberry, S. (1968). The evolution of organizational environments. *Administrative Science Quarterly, 12,* 590-613.
Thomas, G.M., Walker, H.A., & Zelditch, Jr., M. (1986). Legitimacy and collective action. *Social Forces, 65,* 378-404.
Townley, B. (1993). Performance appraisal and the emergence of management. *Journal of Management Studies, 30,* 221-238.
Trice, H.M., Belasco, J., & Alutto, J.A. (1969). The role of ceremonials in organizational behavior. *Industrial and Labor Relations Review, 23,* 40-51.
Tsui, A.S. (1987). Defining the activities and effectiveness of the human resource department: A multiple constituency approach. *Human Resource Management, 26,* 35-69.
Ulrich, D. (1997). *Human resource champions: The next agenda for adding value and delivering results.* Boston: Harvard Business School Press.
Watson, T.J. (1995). In search of HRM: Beyond the rhetoric and reality distinction of the case of the dog that didn't bark. *Personnel Review, 24,* 6-16.
Weaver, G.R., & Gioia, D.A. (1994). Paradigms lost: Incommensurability vs structurationist inquiry. *Organization Studies, 15,* 565-590.
Weber, M. (1978). *Economy and society* (Vol. 1), edited by G. Roth & C. Wittich. Berkeley: University of California Press.
Whittington, R. (1992). Putting Giddens into action: Social systems and managerial agency. *Journal of Management Studies, 29,* 693-712.
Wilkerson, J.L. (1996). Is human resources a dinosaur? *The Human Resources Professional, 9*(4), 3-5.
Wood, D.J. (1991) Corporate social performance revisited. *Academy of Management Review, 16,* 691-718.
Wright, P.M. & McMahan, C. (1992). Theoretical perspectives for strategic human resource management. *Journal of Management, 18,* 295-320.
Wright, P.M., McMahan, G.C., & McWilliams, A. (1994). Human resources and sustained competitive advantage: A resource-based perspective. *International Journal of Human Resource Management, 5,* 301-326.
Zucker, L.G. (1988). Where do institutional patterns come from? Organizations as actors in social systems. In L.G. Zucker (Ed.), *Institutional patterns and organizations: Culture and environment* (pp. 23-49). Cambridge, MA: Ballinger.

CAREER-RELATED CONTINUOUS LEARNING:
DEFINING THE CONSTRUCT AND MAPPING THE PROCESS

Manuel London and James W. Smither

ABSTRACT

Career-related continuous learning (CRCL) is defined as an individual-level process characterized by a self-initiated, discretionary, planned, and proactive pattern of formal or informal activities that are sustained over time for the purpose of applying or transporting knowledge for career development. A theoretical stage model of career-related continuous learning is described. Central components of the model are pre-learning (recognizing the need for CRCL), learning (acquiring new skills and knowledge and monitoring learning), and application of learning (using, evaluating, and reaping the benefits of learning). The model draws on goal setting, control, and self-determination theories to explain learning processes. Research and practical implications of continuous learning for key human resource functions, including job analysis, personnel selection, training and performance appraisal, are discussed.

Research in Personnel and Human Resources Management, Volume 17, pages 81-121.
Copyright © 1999 by JAI Press Inc.
All rights of reproduction in any form reserved.
ISBN: 0-7623-0489-8

INTRODUCTION

Continuous learning has become central to employee career development for several reasons. Organizations increasingly expect employees to be responsible for their own development. The pace of change and the level of global competition have increased. Hesketh and Neal (1999) pointed out that in the past, jobs were assumed to be relatively static. As a consequence, once a job was mastered, employees continued to perform competently as long as they were motivated. Now, however, the increasingly rapid pace of change in job requirements due to organizational, competitive, and technological development means that employees must constantly show the capacity to engage in new learning as they cope with change (London & Mone, 1999). Also, the long-term psychological contract between firm and employee has largely disappeared, thereby creating the need for employees to prepare themselves for new jobs, employers, or careers in order to maintain employment security (Arthur & Rousseau, 1996; Bridges, 1994; Rousseau, 1995). Hall (1976, 1996; Hall & Mirvis, 1995, 1996) called this a "protean" career pattern based on the Greek god Proteus who could change shape as needed. Cianni and Wnuck (1997) noted, "Employees in the 21st century will periodically backtrack in their careers, moving from expert back to novice as they are required to have new competencies that may very well be in areas unconnected to their personal preferences" (p. 105).

There is a need for solid theoretical grounding in this area because (a) the term continuous learning is widely used in the popular literature and among HR practitioners, but like other terms that are widely used (e.g., empowerment), its meaning is ambiguous at best, (b) there is some confusion about whether continuous learning refers to a property of organizations (e.g., the term "learning organization;" Nonaka, 1994; Senge, 1990) or the activity of individuals (or both), and (c) the few studies that have been published to date (e.g., Holt, Noe, & Cavanaugh, 1996; Tracey, Tannenbaum, & Kavanagh, 1995) do not use a consistent theoretical framework to guide their selection of variables or methodologies. As a result, the area is likely to move forward only slowly and is not likely to capitalize on what is already known from years of research that has identified central psychological processes (e.g., goal theory) that can be used to understand and direct research about continuous learning.

This paper goes beyond prior conceptualizations of continuous learning by focusing especially on the elements that comprise continuous learning from an individual perspective, and the theoretical mechanisms that explain the relationships among these elements. We examine extant definitions of continuous learning and then try to state clearly what the construct is and what it is not. We propose a stage model of continuous learning and suggest the theoretical dynamics that underlie the model. In particular, we draw on goal setting, control, and self-determination theories to understand the processes associated with continuous learning. We show how these processes form different patterns of continuous learning.

We then examine the research and practical implications of continuous learning processes for key human resource functions, such as job analysis, personnel selection, training, performance appraisal, compensation, and human resource planning.

DEFINING CONTINUOUS LEARNING

The term "continuous learning" may be viewed on different levels. On a superficial level, it merely describes what everyone does on a daily basis. That is, we are all constantly processing new information and ideas. On a deeper level, the term, as we use it here, implies a deliberate and sustained effort to learn, a readiness and desire to acquire new knowledge and skills, actually engaging in activities that allow us to learn, and applying our increased knowledge and new and improved skills.

We define career-related continuous learning (CRCL) as an individual-level process characterized by a self-initiated, discretionary, planned, and proactive pattern of formal (e.g., institutional) or informal activities that are sustained over time and have the goal of applying or transporting knowledge (including tacit knowledge) for career development today or in the future.

Defining what career-related continuous learning is not will help characterize what we mean by the concept. According to our conceptualization, CRCL is not merely the constant accumulation of new information, nor is it learning for its own sake (e.g., for personal growth or self-actualization). This may characterize continuous learning in general or for other reasons, but not career-related continuous learning per se. CRCL is not an organizational phenomenon, although organizations may be able to leverage individual continuous learning to create a learning organization (Senge, 1990) or use core competencies (Hamel & Prahalad, 1994; Prahalad & Hamel, 1990) to their strategic advantage.

CRCL may vary along a several dimensions. Sometimes it is instrumental, meaning that it has a specific goal or relatively immediate applicability, targeted to specific problems as they arise. Other times the goal is distal and fairly general (e.g., "I want to be ready when my company goes global"). Sometimes the learning is over the long-term, as it is when a person works toward an advanced degree. Other times it is short-term, as it is when a person learns a series of new techniques, practicing and applying each one as it is learned. CRCL may be formal (as in an off-site training program), informal (conscientiously reading, discussing issues, and trying new ideas), or a combination of formal and informal activities. It may involve acquiring formal knowledge (e.g., new concepts in accounting or engineering) as well as tacit knowledge (e.g., ways to work effectively within the organizational culture to influence others) (Wagner & Sternberg, 1985), both of which may make an important contribution to career success.

Continuous learning is usually incremental in that continuous learners obtain and apply bits of new knowledge or skills over time. However, continuous learners may also engage in frame-breaking learning, for instance, to embark on a new occupation (e.g., a teacher who becomes an attorney, a finance manager who moves into marketing, or a news reporter who moves to creating internet Web sites). However, frame-breaking learning is not necessarily part of a pattern of continuous learning. A person may engage in training to switch occupations and then limit his or her learning activities. However, continuous learners are probably more open to exploring possible frame-breaking changes and ultimately tackling the challenge than people who are not continuous learners.

Some Examples

Some examples will help articulate what we mean by career-related continuous learning.

Consider the continuous technical learner in challenging, rapidly changing, competitive fields. The individual might be an engineer in the semiconductor industry who is always learning new techniques and applying them to the development of new products and making operational improvements (e.g., reducing cycle time).

Now, consider continuous learning in other occupations: A veteran actor continues to take lessons, is open to direction (and seeks out working with different directors), seeks different opportunities to apply her craft (different media and vehicles for expression), and shows improvement and versatility.

The director of international human resources for a telecommunications company travels the world to observe ('benchmark') the operations of other firms, actively works on learning new languages, acquires information about different cultures (their economies, values, educational programs, religious practices, and behavioral norms such as modes of negotiation and conflict resolution), and uses the information to shape the company's management practices as it establishes new operations and ventures.

A manufacturing worker on the shop floor becomes fascinated by the firm's continuous quality initiatives, identifies skill and knowledge gaps (e.g., difficulty working with control charts, statistics, and process flow diagrams) by proactively seeking feedback from co-workers, attends evening courses at a community college to acquire the needed skills and knowledge, volunteers for cross-functional teams (hoping to learn more by working with people from other areas), tries new manufacturing techniques, and is continuously on the lookout in book stores for new ideas.

A college professor keeps up with developments in her field and reads widely in a fairly broad range of disciplines, including general literature, and incorporates ideas from these diverse sources into her research and teaching. The academic maintains this level of learning in a new administrative position that provides opportunities for experimenting with new ideas for developing academic departments and improving students' educational opportunities.

A retail branch manager acquires increasingly deeper knowledge of marketing, personal selling, finance, economics, and human resources management by working toward an MBA degree, talking with an informal mentor, reading trade and business publications, and attending short courses funded by the store. The manager is known for trying new approaches to enhance rapid restocking of inventory and to improve the sales staff's responsiveness to customers. When the store merges with a large national retailer, the manager evaluates opportunities and has several excellent offers to move to different positions that promise further career growth.

A middle manager who has been with her firm for ten years begins to think of new career opportunities. She spends several years exploring the feasibility of purchasing a franchise as a side business, and possibly as a full-time occupation. She attends classes in small business development, works with an accountant and financial advisor to set goals for, and monitor, her investments, takes a part-time job in the type of franchise that seems promising, and eventually purchases a business.

A young podiatrist gets fed up with the red tape, standards, and low pay that health maintenance organizations are imposing on allied health professionals and begins to explore new career alternatives. She investigates the labor market in different fields by looking at classified job ads, selects several areas of interest to her that seem to have expanding career possibilities, and then talks to professors and professionals in these fields to determine what the work and rewards are like. She then enters a university program that will give her the credentials for initial entry into the field of human resources management and allow her to expand her education and career possibilities further in various areas of business if she decides to go on for an MBA.

A mechanic whose hobby is investing reads all he can about tracking the market and analyzing a firm's performance and growth and earnings potential. He applies some of the ideas he learned about cash flow, return on investment, and marketing to his garage and finds that competing on price and quality brings him more business.

Now consider some examples of continuous learning that are not career related and don't fall within our definition of CRCL.

A person reads one novel after another for pure enjoyment, and enjoys telling friends about his latest reading.

An individual is always trying new things and continuously pushes herself further, traveling to exotic places, trying new recipes, advancing in the martial arts, and/or reading popular self-development books.

A garment manufacturing worker is laid off and sent to a government supported retraining program to learn data entry skills and secures a job doing that work.

Some people may avoid continuous learning. For instance: A design engineer is assigned to a training course to learn a new software program, attends the class, and uses the material, but does not seek new ways of applying the knowledge and does not seek further training unless it's required.

A medical technician, doing reasonably well and satisfied with his progress, does not have the drive to explore new disciplines or ideas (i.e., he "sticks to the knitting"). He attends continuing education classes only when necessary to maintain his license.

Finally, the proverbial "couch potato" turns off his mind at the end of a stressful day by watching television, reading a mystery novel, or going to bed early.

EXPLORING THE COMPONENTS OF CRCL

In this section, we review the key components that comprise our definition of career-related continuous learning.

An Individual-Level Phenomenon

Although one may question whether organizations can learn, we focus here on the learning of individuals rather than organizations.

CRCL is the process by which one acquires knowledge, skills, and abilities throughout one's career in reaction to, and anticipation of, changing performance requirements (London & Mone, 1999). It may be manifest by training to do the current job better, development in anticipation of tomorrow's job requirements, or retraining for other job/career opportunities inside and outside the organization. Continuous learners are constantly on the lookout for new information about themselves and performance requirements that suggest learning gaps, and they are willing to devote resources (time, energy, and finances) to gain the education needed to close the gap and improve their performance. They monitor their own behaviors and recognize which behaviors and resulting outcomes are most favorable and desirable (London & Smither, 1999). Thus, CRCL is not a property of an organization; instead it is observed in the behavior of individuals. Although it may support organizational goals and be enhanced (or inhibited) by organizational resources or culture, it may also occur independently of the organization. The goals associated with CRCL may or may not coincide with organizational interests.

Career-Related

CRCL is not transcendental, meaning that it does not transcend job and career experiences. The intention is to apply new knowledge or transport existing knowledge to new settings. In all cases, it is job or career-related. However, CRCL is not learning directed purely at solving a problem at hand or participating in one-time, mandatory training provided by organizations. The learning may not be immediately applicable, but it must deal with potentially relevant (applicable) knowledge, skills, and abilities. The career-related continuous learning process contributes to career development, or has the potential (or intention/hope) to do so. As such, it has an imminent quality, one that is real-world based, tied to career or job requirements or opportunities.

Self-Initiated, Discretionary, Planned, and Proactive

A central element of continuous learning is the assumption that learning is self-initiated. People who engage in self-development seek and use feedback, set development goals, engage in developmental activities such as information acquisition and practice, and track their progress on their own (London & Smither, 1999). Some people may be more prone to self-development than others such as those who are high in self-esteem or sensitive to others' views of them (e.g., those who are high in public self-consciousness and self-monitoring). People can learn to be self-developers through cognitive strategies (e.g., mental imagery), self-management training (Frayne & Latham, 1987; Latham & Frayne, 1989), and through reinforcement for feedback seeking and other self-determined behavior.

Sustained Over Time, Cumulative, and Integrated

One cannot determine if a learning activity is continuous learning (CL) merely by looking at a snapshot view of it; instead, it must be seen in context and over time. As such, learning is evolutionary. The learning is cumulative and integrative, meaning that people build on their knowledge and combine their skills and knowledge for new, often innovative applications. For example, when we observe a group of people attending a single educational or training course, we cannot determine who is and who is not engaged in continuous learning. Some people may attend only the one course (and hence are not continuous learners); others may attend several unrelated courses merely because of their intrinsic interest in the various topics (and hence are not engaged in CRCL). The behavior of others in the class (when observed over time and in many settings) may reflect an interest in purposeful, sustained learning directed at career-related interests. These learners exhibit CRCL. Thus, CRCL cannot be measured by taking a snapshot of a learner in a single setting on a single occasion. Instead, we need to observe behavior over time and across settings to determine whether CRCL is present. (This component of CRCL has obvious implications for research methods.)

Pattern of Activities

CRCL is evident from a distinct, nonrandom pattern of activities. However, the pattern will vary from one person to another and from one time to another for a given person. Consider some of the following possible patterns.

Sequential

Cognitive, decision, and behavioral processes occur in a linear sequence. Individuals who exhibit this pattern assess environmental demands, evaluate their skills, consider goals and learning gaps that need to be filled to achieve these goals, establish plans for development, and carry out these plans while evaluating their achievements (both learning acquisition and goal accomplishments). People vary as to the speed with which any given sequence occurs depending on the time span of learning acquisition and application. (For instance, the process of deciding to get another degree, working on the degree, and then applying the new knowledge may take years.)

Iterative and Cyclical

Sequential development may be repeated creating a learning cycle. An individual may enact overlapping cycles for different learning and outcome objectives. Learning may lead to recognizing additional skills and knowledge desired or needed. Experimenting with new behaviors may lead to recognition of new directions for learning. This cycle may branch in new directions at different stages of the continuous learning process (e.g., learning gap recognition, learning, and evaluation), leading the individual to new directions and starting new cycles.

Serendipity

Unplanned events contribute to learning and may set off a more deliberate CL cycle. Unexpected events capture our attention and cause us to think more deliberately and deeply about our readiness to meet the demands of the situation. The unexpected event may prompt us to compare our abilities to the standards or demands of the situation (Carver & Scheier, 1990). This may lead to an understanding of our skill and knowledge gaps and motivate us to embark on learning that we wouldn't have engaged in if the event had not occurred.

Also, an unplanned event is a basis for new knowledge and skills. We find ourselves doing things that we never expected to do or thought we could do. We may even have avoided the event if we had the opportunity to do so because we felt unprepared. Thus, the serendipitous event forces us to learn and may encourage us to establish a more systematic learning process. So, for instance, the human resource manager who is suddenly assigned to an important labor negotiations may need to acquire knowledge in labor relations by reading as much as possible. This may motivate the individual to take a course or perhaps even pursue an advanced degree in labor relations, which in turn would lead to more job responsibilities involving labor/management issues and maybe even a career shift into the labor relations area.

Unorganized Learning

This "garbage can" model of learning may occur as individuals acquire new skills and knowledge, sometimes unwittingly, and then look for applications. Continuing the above example, the human resource manager who now has participated in a successful labor negotiations may seek other opportunities to negotiate—perhaps unrelated to labor relations. For example, the individual may get involved with negotiating contracts with clients to provide human resource services or perhaps designing a training program to educate managers in any are about how to represent management in a labor negotiations. This in turn could set off new career goals and associated learning goals.

In sum, self-determined, continuous learners take responsibility for their own development (London & Smither, 1999). They regularly assess their work environments to identify what they need to know. They develop alternative visions of the future to anticipate what they will need to know tomorrow. They seek feedback and assess their current skills and knowledge to determine learning gaps. They search for development opportunities and set learning objectives. As they learn, begin to apply their new knowledge and skills, and continue to assess job and career requirements, they appraise their progress and recalibrate or change their goals and learning behaviors. Hopefully, as they do so, the organization provides support (e.g., resources, encouragement, empathy, feedback) and opportunities for learning and applying the knowledge and skills.

CRCL is closely related to self-management. Self-managers define goals for themselves and then use self-directed strategies to accomplish them (Williams, Verble, Price, & Layne, 1995). They maintain focus on important tasks until they are completed, base their actions on clearly defined goals, and start and complete tasks in a timely fashion (Williams, Pettibone, & Thomas, 1991). Self-managers monitor and evaluate progress toward their goals on their own (Bandura, 1986).

In CRCL, people judge the gap between their goals and their current skills and accomplishments. People evaluate their goal discrepancies by comparing their behavior to standards. The standards may be idealized images of what they hope to be, the behaviors they need to engage in to achieve their idealized image, or the expectations and daily behaviors of others (Lavalee & Campbell, 1995). (This is the basis of control theory; Carver & Scheier, 1990.) People usually evaluate themselves on the basis of their daily behavior. However, disruptions in their daily lives and expectations direct their attention to higher level goals (e.g., asking themselves, "Am I following all the steps I need to achieve my goal" or

"Maybe my goal is not realistic"). Not meeting the standard induces people to try harder until the discrepancy is reduced significantly or eliminated, the goal or standard is changed, or they focus their attention elsewhere (Pyszczynski & Greenberg, 1987). People are reluctant to give up long-desired goals. They may intensify their efforts even in the face of continued failures, and they may feel considerable frustration and even depression before they relinquish their cherished goals.

Summary

So far, we have defined CRCL, considered some examples, and explored the meaning of each key component in the definition. We have suggested that CRCL is a process with several distinct cognitive, decision, and behavioral stages from pre-learning (i.e., recognizing the need for development and setting learning goals), to learning (i.e., actually acquiring the knowledge and skills), and finally applying the learning and reaping the benefits. The next section maps this process more closely by considering individual and situational antecedents of pre-learning, steps in the learning process, and performance outcomes.

Environmental and Organizational Factors	Personal Characteristics	Organizational Culture and Practices
Value migration Technological change Deregulation Contingent employment Reengineering Quality efforts Decentralization Empowerment Global expansion Downsizing and restructuring Mergers and acquisitions	Self-monitoring Public self-consciousness Feedback seeking Openness to experience Locus of control Self-efficacy Extraversion Proactivity Mastery orientation Cognitive ability Conscientiousness	Empathy Informational feedback Behavioral choices and their consequences Learning resources Learning climate Continuous learning is appraised Reinforcement for learning/innovation

Pre-Learning	Learning	Application of Learning
Recognizing the need for CRCL and setting learning goals	Acquiring new skills and knowledge and monitoring learning	Applying learning in a work context, evaluating learning, reaping benefits from learning

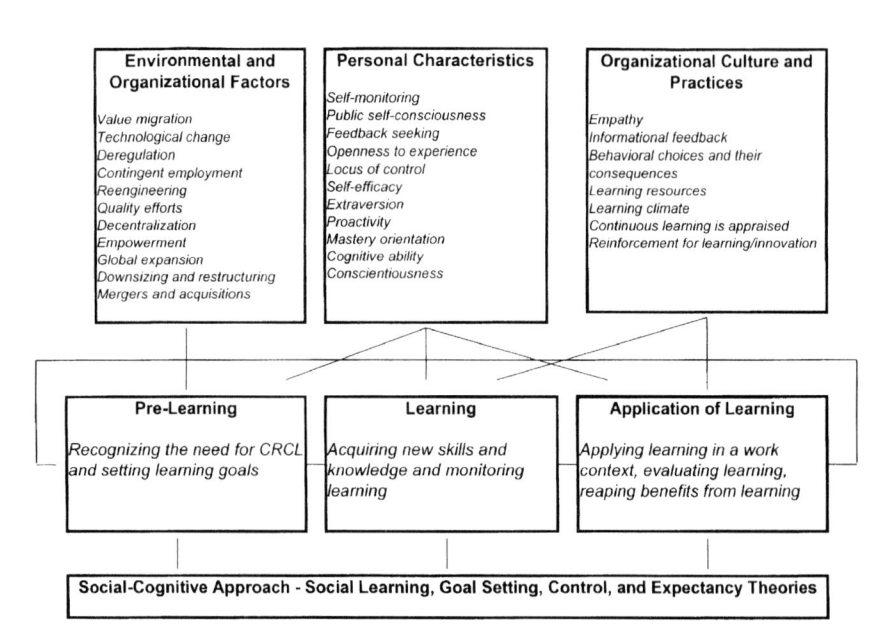

Social-Cognitive Approach - Social Learning, Goal Setting, Control, and Expectancy Theories

Figure 1. A model of career-related continuous learning.

MAPPING THE PROCESS

This section focuses on how people recognize the need for CRCL, set goals for learning, determine how they will learn, engage in the actual process of learning, monitor their learning, apply the learning in the short and long run, and evaluate the effects of learning (i.e., the outcomes of the application(s) over time). Recognition and expectancy give rise to goal setting and commitment to the learning process. This process is outlined in Figure 1. As we review each component of the learning process, we propose measures or indicators of each element. We begin in this section by examining the psychological theories that form the foundation for CRCL.

We adopt a multi-theoretical, integrative approach that draws on several theories of cognition and motivation underlying career-related continuous learning and multiple individual difference variables that may influence the cognition and motivation processes. The specific cognitive and motivational processes and individual propensities may vary from one person to another, emphasizing the equifinality and integrative nature of the sequence of events leading to, and maintaining, continuous learning. That is, people become continuous learners in more than one way, just as the way they engage in continuous learning may vary from one person to the next.

Expectancy theory (Vroom, 1964), control theory (Carver & Scheier, 1990), goal setting theory (Locke & Latham, 1990), and social learning theory (Bandura, 1986) provide useful conceptual frameworks for developing propositions about the nature of CRCL. For example, the expectancy that effort will lead to learning and that learning will have positive (intrinsic or extrinsic) consequences will be an important antecedent of CRCL. Such expectancies may be formed from prior experiences, supervisory promises and support, information about career opportunities and job requirements, role models (i.e., observations of others' experiences), or standards of the discipline (e.g., the requirement or expectation of participation in continuing education). That is, people need to feel that they can learn and that learning will be valuable.

Both goal and control theories emphasize the role of comparisons between standards or goals and current behavior or capabilities. Behavior is directed to reducing gaps between goals and current behavior. Goal theory tells us that CRCL will be most effective when it is conducted in the context of specific, challenging goals to which the learner will remain committed. Goal theory also directs our attention to the role of feedback, not only to identify skill gaps but also to guide CRCL after it is initiated.

Bandura's (1986) social learning theory emphasizes the importance of observation and modeling in learning and behavior. We often learn by observing others and the consequences of their behavior. Thus, if we observe a co-worker's behavior and see the co-worker reinforced for the behavior, then the co-worker's behavior can become part of our repertoire of potential behavior, even though we have not overtly performed the behavior ourselves. Social learning theory also emphasizes the importance of mastery experiences. That is, learners need opportunities to achieve success, overcome failure, and receive realistic encouragement to help them exert greater effort and overcome self-doubts.

In this sense, CRCL can be understood in the context of well-established conceptual models that explain learning, motivation, and performance. Central elements of these theories (i.e., discrepancies between current capabilities and goals, expectancies, instrumentality, goals, commitment, observational learning, feedback, mastery experiences) underlie much of the CRCL model we describe below.

A useful framework for conceptualizing these processes and related individual difference variables is the social-cognitive approach (Cervone, 1997). The social-cognitive approach, in contrast to the trait approach, posits that for a given construct, such as CRCL, there is no single psychological process that is CRCL, nor is CRCL an ingrained aspect of personality. Rather, stable patterns of CRCL emerge from interactions among multiple underlying mechanisms. These include skills, cognitions, and affective mechanisms through which people regulate behavior. So consistent, stable applications of CL to job and career situations are the result of interrelations among the multiple underlying mechanisms. We don't have to posit an independent individual trait or propensity called continuous learning.

In viewing themselves as continuous learners and in fact engaging in continuous learning behaviors related to the job, people consider both information about the situation and information about themselves. If their conception of their capabilities, emotions, and attitudes coincide with their conception of the requirements of the situation, then a consistent pattern of behavior may emerge. Thus, if employees believe that they are capable of learning and that they enjoy learning and they recognize that learning is required or desirable to keeping their job and the organization values learning and provides learning resources, then they will view themselves as having a propensity for continuous learning and, hopefully, actually become continuous learners.

We develop these ideas below as we explore potentially relevant individual difference variables and cognitive and motivational processes at each stage in the continuous learning process. We examine three stages (or processes): prelearning, the learning process itself, and using the learning. We do not argue here that these stages are necessarily discrete or sequential. Instead, we use the three stages as a way to organize our presentation of the central elements and processes associated with CRCL.

The concept of career motivation also provides an underlying foundation for CRCL. London (cf. 1983, 1985; 1993) proposed an integrative model of career motivation that organizes needs, interests, and personality characteristics into three domains associated with career development: insight, identity, and resilience. Insight is the spark that kindles career motivation, identity provides direction, and resilience sustains career motivation over time. People high in these three concepts would be considered high in career motivation and ready to be continuous learners. Specifically in relation to continuous learning, people need insight into their learning needs, identity in terms of a vision for future career movement and development, and resilience in the face of slow progress and career barriers. In the context of the model we describe here, insight and identity are associated primarily with the pre-learning stage, while resilience is associated with the later stages. As noted below, the constellation of variables that makes one high in insight, identity, and resilience may differ for each individual.

Pre-Learning

The first step in CRCL is readiness. At this stage, individuals recognize a gap between their capabilities and requirements of the current job or future career goals. There are number of individual characteristics, psychological processes, and environmental factors that create readiness for CRCL. We describe these next.

Pre-Learning: Personal Characteristics

Several related behavioral processes help people develop insight into themselves and their environment, and recognize the need for or value of continuous learning. These include self-monitoring, public self-consciousness, feedback seeking, and openness to experience.

Snyder (1974, 1979) noted that people differ in the extent to which their behavior is susceptible to situational or interpersonal cues, as opposed to inner states or dispositions. Self-monitoring theory (Snyder, 1974) distinguishes between high self-monitors, who are very sensitive and responsive to social and interpersonal cues about situationally appropriate behavior (e.g., role expectations), and low self-monitors, who are less responsive to such cues. Warech, Smither, Reilly, Millsap, and Reilly (1998) developed a scale that measures both ability and motivational aspects to self-monitoring. For example, self-monitoring ability is represented by items such as, 'When I feel that the image I am portraying isn't working, I can readily change it to something that does'. Self-monitoring motivation is represented by items such as, 'I am highly motivated to control how others see me'. Extending this concept to CRCL, high self-monitors may be more likely than low self-monitors to attend to and act on organizational or group expectations concerning the need for behavior change and learning.

Another process that contributes to self-insight is feedback seeking. People seek feedback by monitoring the environment and by inquiring about their performance (Ashford & Cummings, 1983). Levy, Albright, Cawley, and Williams (1995) pointed to three motives that influence the feedback seeking process: (1) the desire for feedback, resulting from the need to reduce uncertainty (i.e., consistent with social comparison theory; Festinger, 1954); (2) the desire to protect one's ego or maintain a certain level of self-esteem; and (3) the desire to manage the impression one makes on others, that is, to make a positive impression via the way that feedback is sought (e.g., poor performing employees will be motivated to avoid asking for feedback; high performing employees may phrase feedback requests in ways that focus the source's attention on favorable aspects of performance).

Levy et al. (1995) noted that these motives affect different stages of feedback seeking. For example, the desire for feedback affects a person's initial intent to seek feedback, while the desire to enhance one's ego and manage others' impressions affect when (and from whom) the person elects to seek feedback. Ashford and Tsui (1991) found that asking for negative feedback enhanced others' opinions of the manager's overall effectiveness, whereas seeking only positive feedback decreased others' opinions. This is consistent with the idea that self-regulation (including the identification of areas where behavior needs to change and new knowledge is required) depends on adequate and accurate feedback. One note of caution is made by Williams, Bublitz, and Melner (1997) who found (in a laboratory study) that high levels of feedback seeking negatively influenced raters' impressions of the feedback seeker's performance.

Public self-consciousness is another variable that may influence whether people recognize the need for learning. As noted above, one way we recognize the need for learning is to seek feedback from knowledgeable sources (both inside and outside the organization). But some people avoid seeking feedback in public to maintain a favorable, confident public image (Levy et al., 1995). For example, they may intend to seek feedback until they realize the public nature of the setting (e.g., coworkers may overhear the supervisor's response), and then reconsider the wisdom of asking for feedback. People who are high in public self consciousness are more susceptible to feelings of being observed when in the presence of others (Fenigstein, Scheier, & Buss, 1975), and may therefore feel more aware of the public social costs of asking for feedback than those who are low in public self-consciousness. Thus, people who are high in public self-consciousness may be unlikely to seek feedback in public contexts, especially if they are socially anxious. This points to the need for organizations to develop mechanisms by which employees can seek and receive private, non-threatening feedback about current and future job requirements.

Openness to experience, an important personality characteristic, is associated with other characteristics such as being curious, broadminded, intelligent, original, and imaginative. Barrick and Mount's (1991) meta-analysis found that openness to experience was positively related to training proficiency (rho = .25). They suggested that individuals high in openness to experience will be more motivated to learn and hold more positive attitudes about learning experiences.

Pre-Learning: Environmental and Organizational Factors

Experiential learning theories suggest that learning is most likely to take place when people experience a demanding task that requires action, there is an element of surprise so the problem cannot be easily ignored or overlooked, an immediate solution is not evident, and the individual feels capable of taking action (Nonaka, 1994; Sims & McAulay, 1995). Such situational characteristics capture the individual's attention and impose a demand for action.

Some environmental factors create the need for CRCL even for employees who intend to stay in the same job in the same organization. Other environmental factors create conditions that direct employees to change jobs, organizations, or careers. We discuss examples of each of these below.

Vaill (1996) suggested that our work lives are in the midst of "permanent white water." Several well documented factors contribute to permanent white water. These include decentralization and empowerment (change is being initiated from many points, not just the top), an emphasis on service (as the customer's subjective experience, not merely product specifications - "we are not done until the customer says we are done"), emphasis on corporate social responsibility, and organizational multiculturalism and internationalizing the corporate culture. Graddick and Lane (1998) described other external marketplace trends that drive change; these include the increasing rate of technological change, global expansion (which opens up new markets and intensifies competition), and regulatory changes that have brought about an end to some monopolies. These external changes lead to internal responses and initiatives such as reengineering, restructuring, mergers, acquisitions and alliances, downsizing, quality efforts, and cultural renewal. Hall and Mirvis (1996) cited one survey of 400 employers that found 80% had downsized over a recent five-year period, shrinking their work forces by an average of 12%. In such an environment, "learning how to learn and continuous learning have become core career competencies" (Hall & Mirvis, 1996, p.24).

Cascio (1995) and Pearlman, Campbell, Gottfredson, Kehoe, DeVries, and Hackman (1995) pointed out that it is increasingly difficult to view a 'job' as a stable collection of similar tasks performed by a group of employees. Instead, tasks are being reconfigured into broader clusters of work that change over time. The fact that jobs, and the tasks associated with them, are unstable contributes to the need for CRCL.

Vaill (1996) noted that permanent white water conditions are full of surprises because complex systems with their delicate interdependencies and closely calibrated operating specifications tend to produce novel problems that are messy and ill-structured. These white water events are also costly (in monetary terms) and obtrusive (time taken to deal with them is diverted from other pressing issues and is felt as a diversion of time and resources). He concluded that "beyond all of the other new skills and attitudes that permanent white water requires, people have to be (or become) extremely effective learners" (p. 20).

The changing nature of the psychological contract between organizations and employees has destroyed the sense of security that could previously be achieved by long-term organizational memberships, leading to what Jack Welch, CEO of General Electric, has called a one-day contract in which the current contributions of each party to the relationship are all that matters (Hall, 1996). This creates a situation where security must be created by employees through their ongoing investments in CRCL.

This leads to the observation that CRCL can sometimes be 'forced' on employees who would prefer to 'stand still.' As DeVries observed (in Pearlman et al., 1995), workers must beware of thinking 'I have the same job title now as I had a few years ago, so I must have the same job.'

An executive who was viewed as successful a few years ago by focusing solely on satisfying shareholders' financial interests may now be seen as unsuccessful because of a failure to satisfy the interests of other stakeholders, say consumers (Graddick & Lane, 1998). Kaplan and Norton's (1996) balanced scorecard approach considers multiple factors that contribute to organizational success (e.g., financial, customer, internal business processes, learning and growth). As organizations consider short- and long-term measures, financial and non-financial indicators, and internal and external perspectives, the skills required for effective performance (throughout the organization) are likely to change. Such changes contribute to the need for CRCL.

As mentioned above, other environmental factors create the need for people to engage in CRCL so that they may change jobs, organizations, or careers. For example, value migration (Shapiro, Slywotzky, & Tedlow, 1997) can be triggered by evolving customer needs and new competitive options becoming available to customers. It is generally experienced as discontinuous change and may occur rapidly (e.g., IBM made $6 billion in 1990 and lost $5 billion in 1992). One example is the computing industry where value migrated from mainframes to minis to PCs to processor chips and operating systems. Where I.B.M., D.E.C., Wang, and Apple were once dominant competitors, other firms such as Microsoft, Novell, and Intel are now seen as market leaders. In retailing, traditional competitors such as Sears, Penney, and Ward focused on each other while value migrated to once unnoticed competitors such as Wal-Mart, Target, K-Mart, Toys-R-Us, and The Gap. Value migration limits employee growth and often leads to downsizing (in firms that are losing market dominance). Moreover, there is usually not enough room in the new value 'space' to accommodate all the traditional competitors plus the new competitors (Shapiro et al., 1997).

The net effect of value migration is to reshape industries and thereby change the organizations where jobs are available and the skills required to perform those jobs. By doing so, value migration creates a need for CRCL. Similarly, because the life cycle of many products and technologies is becoming shorter (Handy, 1994), employees are forced to repeatedly engage in cycles of CRCL.

Another environmental factor that contributes to CRCL is the increasing emphasis on contingent work (Smither, 1995). The *Wall Street Journal* (Ansberry, 1993) recently wrote that "contingent workers...are the medium for a corporate America preaching a message of flexibility and cost cutting." The number of people working for temporary-employment agencies has increased from 470,000 to 1.6 million over the past ten years (Fierman, 1994). Moreover, a recent poll of CEOs indicated that 44% had increased their use of contingent workers over the past five years, whereas only 13% had decreased use of such workers (Fierman, 1994).

Contingent workers are not confined to simple clerical positions; they can be found in technical areas, such as engineering, programming, and nursing. At least one firm provides companies with temporary senior executives. A recent study by the Economic Policy Institute and the Women's Research & Education Institute found that, among professionals, 31% of women and 25% of men held nonstandard work arrangements (e.g., temporary, business start up, or part-time work). The study also found that about 40% of these jobs pay more than comparable full-time positions ("Nonstandard jobs," 1997).

The increasing prevalence of contingent work, especially among technical and professional employees, creates a need for continuous learning so that these employees can remain competitive and attractive to employers. In sum, contingent work escalates the need for CRCL.

Summary

Personal characteristics such as self-monitoring, feedback seeking, public self-consciousness, and openness to experience influence the extent to which people seek feedback and are sensitive to environmental factors (e.g., technological change, global expansion, deregulation, value migration, contingent employment) that create the need for CRCL. As Hamel and Prahalad (1989, p. 69) have said, "An organization's capacity to improve existing skills and learn new ones is the most defensible competitive advantage of all."

Once the need for learning is identified, other personal characteristics, such as proactivity, locus of control, self-efficacy, extraversion, mastery orientation, and cognitive ability, influence whether people will identify and benefit from formal or informal learning opportunities. At the same time, organizational and other environmental characteristics (e.g., culture and learning climate, availability of developmental resources, staffing and compensation policies) influence the opportunities and rewards for CRCL. We describe these next.

Learning

In this stage, people acquire knowledge, skills, and experience to reduce the gap between their capabilities and current or anticipated job requirements. A central feature of CRCL is that, although it may occur in institutional (e.g., formal, classroom) settings, it is also likely to occur outside of such formal settings.

Vaill (1996) described a number of important differences between institutional learning and what he calls "learning as a way of being." For example, in institutional learning, the goal and material relevant to achieving it can be clearly specified. In contrast, CRCL may sometimes be more creative or exploratory (i.e., the learner is not exactly clear what is to be learned or what the learning will look like). In institutional learning, goals are decided by others (e.g., trainers, teachers, supervisors), people learn from an authority about something that is already known, learning tends to be answer oriented (i.e., getting the "right" answer), and everyone in the group is expected to learn pretty much the same thing. In contrast, much CRCL is self-directed; no textbooks or learning materials have been specifically designed to guide the learning. Institutional learning generally occurs off-line; skills are first learned and later applied at work. In contrast, CRCL often occurs on-line; learners learn as they go along. The goal of institutional learning tends to be personal mastery. But in CRCL, people may be faced with a continual stream of novel problems so the learner always feels like a beginner. Personal mastery is almost a contradiction in terms.

In sum, in CRCL learners often have substantial control over the purposes, content, form, and pace of learning, and they are the primary judges of when sufficient learning has occurred. Also, because CRCL is often self-directed, occurs in a context where others react to the learner's new behaviors, and the learner can, at best, have only partial knowledge of the system or environment in which the learning will be applied, it is clear that the optimal methods and outcomes of learning cannot always be precisely defined (i.e., there is no cookbook).

Experiential learning and role modeling are likely to be especially important at this stage. It is important to distinguish between experiential learning in natural (e.g., on-the-job) versus institutional learning contexts. For example, in institutional contexts, experiential learning is sometimes seen as merely a change of pace or as a way of enriching traditional (e.g., lecture) instructional approaches. Vaill (1996) noted that, in such a setting, "experts decide what learnings learners need; experiences are then designed for learners so that they will all have the same experiences at the same time and same place and from the same teaching material and can then be graded on their attainment of the principles and practices" (p. 122). In contrast, experiential learning in work contexts is much less structured and more ambiguous. Observational learning and role modeling play a critical role. Learners observe others, practice, and receive feedback. Such learning can also be guided or enhanced by various relationships (e.g., boss/employee, mentor/mentee).

Survey evidence indicates that formal, institutional learning may represent only a modest portion of CRCL. For example, only about 25% of the employed U.S. workforce participated in some form of employer-sponsored formal training activity in 1991 (Olson, 1997). Also, formal training was unevenly distributed, being concentrated among more skilled and educated workers in the upper portions of the wage distribution. The relationship between age and formal training followed an inverted-U shape, with the probability of receiving training peaking at about age 40. Kozlowski and Hults (1987) cite Kaufman (1974) who found that fewer than 25% of engineers engage in formal technical development activities. Much more common was development via the work itself, reading, peer contacts, or other informal activities.

The fact that CRCL can occur in less formal settings, as well as in institutional settings, has a number of implications for the design of research about CRCL and organizational practices and interventions that may enhance CRCL (discussed later).

Learning: Personal Characteristics

Having recognized a gap between their capabilities and the requirements of the current or anticipated situation, some individuals are more likely than others to take action and engage successfully in CRCL. Some of the same variables that affect pre-learning also affect the learning process itself. These include locus of control, self-efficacy, extraversion, mastery orientation, and cognitive ability.

Proactivity (Bateman & Crant, 1993; Crant, 1995) refers to the tendency of some people to take action to influence their environments. Proactive people are relatively unconstrained by situational forces. They identify opportunities, show initiative, take action, and persevere until they bring about meaningful change. Proactive people appear to accept and act on the maxim that "it is better to beg forgiveness than to ask permission." The concept of proactivity is grounded in an interactionist approach and is consistent with the idea that people create (i.e., directly influence) their environments, thereby making successful performance more likely. Proactive people are thus more likely to take advantage of opportunities for development than others.

People with an internal locus of control see the reinforcements (positive and negative) they receive as being the result of their own behavior, whereas those with an external locus of control tend to see the outcomes they receive as being largely influenced by outside forces (Rotter, 1966). That is, people with an internal locus of control believe that events and outcomes at work are strongly affected by their behavior and are therefore under their control. People with an external locus of control believe that events and outcomes at work are beyond their control and are instead controlled by others (or luck). Because internals are more likely to accept feedback and take action to correct and improve performance, internals are more likely to benefit from learning opportunities. Research has found that, relative to those with an external locus of control, those with an internal locus of control experience greater position mobility (i.e., promotions and job changes), engage in more entrepreneurial activity, are more active in managing their own careers, and (as executives) engage in more innovation, risk taking, environmental scanning, and have longer planning horizons (Whetten & Cameron, 1995). This pattern of behavior should be positively related to CRCL.

Self-efficacy refers to a person's appraisal of his or her ability to execute the actions or behaviors required for successful learning and performance on a specific task. Because self-efficacy influences motivation (e.g., via the difficulty of self-set goals and commitment to goals), it guides the actions a person elects to undertake, the level of effort the person is likely to exert, and the extent to which the person will persist when faced with obstacles (Bandura, 1997). Of course, a person's previous task accomplishments affect self-efficacy; thus people with a history of successful CRCL may therefore be more likely than others to subsequently engage in CRCL. Self-efficacy reflects employees' beliefs in their own capacity to improve (e.g., will acting on developmental feedback actually result in successful learning?).

Self-esteem may also affect the goals people set for themselves, although the effect of self-esteem may depend on the extent to which the employee identifies with his or her work group. For example, in a laboratory study, Pilegge and Holtz (1997) found that high self-esteem individuals who strongly identified with their work group set higher goals and achieved more than high self-esteem individuals who did not identify with their work group or low self-esteem persons.

Extraversion, one of the big five dimensions of personality, may also offer an advantage in that CRCL is likely to be enhanced by relationships (e.g., teams, mentors, networks) that offer the opportunity for mutual learning (Hall, 1996). Loners and those who are uncomfortable with being dependent on others are likely to be at a disadvantage. Supportive relationships create the possibility of learning in an atmosphere where people can experiment, risk failure, and overcome setbacks with the help of others who they trust and who care about their success.

Dweck (1986) has described two orientations to learning: mastery orientation and performance orientation. For those with a mastery orientation, the learner's attentional focus is on developing competence. They want to acquire knowledge and skill until they reach a level of mastery that reflects a deep (expert) understanding, and they view feedback about skill deficits as an opportunity for improvement. In contrast, people with a performance orientation often compare their performance to that of others and tend to focus on doing better than others (e.g., on a test or on the job). They tend to see failure as indicating a lack of ability, and therefore view performance feedback as threatening. A competitive, performance orientation appears to be less effective for acquiring, retaining, and transferring material than a mastery orientation. Focusing on competition apparently interferes with the deep cognitive processing that leads to mastery. Moreover, learners with a mastery orientation are more likely to be receptive to and make constructive use of feedback and thus persist until they attain competence. Because development naturally involves failure and frustration as the employee progresses from incompetence to competence, a mastery orientation should be important in sustaining CRCL (Squires & Adler, 1998).

Among those who decide to engage in CRCL, success will certainly be influenced by cognitive ability. As noted above, CRCL is often driven by the increasing complexity of work (i.e., as organizations empower employees, drive responsibilities to lower levels, and compete in increasingly competitive global markets). Gottfredson (in Pearlman et al., 1995) has pointed out that dealing with complexity is the essence of intelligence. She also argued that, because some level of intelligence appears to be necessary to handle complexity (and because we do not, at least as yet, know how to develop intelligence) only about half of the workforce is likely to be able to function successfully in jobs that routinely require independent learning (or, in our language, CRCL).

Kanfer and Ackerman (1989) showed that cognitive ability is especially important early in task acquisition, and that the role of cognitive ability in influencing performance diminishes as performing the task becomes more automatic. However, if work requires continuous learning of new tasks, then workers will nearly always be engaged in controlled processing oriented to learning one task or another, and cognitive ability will remain continually important for CRCL and performance.

Learning: Organizational Culture and Practices

Implicit in the process of continuous learning is support from the environment, the idea of "empowered self-development" (London & Smither, 1999). CRCL will be shaped by the extent to which there is "…an organization-wide concern, value, belief, and expectation that general knowledge acquisition and application is important" (Tracey et al., 1995, p. 245). The organization can offer the resources and encouragement that promote continuous learning for performance improvement and prepare employees to meet future performance requirements (Holt et al., 1996).

In a study that considered both individual characteristics and perceptions of organizational support, London, Larsen, and Thisted (1999) examined the relationships between support for development (i.e., self-perceptions of empowerment and feedback), self-managing behaviors (i.e., supervisors' perceptions of employees' feedback seeking and self-management for development), supervisors' ratings of job performance, and individual differences (i.e., career motivation, public self-consciousness, gender, and age). Respondents were 115 employees and their supervisors in a large Danish bank. Self-management for career development was higher for individuals who perceived receiving more feedback, felt more empowered, and were higher on career motivation. Positive reinforcement contributed to feelings of empowerment.

Deci and his colleagues (cf. Deci, Connell, & Ryan, 1989; Deci, Eghrari, Patrick, & Leone, 1994) identified factors in the environment that influence self-determination. These include constructive (non-threatening) performance feedback, behavioral choices and their likely consequences, empathy (the knowledge that others understand one's perspective), a reason to act, and reinforcement of self-competence. When managers create an environment that has these elements, employees will be more motivated to engage in behaviors that satisfy their own needs. Such an environment supports their sense of self-competence and the belief that they can control their own behaviors and outcomes.

Deci et al. (1989) found that managers can create an environment that empowers self-development by engaging subordinates in discussion and encouraging them to be self-reflective and try new behaviors. To do so, managers were taught to avoid giving harsh feedback, use a minimum of controlling language, treat poor performance as a problem to be solved rather than a focus for criticism, recognize and accept the subordinates' perspective (the limits under which the subordinates' operate), and acknowledge the needs and feelings of subordinates.

Tracey et al. (1995) distinguished four elements of a continuous learning work environment based on the work of Dubin (1990), Noe and Ford (1992), and Rosow and Zager (1988): (1) Knowledge and skill acquisition should be built into all jobs as part of the job requirement and performance evaluation. (2) Learning is supported by social interaction and work relationships within and between groups such that team members can encourage and reinforce each other's learning. (3) The organization provides formal training resources that enable self-development and performance appraisal criteria and rewards that inform employees about the importance of continuous learning for performance improvement. (4) Innovation is expected and rewarded as the organization strives to enhance its competitive position. The result is a continuous learning environment in which employees share perceptions and expectations that learning is an essential part of the way the organization does business. "These perceptions and expectations constitute an organizational value or belief and are influenced by a variety of factors, including challenging jobs; supportive social, reward, and development systems; and an innovative and competitive work setting" (Tracey et al., 1995, p. 241).

Linking feedback to learning resources is an essential feature of organizations that support CRCL. In fact, organizations run a risk in providing developmental feedback without also providing appropriate developmental resources. In one field study, Maurer and Tarulli (1996) found that the degree to which employees felt that such resources were available to them directly affected the employee beliefs in their own capacity to improve. In turn, this "developmental self-efficacy" influenced the attitudes of employees towards the entire appraisal process. Similarly, research by Hazucha, Hezlett, and Schneider (1993) indicated that self ratings and others' ratings of effort to change following the implementation of multi-source feedback were related to whether the organization was reported as having a formal and active career development program.

All of this points to the importance of the organizational climate concerning CRCL. Climate can be thought of as a set of individual perceptions about the work or organizational context. But organizational context and practices can affect individual-level perceptions in common ways. Social interactions among co-workers facilitate such shared perceptions. To the extent that these perceptions are consensual they provide a basis for collective response tendencies among employees. Climate thereby offers a framework for conceptualizing motivation at a collective, rather than merely at an individual, level (Kozlowski & Hults, 1987). Kozlowski and Hults (1987) characterized several features of an organizational climate that supports what they called technical updating (e.g., continuing education, professional activities). The features of such a climate included organizational support (e.g., for journal subscriptions and time to read them, financial support for training, time to explore advanced ideas), supervisor support (e.g., mutual goal setting and career counseling, performance feedback), innovativeness (e.g., a competitive, technologically-superior, state-of-the-art organization), job characteristics (e.g., challenging work that stretches employees' limits), and information exchange (free sharing of information). They found that such climates were more prevalent in organizations with more complex technological systems, and less prevalent in organizations with more routine, standardized work. Finally, climate was positively associated with organizational commitment, technical performance, and technical continuing education.

Perceptions of the work environment related to voluntary development were also examined by Maurer and Tarulli (1994). They focused on company policies that make it possible to participate in learning activities, supervisor support, coworker support, and the extent to which the company values learning. They found that these perceptions of the work environment were related to participation in learning activities. They also found that the more employees valued supervisor support, the stronger was the relationship between supervisor support and intended participation in learning activities. The latter finding illustrates the value of considering the role that moderator variables may play in explaining CRCL.

Noe and Wilk (1993) also found that work environment perceptions (i.e., social support and situational constraints), as well as learning attitudes (i.e., motivation to learn) and perceptions of development needs had a significant effect on the amount of developmental activities in which employees participated.

In contrast, some work context factors, such as demands for conformity and obedience or discouragement of risk taking (Porras & Robertson, 1992), are likely to suppress CRCL.

Some Examples

At Intel, manufacturing supervisors are expected to acquire and demonstrate competency in each of four skill and knowledge areas: leadership and team skills, administrative, manufacturing operations, and safety and legal. They have specified four levels of competency within each of the four areas. Supervisors are expected to fully attain one level of competency in an area before moving to the next level of competency in the area. It is recognized that competency can be acquired via formal or informal mechanisms, such as reading, on-the-job experience, and courses. Each supervisor and his or her manager separately diagnose the supervisor's competency level in each of the four areas. Then they sit together, discuss differences, reach agreement, and look at how to build competency. Except for the safety and legal area (where requirements are relatively fixed), there is a menu of choices for developing competencies in each area. For the entry-level of competency (across all four areas), they list 25 courses and 15 readings (e.g., books). Less formal learning approaches are also specified, with an emphasis on learning from peers, managers, and on-the-job tools and resources. Moreover, classroom training is not always the dominant tool for building competencies. This approach creates a culture where there is an expectation of continuous learning, feedback to identify skill gaps, formal and informal resources for learning, and the opportunity for employees to help shape the content and format of their learning.

Another example of support for CRCL is the use of a career center where employees can obtain help with self-evaluation, career planning, and searching for new job and career opportunities within and outside the firm. This approach explicitly acknowledges that employees need to adapt in response to rapid organizational change and sometimes alter their career paths. This concept was extended with the recent creation of The Talent Alliance, a nonprofit consortium of firms such as AT&T, Johnson & Johnson, GTE, Unisys, and UPS. The Talent Alliance supports employees' career management and helps them make job matches between companies (Lancaster, 1997). For example, it helps employees whose skills are no longer needed by their company to find other employment. Also, it helps member firms identify candidates who have the skills needed as organization requirements change. Centered around a Web site, employees at all organization levels and career stages can take tests concerning personality, skills, and interests. The Talent Alliance also offers private sessions with career counselors on the internet or in person to help employees formulate career plans based on the results. It then matches employees with appropriate training and education available through the corporate members' contacts in academia and at training companies and institutes.

Summary

Proactivity, locus of control, self-efficacy, extraversion, and mastery orientation are likely to affect whether people seek out and act on opportunities for CRCL; cognitive ability will influence the extent to which they benefit from such developmental activities. At the same time, a host of organizational features (e.g., climate, norms, resources, feedback, empathy, co-worker support) may encourage or inhibit CRCL.

The effect of any of these antecedent variables on CRCL may be indirect. Because many individual difference and organizational variables are affecting individual goals at the same time, the effect of any one antecedent variable on CRCL is likely to be modest. The model suggests that effects of these different classes of variables may be additive. For example, organizational incentives may enhance or detract from the effect of personality dispositions on CRCL. Thus, researchers should measure individual and contextual variables whenever possible.

Application of Learning

One indispensable element of CRCL is the realization that the learning will matter in terms of actual consequences; it will affect the learner and others (inside or outside the organization). CRCL is not complete until learning is applied. As Vaill (1996) stated when describing successful learners:

> Whatever it is they have become good at is not fragmented for them, not isolated from its environment, not isolated in time or in space. They know the relevant elements and their interrelations intimately. These learners have a feeling for the meaning of the subject beyond its technical details and formal structure. They have operative knowledge about this system, which is to say they know how to get this complex something to work in the way they intend...(p. 111).

Although we discuss application of learning as a separate step, we recognize that such application may occur at the time of learning, especially when CRCL occurs outside of traditional, institutional settings. Critical tasks at this stage are integrating new knowledge with existing knowledge, practice (i.e., applying the knowledge in a setting where behavior has important consequences), and receiving feedback from others or the environment about the effectiveness of the new learning and behavior.

CRCL often begins with declarative learning (e.g., obtaining factual knowledge), progresses to knowledge compilation (e.g., transforming declarative knowledge into procedural knowledge—from knowing 'what' to knowing 'how'), and culminates in proceduralization—the application and use of knowledge to solve a problem or perform a task. The concept of automaticity is also important at this stage of CRCL. Processing that once demanded active attention and control can become more automatic. This frees the learner's attentional capacity and cognitive resources to deal with other learning and tasks. For most tasks, automaticity requires extensive practice. This indicates the importance of creating many opportunities for learners to practice and apply their learning.

Feedback is also critical at this application stage. Feedback will be accepted only if the learner perceives the source of feedback as credible and expert. Also, very frequent feedback will not necessarily be more useful than less frequent feedback. Very frequent feedback can suggest to learners a loss of personal control, and it can lead learners to rely on external feedback and neglect to develop their own abilities to judge their own performance.

Finally, skills and knowledge acquired from CRCL are cumulative and integrative. Moreover, learners may integrate and apply apparently unrelated skills to address complex problems. For example, a manager may combine two new skills—computer skills (e.g., e-mail, internet, and spreadsheets) and conflict resolution skills—to identify underlying concerns and gather, analyze, and share information with geographically dispersed team members to help develop a shared understanding and build consensus among team members.

Application of Learning: Personal Characteristics

It can be expected that people high in conscientiousness should be especially likely to apply skills and knowledge acquired from CRCL. Conscientiousness is associated with being dependable, thorough, organized, planful, achievement-oriented, hard working, and persevering. Barrick and Mount's (1991) meta-analysis found that conscientiousness was positively related to job proficiency (rho = .23) and training proficiency (rho = .23). They also cite evidence that conscientiousness is related to educational achievement.

Applications of Learning: Organizational Culture and Practices

Effective learning is enhanced when there are situational cues on the job (e.g., job aids, performance goals shared with the boss) that remind learners to use their new skills, and learners are reinforced by co-workers (or the boss) for exhibiting new skills. Tracey et al. (1995) found that supervisory training is more likely to transfer to the job when the work setting supports the specific skills learned and, more generally, creates an environment that supports continuous learning. Supervisors who reported more environmental support were viewed by their managers as having changed their behavior in accord with the goals of the course. Cianni and Wnuck (1997) have also posited that a supportive team environment can promote adaptation to career shifts and movements into unfamiliar roles and responsibilities. Team members will help each other be continuous learners.

The level of organizational support for CRCL will affect outcome expectancies. Outcome expectancies refer to an individual's beliefs about the consequences of his or her behavior. Outcome expectancies are task- and situation-specific. I may expect rewards for successful performance in one setting (or for one task) but I may expect no rewards for another task/setting. Several studies have found a relationship between outcome expectancies and performance (Frayne & Latham, 1987). Of course, outcome expectancies play a central role in several theories of human behavior such as social cognitive theory (Bandura, 1986) and expectancy theory (Vroom, 1964). Thus, organizational practices that link the application of CRCL to formal recognition, job security, salary increases, or merit pay are likely to increase the probability that people will engage in CRCL and apply the learning in the organizational setting.

Moreover, results of research are consistent with the comments of a training manager from one firm who emphasized that all the training in the world will be unsuccessful unless "management is walking the talk from the top to the bottom." For learning to be successful, it must be embedded in a broader culture change.

Summary

Individual characteristics, such as conscientiousness, and situational support factors, such as feedback and rewards from the boss, affect the extent to which learning is applied. These factors also affect the extent to which the employee is motivated to engage in continuous learning, starting the cycle anew.

This completes our mapping of the CRCL process. We now turn to how CRCL affects human resource management (HRM). As we stated at the outset, CRCL is not an independent phenomenon, but is integrally tied to basic HRM functions.

IMPLICATIONS FOR HUMAN RESOURCE MANAGEMENT PRACTICES

This section discusses the practical and research implications of the model for different areas of HRM. Separate sections deal with job analysis, selection, training, career development, appraisal, compensation, and leader behavior. Each section concludes with practical recommendations (or our best guess about these given the state of research) and suggestions for research (i.e., looking at how CL affects and is affected by HR practices in the area). We then consider more specific directions for CRCL research.

Job Analysis

Traditional HRM views the job as the primary level of analysis. Job analysis is used to identify important tasks as well as the knowledge, skills, abilities, and other characteristics needed to effectively perform the job. But Cascio (1995) has argued that it is becoming more difficult to view a 'job' as a stable collection of similar tasks performed by a group of employees. Reilly and McGourty (1998) pointed out that smaller, flatter organizational structures (and the self-directed work teams often associated with them) require employees to be more flexible and to play a greater role in deciding how work gets done. Also, the growing complexity of many jobs sometimes makes it difficult for one person to perform them, and this contributes to the increasing use of teams as a work unit. This emphasis on teams challenges our common notions of job relatedness, since the 'job' may be a fluid and continually changing construct. Gottfredson (Pearlman et al., 1995) has also argued that work is becoming more complex, and that traditional job analysis is better suited for routine work than for complex and fluid jobs.

Job analysis practices will need to evolve to the extent that jobs, and the tasks associated with them, are becoming less stable. Organizations need to develop job analysis approaches that are more responsive to the fluid, changing nature of work.

The use of learning-based competency models is one approach to addressing the learning requirements associated with the fluid nature of work. Graddick and Lane

(1998) described several approaches for developing competency models. One relatively traditional approach (i.e., grounded in the critical incidents method of job analysis) gathers behavioral descriptions of superior performers and contrasts these with descriptions of average performers to identify important competencies that distinguish these two groups. A value-based approach derives competencies, formally or informally, from the cultural or normative values held by the organization or its leaders. A strategy-based approach involves forecasting competencies that will be needed for the firm to respond proactively to its anticipated future.

In contrast, learning-based competencies emphasize the ability of employees to learn new skills and adapt to changing roles and circumstances. They focus on developing employees' abilities to adapt to quickly changing environments and reflect relatively more enduring personal skills. A learning-based approach to developing competency models is likely to be especially useful in organizations that face rapidly changing environments and therefore want to encourage and harness the benefits of CRCL.

Moreover, organizations need to lessen their use of task-oriented job analysis procedures that by their very nature assume jobs are static and predictable. At the least, worker-oriented job analysis approaches—that focus on the underlying skills, abilities, and characteristics needed to adapt to changing work—would appear to be more desirable. The use of team-based job analysis or learning-based competency models may better focus the organization's attention on the vital role that CRCL and adaptability will play in individual effectiveness and organizational performance.

Researchers need to develop procedures that strike a balance between current work requirements and the dynamic nature of work. Also, some of the methods (interviews, observation) and sources (e.g., incumbents, supervisors) of traditional job analysis may underestimate the importance of CRCL. And when they do recognize the importance of CRCL, they may be ill suited to identifying the particular characteristics that will be required to take advantage of and benefit from CRCL opportunities.

Selection

An organization's selection practices will critically affect its ability to support and enhance CRCL. In traditional HRM, the organization first identifies the knowledge, skills, abilities, and other characteristics (KSAOs) required to effectively perform the 'job,' and then selects candidates who possess these KSAOs. Perhaps the most obvious implication for practice is the need to identify the personal characteristics that are most correlated with the disposition to engage in and the ability to benefit from CRCL, and to select employees who possess such characteristics. This raises questions about the role of many useful and well established selection practices such as work samples, job simulations, and job knowledge tests. These practices are generally developed by specifying the domain of important tasks or knowledge required on the job, and ensuring that the content of the selection measure (e.g., work sample, simulation, or test) is representative of the content of the job. But such approaches may implicitly assume that the content of the job (i.e., the tasks and knowledge associated with the job) is stable.

Thus, it may be desirable to combine the use of work samples, simulations, or knowledge tests (which appear to focus on current requirements) with the use of ability and personality measures that may better serve to predict CRCL. Here, it would be helpful to better understand the most parsimonious combination of measures required to predict current work performance as well as CRCL. Note also that predicting CRCL is not merely the same as predicting learning (as traditionally measured by achievement tests and post-training learning measures). Instead, predicting CRCL will require researchers to identify the characteristics that create readiness for CRCL, acquisition of knowledge and skill, and the application (and monitoring) of new learning to changing real-world problems in the work environment. Thus, we are not asking something as simple as "Does cognitive ability predict training performance?" Understanding the personal characteristics associated with CRCL may require (as our model suggests) identifying a number of personal characteristics, each of which helps explain a different aspect of CRCL. By doing so, researchers can help practitioners understand the linkages between personal characteristics and CRCL, and specify the optimal set of predictors for identifying candidates who can meet immediate work requirements and adapt to changes in the way work is accomplished.

The importance of CRCL also raises a question about the extent to which a close fit between the candidate and organization is always desirable. For example, in an environment that is rapidly changing, it may be undesirable to hire people who fit the current organizational culture or system. Instead, it may be better to seek a slight misfit between the person and the current system. Finding candidates who will challenge but not alienate the system may cultivate CRCL both for the new hire and for those employees already operating in the system. This in turn may contribute to the system's ability to adapt to changing customer, technical, and business requirements.

Training

Vaill (1996, p. 76) quoted Gardner (1964) as saying, "The ultimate goal of the educational system is to shift to the individual the burden of pursuing his own education." This view seems consistent with the recognition that many important aspects of CRCL will occur outside of a formal, institutional learning context that is controlled by the organization. Clearly, traditional training programs have a place in shaping CRCL. But they are insufficient.

Two approaches that organizations can use to foster CRCL are *action learning* and *self-management training*. Vaill (1996) described action learning as creating teams of working managers to work on real organizational problems and structuring the experience in such a way that both useful solutions to these problems emerge and substantial learning occurs for participants, learning that goes beyond the technical details of the particular problem. Raelin (1997) said that action learning is based on the premise that managers learn most effectively from other managers and facilitators while engaged in the solution of real-world problems.

Self-management is a collection of cognitive and behavioral strategies that help individuals to structure their environment, enhance their motivation, set goals for themselves, and facilitate actions required to attain their goals (Frayne & Latham, 1987; Gist, Stevens, & Bavetta, 1991; Latham and Frayne, 1989; Locke and Latham, 1990, p. 278; Neck, Steward, & Manz, 1995). Because self-management training is quite general, it offers the latitude for employees to set goals and use reinforcers that are appropriate for the setting or culture in which they will be learning and working. Self-management training consists of several elements. Trainees are first introduced to self-management principles. They are then taught to use a systematic approach to self-evaluation where they gather data about their behavior. Based on this self-evaluation, they establish self-set (short- and long-term) goals that provide direction for their efforts, thereby avoiding what might otherwise be sporadic activity lacking a clear purpose. They also identify features in the environment that may facilitate or hinder their progress. They then collect data and maintain a record concerning their progress toward the self-set goals. Based on performance relative to self-set goals, they are taught to use self-reinforcement (or self-punishment). Providing rewards for achieving goals should exert a positive influence on future behavior, whereas self-administered punishers should reduce the likelihood of undesired behaviors. Generally, trainees create a written contract with themselves that specifies the goals, action plans, and reinforcers. The contract prompts the individual to follow through on the planned course of action and may enhance goal commitment. Self-management training also includes a relapse prevention component to help the individual recognize common problems, pitfalls, and high-risk situations, and develop procedures for overcoming these circumstances. Self-management has proven successful in changing behavior and increasing performance in organizational settings (Frayne & Latham, 1987; Gist et al., 1991; Latham & Frayne, 1989; Locke & Latham, 1990, p. 278).

Organizations that seek to actively cultivate CRCL may consider self-management training, action learning, and self-leadership training (i.e., training in self-dialogue, mental imagery, and positive patterns of thought and affect). Creating a culture that supports each stage of CRCL also is important. Such a culture will provide opportunities for employees to receive information about shifting organizational expectations, feedback to help identify skill gaps, formal and informal developmental opportunities, and a work environment that encourages and rewards the application of new skills.

For researchers, evaluating the effectiveness of CRCL will be a substantial challenge. Researchers will need to develop ways to assess the reactions, learning, behavior change, and organizational impact of CRCL that occurs outside of traditional institutional settings. (Later, we offer some ideas about criteria that can be used to assess outcomes of CRCL.) The measurement and evaluation issue is complicated by the fact that different employees may have different self-set learning goals, and as noted above, CRCL may not always be aligned with or in response to organizational concerns. As a result, common evaluation practices, such as the use of standardized pre- and post-tests to assess learning or using a standardized behavioral observation scale to assess on-the-job behavior, often are not feasible (e.g., the content to be learned and the behavior to be changed vary depending on each employee's unique, and sometimes unstated, goals).

Career Development

Hall and Mirvis (1996) questioned whether organizations should be in the career management business. For example, the organization may elect to provide challenging work and resources to support employee development, but leave it to individuals to manage their own careers. DeVries (in Pearlman et al., 1995) emphasized that individuals need to become skilled at unlearning and recognizing their limiting assumptions (i.e., we are 'prisoners of our experience'). Carson and Carson (1997) proposed that organizations should help employees avoid the dangers of career entrapment. People who feel they are in relatively secure, steady state jobs are likely to engage in psychological preservation of their careers, perceive few career opportunities, and wind up entrenched and unhappy. Organizations can help employees avoid career entrapment by providing job rotation (Campion, Cheraskin, & Stevens, 1994) or cross-functional moves and sanctioning more diverse career paths.

Organizations should probably not assume that, because the traditional concept of careers (i.e., as progression through a vertical ladder of positions in a single organization) is disappearing, active career management offers little value to the organization. An organization that adopts a completely passive approach to career management risks the likelihood that the skills and operating assumptions of its employees will become obsolete. To promote CRCL, career management practices may focus less on active succession planning (choreographing movement up a vertical career ladder) and more on providing challenging work via job rotation and related interventions.

Performance Appraisal

As Tracey et al. (1995) emphasized, continuous learning needs to become an element of performance in and of itself that is supported and rewarded by the organization. This points to the importance of developing appraisal practices that emphasize the value of CRCL. London and Mone (1999) noted that participation in continuous learning is a measurable dimension of performance. Employees establish goals for development in relation to performance expectations, and they can be appraised and rewarded for their developmental accomplishments, even when they do not give rise to immediate performance improvements. Supervisors can be evaluated and rewarded on the support they provide for their subordinates' continuous learning. As such, rewards for development cannot wait for newly learned behaviors to be applied sometime in the future. Learning must be measured and rewarded now for the organization to ensure it will remain competitive.

London and Mone (1999) suggested performance dimensions that reflect continuous learning. These include (1) anticipating learning requirements by, for instance, identifying future job requirements and implications for needed skill updates, (2) setting development goals that reflect needed knowledge and skill structures, (3) participating in learning activities, (4) asking for feedback to test goal relevance, and (5) tracking progress. In addition, evidence of continuous learning can be obtained from post-training skills and knowledge test scores, performance in trial situations (e.g., simulations), as well as actual changes in behavior and performance on the job. Some of these performance dimensions can be evaluated objectively in that behaviors can be observed and recorded. So for instance, it is fairly easy to determine if an employee established a written development plan and participated in training. However, plans may also be kept in a person's mind, and learning occurs outside of formal classes. So continuous learning may also be assessed subjectively by, for instance, asking supervisors to rate their employees and employees to rate themselves on such survey items as the extent to which they set development goals and the extent to which goals were accomplished.

Organizations can also use competency models as a basis for placing value on employee knowledge and skill instead of relying on traditional methods of appraisal (Reilly & McGourty, 1998). Competency measurement involves verifying that an employee has mastered a specific level of a competency and may be accomplished via peer assessment, supervisory assessment, accredited external certification, or internal board certification.

Including measures of learning in formal appraisal processes is a critical element in creating and sustaining CRCL. Competency certification is one approach to clearly specify the knowledge and skills that employees are expected to acquire and to track the extent of CRCL.

Researchers should examine the effect that competency-based certification requirements have on other CRCL activities. On the one hand, certification programs may focus attention on only a limited set of learning goals (as specified by the organization), and thereby diminish effort devoted to other CRCL activities. Alternatively, certification programs may remind employees of the value of CRCL and thereby encourage employees to actively pursue CRCL in other areas (including areas where there is no formal certification). In sum, organizations will benefit by better understanding how certification programs (which can be thought of as organizationally-driven CRCL) affect more voluntary, self-directed CRCL activity.

Compensation

Compensation policy is another area where organizations can use their human resource practices to encourage CRCL. Heneman and Gresham (1998) have stated that the continual upgrading of employee knowledge, skills, abilities, and competencies will more likely be accomplished with performance-based pay plans than with traditional pay plans. Specifically, they propose that skill-based pay or competency-based pay appear better suited than other pay plans to sustaining CRCL. Skill-based pay plans benefit the organization by enhancing the skill and versatility of its work force, and they create a culture where continuous learning is explicitly valued and rewarded (London & Smither, 1999). The use of such plans appears to be increasing. For example, in a 1990 survey, 51% of companies reported use of competency-based pay as compared with 40% in 1987 (Lawler, Ledford, & Chang, 1993). However, to date, there is too little research to draw conclusions about the effect of skill-based pay or competency-based pay on productivity (Heneman & Gresham, 1998).

The use of skill-based pay is an example of creating a climate that explicitly values and reinforces CRCL. But skills should be clearly and precisely defined and skill assessment needs to be accomplished via valid certification programs such as those described above. Research is needed to understand the effect of skill-based pay plans on the acquisition of skill and knowledge that are not rewarded by the plan. Also, it would be helpful to know more about the effect of skill-based pay on performance (in addition to understanding its effect on learning). Finally, early research (Mericle & Kim, 1997) suggests that some employees may be more likely than others to respond positively to skill-based pay plans. Identifying the characteristics of such employees will also be an important undertaking.

Summary

We believe that it is important to consider the combined effects of the HR practices described above, rather than viewing these practices as independent and unrelated to each other. Three views or perspectives concerning HR effectiveness have recently been described in the literature dealing with the economic impact of strategic human resource practices (Delery & Doty, 1996). A universalistic view argues that a particular HR practice (e.g., providing multi-source feedback to help guide CRCL) should be effective regardless of the firm's business strategy or culture. A contingency view argues that a particular practice will be effective only if it fits the firm's business strategy (e.g., multi-source feedback will prompt CRCL in firms with a prospector strategy but not in firms with a defender strategy; Miles & Snow, 1984). A configurational view argues that a particular practice will be effective only if it fits the firm's business strategy and is reinforced by a host of other HR practices (e.g., multi-source feedback needs to be embedded in a climate that encourages innovation and risk taking, offers support from co-workers, opportunities for challenging work and formal development, and recognition and rewards for learning). Configurational perspectives emphasize the pattern (and internal consistency) of HR practices and suggest that a configuration or pattern of HR activities can have synergistic effects. Configurational perspectives also presume equifinality; that is, there may be several unique (internally consistent) configurations of HR practices that can result in maximal performance.

Ultimately, we hypothesize that a configurational view is likely to be important in explaining CRCL. That is, CRCL is not likely to be encouraged and sustained by implementing only one or two HR practices (e.g., multi-source feedback and formal training opportunities). Instead, organizations will need to consider how a host of HR and management practices combine to shape the firm's culture and thereby provide a context which encourages CRCL and avoids sending mixed messages. Such a configurational view will probably be important to develop core competencies that the organization can use to create and sustain a strategic, competitive advantage.

Testing universalistic, contingency, and configurational hypotheses has implications for research design and analysis. Clearly, it is important to gather information about each firm's business strategy as well as many of the firm's HR practices. Contingency effects can be demonstrated by showing an interaction between HR practice and business strategy (e.g., a specific practice is effective only in a firm with a specific strategy). Delery and Doty (1996) illustrated approaches to test configurational hypotheses.

ADDITIONAL RESEARCH CONCERNS

Measuring CRCL

In this section, we offer some examples of how CRCL could be measured. As emphasized above, measuring CRCL requires longitudinal assessment that considers formal, institutional development activities as well as the informal, experiential learning that occurs in the course of trying to confront and solve real-world problems in organizations. This points to the importance of measuring a wide variety of behaviors that may serve indicators of CRCL.

For example, feedback seeking should measure the extent to which people initiate constructive discussions with supervisors or other co-workers about performance problems, ask them to critique or provide feedback about performance, seek advice or ideas about how to perform more effectively, identify gaps between their current performance and others' expectations, and use feedback from others to set learning or performance improvement goals.

Examples of development activities include initiating a career planning discussion with a supervisor or mentor, reading books or articles to help develop skills or acquire business knowledge, attending training programs, workshops, or conferences (on or off work time), using computer-based training software, videos, or audiotapes, asking for assignments (including task force or committee assignments) that will help develop skills, setting learning goals and making commitments to others to improve specific skills, practicing a new skill, requesting coaching from a supervisor, mentor, or other co-workers, reading business publications and newspapers, keeping up-to-date on industry practices or professional/technical developments, asking others for information about the organization or emerging technologies, and suggesting to others how the job can be performed more efficiently. It will also be important to look at relational measures that reflect mutual learning and reciprocity via teams, projects, and information networks (Hall, 1996).

Although organizational influences are important, personal concerns about CRCL may transcend organization-specific interests. Because of this, it is essential to gather information about CRCL that occurs independently or outside of the organization in which the learner is currently employed. Thus, data gathered from personnel files, while important, will not by itself provide a complete or accurate indication of the level of CRCL. Similarly, measures of traditional career progress (progression through ladder-like career sequences) will also provide limited insight about the consequences of CRCL (Hall, 1996).

To capture CRCL, researchers will need to gather data concerning activities over a realistic period of time (not merely a snapshot of activity at a single time). This points to the value of longitudinal studies (rather than cross-sectional research).

It is also possible that mandatory development activity may be negatively related to voluntary activity because employees have limited time to devote to development (Noe & Wilk, 1993). Therefore, it is important to measure the level of mandatory development activity, even though such activity is not part of our definition of CRCL.

Finally, self-report measures of CRCL may be more complete and accurate than reports from supervisors, peers, or archival records. That is, co-workers are not always in a position to observe the CRCL activities of others (especially when such activities occur outside of the organization or are not related to the person's current work or aspirations within the organization). However, the reliance on only self-report measures creates potential problems (e.g., method variance) when employees provide self-reports of developmental activity and personal characteristics (e.g., personality, self-efficacy), especially when all data is collected in a single session. One obvious solution is to gather data from co-workers as well as self-report data. These two sources can provide a more complete picture than either source alone.

Evaluating the Effects of CRCL

In addition to knowledge and skill acquisition, there are a number of outcomes (criteria) that should be considered when evaluating CRCL. These include seeking and participating in further learning, increasing tacit knowledge (e.g., about how to get things done in the context of the organization's culture), identifying new opportunities in which learning can be applied, the impact on the learner's job performance and career path (e.g., more challenging or less routine assignments, career advancement, increased earnings), affective reactions (e.g., reactions to the learning process, job satisfaction, energy, feelings associated with growth and change), changes in feedback seeking (including the frequency, sources, or kind of feedback), impact on self-efficacy and self-esteem, evidence of creativity and innovativeness (e. g., designing products or delivering services in new ways), the short- and long-term impacts from action learning, multi-source feedback about behavior change, and the extent to which learners become more versatile (i.e., able to use a variety of different skills and approaches to address problems).

Additional Variables

We have identified many of the personal, situational, and organizational variables that are likely to play a key role in guiding and shaping CRCL. We believe it is important to measure both personal and organizational/situational variables in studies of CRCL. Only by doing so will we be able to understand whether the effects of such variables are additive, as well as understanding whether some variables moderate the effects of other variables in the model.

Of course, many other variables may also affect the course of CRCL. Demographic variables of interest may include position (e.g., higher level employees receiving more CRCL opportunities), tenure, job experience, occupation (e.g., requiring updating to avoid obsolescence), education, age, race, and distance from career goal (e.g., Noe, 1996, found that distance from career goal was positively related to developmental activity). Other personality variables of interest may include optimism, need for control, social anxiety, self-esteem, uncertainty orientation, job involvement (i.e., work is a central life concern), and tolerance of ambiguity. Other individual characteristics of interest include awareness of career interests and needs (e.g., the extent to which someone has engaged in self-assessment or career exploration, and agrees with the firm's assessment of their developmental needs), career insight (knowledge of career-related strengths and weaknesses and having a career goal), skills of introspection (e.g., the extent to which someone reflects on or thinks about what works and doesn't work and the skills needed to enhance performance or career options), and employees' perceptions of their 'obligations' (via their implicit contracts) to participate in CRCL. It is also worth noting that some characteristics may be of interest because they may act as moderator variables. For example, some learners may consider co-worker support important while others do not. This, in turn, would affect the correlation between other variables (e.g., the presence of co-worker support) and CRCL.

CONCLUSION

A host of economic and organizational trends have accelerated the need for employees to engage in career-related continuous learning. This is true for employees who intend to remain in the same job, as well as for employees who wish to change jobs or careers. The ability of employers to harness the benefits of CRCL will affect the skill and motivation of their work force, and their ability to harness intellectual capital to create sustained competitive advantage.

Yet, survey data indicates that only a small portion of the work force participates in employer-sponsored training each year. We have therefore emphasized the need to study the way employees pursue and use CRCL outside of traditional, institutional learning settings. We have described personal and organizational variables (e.g., antecedents, consequences, moderators) that are likely to shape CRCL. Employers who are concerned with CRCL need to reformulate many aspects of their human resource management system, including job analysis and design, selection, training, career development, appraisal, and compensation. Researchers who seek to study CRCL need to consider different designs and criteria than those associated with traditional training research. Moreover, CRCL is an area that challenges the scientist-practitioner model. Only by partnerships between scientists and practitioners will we be able to develop a research framework (based on data gathered in applied settings) that can guide the continuous learning that employees increasingly need and employers increasingly seek to harness.

REFERENCES

Ansberry, C. (1993, March 11). Workers are forced to take more jobs with fewer benefits: Firms use contract labor and temps to cut costs and increase flexibility. *Wall Street Journal*, p. 1.

Argyris, C., & Schön, D. (1978). *Organizational learning: A theory of action perspective*. Reading, MA: Addison-Wesley.

Arthur, M.B., & Rousseau, D.M. (1996), Introduction: The boundaryless career as a new employment principle. In M.B. Arthur & D.M. Rousseau (Eds.), *Boundaryless careers* (pp. 3-20). Oxford: Oxford University Press.

Ashford, S.J., & Cummings, L.L. (1983). Feedback as an individual resource: Personal strategies of creating information. *Organizational Behavior and Human Performance, 32*, 370-398.

Ashford, S.J., & Tsui, A.S. (1991). Self-regulation for managerial effectiveness: The role of active feedback seeking. *Academy of Management Journal, 32*(2), 251-280.

Bandura, A. (1986). *Social foundations of thought and action: A social cognitive theory*. Englewood Cliffs, NJ: Prentice-Hall.

Bandura, A. (1997). *Self-efficacy: The exercise of control*. New York: W.H. Freeman.

Barrick, M.R., & Mount, M.K. (1991). The big five personality dimensions and job performance: A meta-analysis. *Personnel Psychology, 44*, 1–26.

Bateman, T.S., & Crant, J.M. (1993). The proactive component of organizational behavior: A measure and correlates. *Journal of Organizational Behavior, 14*, 103-118.

Bridges, W. (1994). *Jobshift*. Reading, MA: Addison-Wesley.

Campion, M.A., Cheraskin, L., & Stevens, M.J. (1994). Career-related antecedents and outcomes of job rotation. *Academy of Management Journal, 37*(6), 1518-1542.

Carson, K.D., & Carson, P.P. (1997). Career entrenchment: A quiet march toward occupational death? *Academy of Management Executive, 21*, 62-75.

Carver, C.S., & Scheier, M.F. (1990). Origins and functions of positive and negative affect: A control-process view. *Psychological Review, 97*, 19-35.

Cascio, W.F. (1995). Whither industrial and organizational psychology in a changing world of work? *American Psychologist, 50*, 928-939.

Cervone, D. (1997). Social-cognitive mechanisms and personality coherence: Self-knowledge, situational beliefs, and cross-situational coherence in perceived self-efficacy. *Psychological Science, 8*, 43-51.

Cianni, M., & Wnuck, D. (1997). Individual growth and team enhancement: Moving toward a new model of career development. *Academy of Management Executive, 21*, 105-115.

Compeau, D.R., & Higgins, C.A. (1995). Computer self-efficacy: Development of a measure and initial test. *MIS Quarterly, 19*(2), 189-206.

Crant, J.M. (1995). The proactive personality scale and objective job performance among real estate agents. *Journal of Applied Psychology, 80*, 532-537.

Deci, E.L., Connell, J.P., & Ryan, R.M. (1989). Self-determination in a work organization. *Journal of Applied Psychology, 74*, 580-590.

Deci, E.L., Eghrari, H., Patrick, B.C., & Leone, D.R. (1994). Facilitating internalization: The self-determination theory perspective. *Journal of Personality, 62*, 119-142.

Delery, J.E., & Doty, D.H. (1996). Modes of theorizing in strategic human resource management: Tests of universalistic, contingency, and configurational performance predictions. *Academy of Management Journal, 39*, 802-835.

Driver, M.J. (1979). Career concepts and career management in organizations. In C.L. Cooper (Ed.), *Behavioral problems in organizations* (pp. 79-139). Englewood Cliffs, NJ: Prentice-Hall.

Dubin, S.S. (1990). Maintaining competence through updating. In S.L. Willis & S.S. Dubin (Eds.), *Maintaining professional competence* (pp. 9-43). San Francisco: Jossey-Bass.

Fenigstein, A., Scheier, M.F., & Bass, A.H. (1975). Public and private self-consciousness: Assessment and theory. *Journal of Consulting and Clinical Psychology, 43*, 522-527.

Festinger, L. (1954). A theory of social comparison processes. *Human Relations, 7*, 117-140.

Fierman, J. (1994, January 24). The contingency workforce. *Fortune*, pp. 30-36.

Frayne, C.A., & Latham, G.P. (1987). The application of social learning theory to employee self-management of attendance. *Journal of Applied Psychology, 72*, 387-392.

Gist, M.E., Stevens, C.K., & Bavetta, A.G. (1991). Effects of self-efficacy and post-training intervention on the acquisition and maintenance of complex interpersonal skills. *Personnel Psychology, 44*, 837-861.

Graddick, M.M., & Lane, P. (1998). Evaluating executive performance. In J.W. Smither (Ed.), *Performance appraisal: State-of-the-art in practice* (pp. 370-403). San Francisco: Jossey-Bass.

Hall, D.T. (1996). Protean careers of the 21st century. *Academy of Management Executive, 10*(4), 9-16.

Hall, D.T. (1996). Long live the career: A relational approach. In D.T. Hall & Associates (Eds.), *The career is dead, Long live the career* (pp. 1-11). San Francisco: Jossey-Bass.

Hall, D.T. (1976). *Careers in organizations*. Santa Monica, CA: Goodyear.

Hall, D.T., & Mirvis, P.H. (1996). The new protean career: Psychological success and the path with a heart. In D.T. Hall & Associates (Eds.), *The career is dead, Long live the career* (pp. 15-45) San Francisco: Jossey-Bass.

Hall, T.D., & Mirvis, P.H. (1995). Careers as lifelong learning. In A. Howard (Ed.), *The changing nature of work* (pp. 323-362). San Francisco: Jossey-Bass.

Hamel, G., & Prahalad, C.K. (1989, May-June). Strategic intent. *Harvard Business Review*, pp. 63-76.
Hamel, G., & Prahalad, C.K. (1994). *Competing for the future*. Boston: Harvard Business School Press.
Handy, C. (1994). *The age of unreason*. Boston: Harvard Business School Press.
Hazucha, J.F., Hezlett, S.A., & Schneider, R.J. (1993). The impact of 360-degree feedback on management skills development. *Human Resource Management, 32*, 325-351.
Heneman, R.L., & Gresham, M.T. (1998). Performance-based pay plans. In J.W. Smither (Ed.), *Performance appraisal: State-of-the-art in practice* (pp. 496-536). San Francisco: Jossey-Bass.
Hesketh, B., & Neal, A. (1999). Technology and performance. In D.R. Ilgen & E.D. Pulakos (Eds.), *The changing nature of work performance: Implications for staffing, personnel actions, and development* (pp. 21-55). San Francisco: Jossey-Bass.
Holt, K., Noe, R.A., & Cavanaugh, M. (1996). *Managers' developmental responses to 360-degree feedback*. Paper presented at the Annual Meeting of the Society for Industrial and Organizational Psychology, San Diego, CA.
Kanfer, R., & Ackerman, P.L. (1989). Motivation and cognitive abilities: An integrative/aptitude treatment interaction approach to skill acquisition. *Journal of Applied Psychology, 74*, 657-690.
Kaplan, R.S., & Norton, D.P. (1996). Using the balanced scorecard as a strategic management system. *Harvard Business Review, 74*(1), 75-85.
Kozlowski, S.W.J., & Hults, B.M. (1987). An exploration of climates for technical updating and performance. *Personnel Psychology, 40*, 539-563.
Lancaster, H. (1997, March 11). Managing your career: Companies promise to help employees plot their careers. *The Wall Street Journal*, p. B1.
Latham, G.P., & Frayne, C.A. (1989). Self-management training for increasing job attendance: A follow-up and replication. *Journal of Applied Psychology, 72*, 411-416.
Lawler, E.E., Ledford, G.E., & Chang, L. (1993). Who uses skill-based pay and why? *Compensation & Benefits Review*.
Lavalee, L.F., & Campbell, J.D. (1995). Impact of personal goals on self-regulation processes elicited by daily negative events. *Journal of Personality and Social Psychology, 69*, 341-352.
Levy, P.E., Albright, M.D., Cawley, B.D., & Williams, J.R. (1995). Situational and individual determinants of feedback seeking: A closer look at the process. *Organizational Behavior and Human Decision Processes, 62*, 23-37.
Locke E.A., & Latham, G.P. (1990). *A theory of goal setting and task performance*. Englewood Cliffs, NJ: Prentice Hall.
London, M., Larsen, H.H., & Thisted, L.N. (1999). Relationships between empowerment, feedback, and self-management for career development. *Group and Organization Management, 24*(1), 5-27.
London, M., & Mone, E.M. (1999). Continuous learning. In D. R. Ilgen & E. D. Pulakos (Eds), *The changing nature of work performance: Implications for staffing, personnel actions, and development* (pp. 119-153). San Francisco: Jossey-Bass.
London, M., & Smither, J.W. (1999). Empowered self-development and continuous learning. *Human Resource Management Journal, 38*(1), 3-16.
Maurer, T.J., & Tarulli, B.A. (1994). Investigation of perceived environment, perceived outcome, and person variables in relationship to voluntary development activity by employees. *Journal of Applied Psychology, 79*, 3-14.
Maurer, T.J., & Tarulli, B.A. (1996). Acceptance of peer/upward performance appraisal systems: Role of work context factors and beliefs about managers' development capability. *Human Resource Management, 35*, 217-241.
Miles, R., & Snow, C.C. (1984). Designing strategic human resource systems. *Organizational Dynamics, 13*(1), 36-52.

Neck, C.P., Steward, G.L., & Manz, C.C. (1995). Thought self-leadership as a framework for enhancing the performance of performance appraisers. *Journal of Applied Behavioral Science, 31*, 278-302.

Noe, R.A. (1996). Is career management related to employee development and performance? *Journal of Organizational Behavior, 17*, 119-133.

Noe, R.A., & Ford, J.K. (1992). Emerging issues and new directions for training research. In G.R. Ferris & K.M. Rowland (Eds.), *Research in personnel and human resources management* (Vol. 10, pp. 345-384). Greenwich, CT: JAI Press.

Noe, R.A., & Wilk, S.L. (1993). Investigation of the factors that influence employees' participation in development activities. *Journal of Applied Psychology, 78*, 291-302.

Nonaka, I. (1994). A dynamic theory of organizational knowledge creation. *Organization Science, 5*(1), 14-36.

Nonstandard jobs: A new look. (1997, September 15). *Business Week*.

Olson, C.A. (1997) *Who receives formal firm sponsored training in the U.S.?* Unpublished manuscript. School of Business & Industrial Relations Research Institute, University of Wisconsin-Madison.

Pearlman, K., Campbell, J., Gottfredson, L., Kehoe, J., DeVries, D., & Hackman (1995). *Is 'job' dead? Implications of changing concepts of work for I/O science and practice*. Panel discussion (K. Pearlman, Chair) at the Tenth Annual Conference of the Society for Industrial and Organizational Psychology, Orlando, Florida.

Pilegge, A.J., & Holtz, R. (1997). The effects of social identity on the self-set goals and task performance of high and low self-esteem individuals. *Organizational Behavior and Human Decision Processes, 70*, 17-26.

Porras, J.I., & Robertson, P.J. (1992). Organization development: Theory, practice, and research. In M.D. Dunnette & L.M. Hough (Eds.), *Handbook of industrial and organizational psychology* (2nd ed., Vol. 3, pp. 719-822). Palo Alto, CA: Consulting Psychologists Press, Inc.

Prahalad, C.K., & Hamel, G. (1990). The core competence of the corporation. *Harvard Business Review, 68*(3), 79-91.

Pyszczynski, T., & Greenberg, J. (1987). Self-regulatory perseveration and the depressive self-focusing style: A self-awareness theory of reactive depression. *Psychological Bulletin, 102*, 122-128.

Raelin, J.A. (1997). Individual and situational precursors of successful action learning. *Journal of Management Education*, pp. 368-394.

Reilly, R.R., & McGourty, J. (1998). Performance appraisal in team settings. In J.W. Smither (Ed.), *Performance appraisal: State-of-the-art in practice* (pp. 244-277). San Francisco: Jossey-Bass.

Rosow, J.M., & Zager, R. (1988). *Training: The competitive edge*. San Francisco: Jossey-Bass.

Rotter, J.B. (1966). Generalized expectancies for internal vs. external control of reinforcement. *Psychological Monographs, 80*, 1-28.

Rousseau, D.M. (1995). *Psychological contracts in organizations*. Thousand Oaks, CA: Sage.

Senge, P.M. (1990). *The fifth discipline: The art and practice*. New York: Doubleday.

Shapiro, B.P., Slywotzky, A.J., & Tedlow, R.S. (1997). How to stop bad things from happening to good companies. *Strategy and Business*, first quarter (6), 25-41.

Sims, D., & McAulay, L. (1995). Management learning as a learning process: An invitation. *Management Learning, 26*(1), 5-20.

Smither, J.W. (1995). Creating an internal contingent workforce: Managing the resource link. In M. London (Ed.), *Employees, careers, and job creation* (pp. 142-164). San Francisco: Jossey-Bass.

Snyder, M. (1979). Self-monitoring processes. In L. Berkowitz (Ed.), *Advances in experimental social psychology* (Vol. 12, pp. 85-128). New York: Academic Press.

Snyder, M. (1974). The self-monitoring of expressive behavior. *Journal of Personality and Social Psychology, 30*, 526-537.

Squires, P., & Adler, S. (1998). Linking appraisals to individual development and training. In J.W. Smither (Ed.), *Performance appraisal: State-of-the-art in practice* (pp. 445-495). San Francisco: Jossey-Bass.

Tracey, J.B., Tannenbaum, S.I., & Kavanagh, M.J. (1995). Applying trained skills on the job: The importance of the work environment. *Journal of Applied Psychology, 80,* 239-252.

Vaill, P.B. (1996). *Learning as a way of being.* San Francisco: Jossey-Bass.

Vroom, V. (1964). *Work and motivation.* New York: Wiley.

Wagner, R.K., & Sternberg, R.J. (1985). Practical intelligence in real-world pursuits: The role of tacit knowledge. *Journal of Personality and Social Psychology, 49,* 436-458.

Warech, M.A., Smither, J.W., Reilly, R.R., Millsap, R.E., & Reilly, S.P. (1998). Self-monitoring and 360-degree ratings. *Leadership Quarterly, 9,* 449-473.

Williams, J.R., Bublitz, S.T., & Melner, S.B. (1997, April). *Feedback seeking: The costs may in fact be real.* Paper presented at the 1997 Convention of the Society for Industrial and Organizational Psychology, St. Louis, MO.

Williams, R.L., Pettibone, T.J., & Thomas, S.P. (1991). Naturalistic application of self-change practices. *Journal of Research in Personality, 25,* 167-176.

Williams, R.L., Verble, J.S., Price, D.E., & Layne, B.H. (1995). Relationship of self-management to personality types and indices. *Journal of Personality Assessment, 64,* 494-506.

Whetten, D.A., & Cameron, K.S. (1995). *Developing management skills.* New York: Harper Collins.

TRAINING IN ORGANIZATIONS:
MYTHS, MISCONCEPTIONS, AND MISTAKEN ASSUMPTIONS

Eduardo Salas, Janis A. Cannon-Bowers,
Lori Rhodenizer, and Clint A. Bowers

ABSTRACT

Despite scientific advances in the field of training effectiveness research and millions of dollars invested in the field of training, the crucial question of why training research has not had as great an impact on practice as it could remains. One answer to this question is that many myths, misconceptions, mistaken assumptions and misnomers about the science and practice of training exist. This paper critically examines some of these myths with the goal of highlighting, examining and dispelling common myths, misconceptions and mistaken assumptions that may be driving the manner in which organizations approach training. Where applicable, research needs that are required to address and more fully remedy problems associated with the myth are suggested. It is our hope that this paper challenges the training community to continue to adopt a broader and deeper view of training in order to advance the science and practice of training.

INTRODUCTION

There is no doubt that almost every employee in any organization will go through some form of training during their careers. Clearly, organizations believe that training can be a crucial factor in enhancing organizational effectiveness. This is evidenced by the fact that collectively, American industry invests 200 million dollars in this activity each year (Carnevale, Gainer, & Villet, 1990). To maximize the return on this massive investment, we must understand what training effectiveness is and which factors influence it. Fortunately, in recent years, there has been a resurgence of research aimed at understanding the nature of training effectiveness. We have witnessed a plethora of theoretical and empirical work that begins to suggest how organizations can achieve training effectiveness (see Goldstein, 1989; 1993; Quiñones & Ehrenstein, 1997; Smith, Ford, & Kozlowski, 1997; Tannenbaum & Yukl, 1992). A number of trends have contributed to this state of affairs.

First, human resource specialists have increasingly demanded answers to **how** and **why** training interventions work. Therefore, scientists and practitioners have sought to establish general principles regarding how to achieve training effectiveness (see Salas, Cannon-Bowers, & Blickensderfer, 1997). Second, the simplistic view of training (that the training program itself is solely responsible for skill acquisition) no longer holds. Training research now demonstrates the importance of variables such as pre-training and post-training conditions, organizational climate, and motivation (Cannon-Bowers, Salas, Tannenbaum, & Mathieu, 1995; Kozlowski & Salas, 1997; Rouiller & Goldstein, 1997; Tannenbaum, Mathieu, Salas, & Cannon-Bowers, 1991; Tracey, Tannenbaum, & Kavanagh, 1995). Third, the cognitive revolution in psychology has also had an impact on training research. Cognitive theories, methods, and tools have driven (and will continue to drive) much of the recent training research (e.g., Kraiger, Ford, & Salas, 1993; Smith et al., 1997; Tannenbaum & Yukl, 1992). Fourth, new technology (e.g., intelligent systems) has made it possible to explore innovative learning strategies such as distance learning, distributed training, just-in-time training, and collaborative learning (Cannon-Bowers, Burns, Salas, & Pruitt, 1998). Finally, the continuous learning "movement" in organizations (i.e., the notion that training should not be viewed as a single event, but as a continuous process) has also generated new ways of looking at training with a focus on performance assessment and diagnosis in the work environment itself (Tannenbaum, 1997).

Despite these advances and all of what we know about training effectiveness, a crucial question remains: Why does training research not have as great an impact on practice as it could? Dipboye (1997) suggests that organizational barriers, such as negative attitudes concerning the scientific approach to training, the trainee's acculturation in the organization, and the lack of fairness with which training is implemented, prohibit the implementation of a rational instructional systems approach to training within many organizations. Others have argued that it is a lack of a "translation mechanism" in the training community. In fact, some have recently called for the need for more reciprocity between scientific findings and the practical implications of these (Salas et al., 1997) as a way to impact training practice. Salas et al. (1997) suggested a framework for this and illustrated how it might work. However, much needs to be done here if human resource researchers are ever to influence organizational practices and effectiveness.

Another answer to the question of why training research is not implemented is that many myths, misconceptions, mistaken assumptions and misnomers about the science and practice of training exist. This paper critically examines some of these myths with the objective of dispelling them and suggesting new opportunities for research. It is our hope that this paper challenges the training community to continue to adopt a broader and deeper view of training in order to advance the science and practice of training.

FRAMEWORK FOR CONCEPTUALIZING TRAINING

Before addressing the myths in training, we first describe a framework that will help organize our thoughts. Recently, training has been conceptualized as a multifaceted framework that includes factors that occur around, before, during, and after training (e.g., Cannon-Bowers et al., 1995; Noe, 1986; Noe & Schmitt, 1986; Tannenbaum et al., 1991; Thayer & Teachout, 1995; Tannenbaum & Yukl, 1992). This new framework of training effectiveness transcends traditional instructional systems approaches that may be viewed as inflexible because it also considers "characteristics of the organization and work environment and characteristics of the individual trainee as crucial input factors" (Cannon-Bowers et al., 1995, p. 143). The major features of this training effectiveness framework are summarized in Table 1 (for a thorough review of training effectiveness models see Ford, Kozlowski, Kraiger, Salas, & Teachout, 1997; Tannenbaum, Cannon-Bowers, Salas, & Mathieu, 1993). As indicated in the table, the variables that influence training effectiveness can be categorized into organizational influences, pre-training, training, and post-training factors.

Organizational influences are the factors that shape how training will be implemented in organizations as well as how it will be perceived by employees. These factors set the tone or climate for the design and delivery of training. For example, the kind of support supervisors give employees to apply the newly acquired skills or the opportunities to practice will dictate transfer of training (see Ford, Quiñones, Sego, & Sorra, 1992). Research clearly indicates variables such as the context, the organizational climate, the notification process and the support for learning have an influence on training effectiveness (e.g., Tannenbaum & Yukl, 1992; Quiñones, 1997).

Pre-training factors (see Table 1) include individual characteristics, needs assessment issues, and motivational characteristics that exist prior to the conduct of training. For example, pre-training factors include the trainee's motivation for training, prior knowledge and abilities, and self-efficacy. Research has indicated that these factors can influence training effectiveness (Cannon-Bowers et al., 1995; Smith-Jentsch, Jentsch, Payne, & Salas, 1996; Tannenbaum et al., 1991).

Training design and delivery factors include the components considered in traditional instructional systems approaches to training. These factors follow a methodological and rational approach to training that begins with the identification of targeted knowledge, skills, and attitudes (KSAs) and development of training objectives. Instructional design principles (e.g., learning, practice, feedback), training objectives, and media required to train the targeted KSAs are delineated and specified here. An abundance of research exists which indicates that these factors contribute to the overall effectiveness of training (e.g., Goldstein, 1993;

Table 1. Variables in Typical Training Effectiveness Models

Organizational Influences	Pre-Training Factors	Training Design & Delivery Factors	Post-Training Factors
Organizational Climate: Situational constraints Trainee notification process	*Individual Characteristics:* Abilities Attitudes Self-efficacy	*Training Objectives:* Specificity Relevance Measurable	*Individual Characteristics:* Abilities Attitudes Self-efficacy
Organizational Support: • Opportunity to practice • Supervisor support • Cultural factors	*Competencies (based on needs assessment):* • Knowledge • Skills • Attitudes	*Application of Principles:* • Learning • Feedback • Practice • Diagnosis	*Climate for Transfer:* • Cues to perform • Consequences of performance • Follow Up
Philosophy on Training: Support for learning • Resources allocated	*Motivation to Learn:* • Reactions to previous training • Instrumentality • Expectations	*Instructional Strategies:* • Team training • Self-correction • Simulation-based • Event-based *Instructional components:* • Information presentation • Demonstration of KSAs • Practice of KSAs • Feedback on KSAs	*Motivation to Transfer:* • Support to practice • Rewards • Expectations *Maintenance Interventions:* • Relapse Prevention • Refresher Training

Kraiger & Jung, 1997; Salas & Cannon-Bowers, 1997; Smith et al., 1997; Tannenbaum & Yukl, 1992).

Post-training factors influence the transfer of training to the work environment. Similar to the pre-training factors, post-training factors include individual characteristics (e.g., ability, attitudes, and self-efficacy) in addition to the climate for transfer and maintenance interventions. These factors determine whether the skills learned in training will be used on the job. Researchers have found considerable support for the contribution of these factors to training effectiveness (e.g., Baldwin & Ford, 1988; Rouiller & Goldstein, 1997; Tannenbaum & Yukl, 1992; Thayer & Teachout, 1995).

As has been argued by Dipboye (1997), most organizations do not follow a systematic approach to training, thus ignoring the impact of many of the organizational influences on all of the pre-training, training, and post-training factors described above. For instance, the needs assessment, evaluation, and validation steps are often disregarded or considered "non-value" added. Further, the design phase is often presented without careful consideration of trainee characteristics, learning principles, and media selection.

MYTHS, MISCONCEPTIONS AND MISTAKEN ASSUMPTIONS

Stories in Greek mythology have been repeated countless times as they have been passed down through generations. As historians seek to reveal the realities mixed amongst the fiction in myths, many believe that the substance of much Greek mythology lies in the deeper and broader interpretation of the stories. It is in this sense that myths "supply models for human behavior" (Eliade, 1963). However, when myths are accepted unchallenged, they can lead us astray. In training, unchallenged myths and beliefs may serve as a basis for erroneous models upon which to base the design and execution of training programs. Further, some common assumptions about the nature of training, and misconceptions about how training should be designed also drive the way organizations view and conduct training. Some of these are oversimplifications of truisms (e.g., "practice makes perfect"); others are merely unchallenged assumptions (e.g., "organizations value training"); still others are mistaken beliefs (e.g., "attitude change indicates behavioral change"). Our goal here is to highlight, examine and dispel some common myths, misconceptions and mistaken assumptions that may be driving the manner in which organizations approach training. To do this, we provide an explanation of each of the myths, misconceptions, and mistaken assumptions (which we will refer to simply as "myths" from here on for the sake of simplicity) that we believe exist in many organizations. These are organized around the broad categories laid out in Table 1 (i.e., those associated with organizational factors, pre-training practices, training design and delivery and post-training factors). Where applicable, we also highlight research needs that are required to address and more fully remedy problems associated with the myth.

Myths Associated with Organizational Factors

As noted, several organization-level factors affect the way training is conducted, and hence its effectiveness. Some of the myths and misconceptions that exist at this level are relatively subtle and indirect. However, their impact on training effectiveness can be potent. According to Dipboye (1997), organizational characteristics, specifically, the organization's attitude toward training and the organizational context, can influence training effectiveness. The following myths apply:

Myth 1: Organizations Use Training to Train People

Do *organizations use training to train people*? A critical look at the reasons many organizations provide training reveals that they may use "training programs" to accomplish organizational objectives other than training their employees. For instance, organizations use training programs to show their employees the organization's concern for a controversial topic, such as sexual harassment or workman's compensation. In these instances, organizations are using training as a means for communicating to employees or providing awareness concerning a topic. By providing training, the organization appears to be responding to an organizational need. Instead, the organization may be attempting to protect itself from potential litigation or appear "politically correct." They may also engage in training for purposes of image or appearance. That is, in an attempt to improve its reputation in the community or appear "legitimate," organizations invest in training activities. In addition, organizations frequently use training as a perk or to reward an employee. Alternatively, the organization may be using the training program as a "quick fix" to a problem rather than as a real opportunity to train employees. Organizations (and individual managers) may also be "checking a box" by offering training that is perceived to be important, but that accomplishes little when it comes to actual performance change.

The problem with this practice is that it may sour employees to real training. After being forced to sit through sometimes hours of boring and seemingly worthless "training," employees may be reluctant to attend training that has real potential value. Hence, labeling everything as training may be detrimental to trainee motivation in the long run.

Research needs. Organizations send many unintentional messages to their employees regarding training. Yet, there is virtually no systematic research to understand these messages and capture the many implicit goals training may serve. For example, we need to know through what specific mechanisms (formal or informal) organizations communicate to employees how training activities are perceived.

Myth 2: Organizations Send a Message that Training is Valued

Whether intentional or not, many organizations *fail to send a message that training is valued*. According to a recent survey, this message is heard by many workers who report that they "don't have high regard for their" job training (Bassi, Benson, & Cheney, 1996, p. 6). It can be voiced in numerous ways. For instance, research indicates that organizations on average allocated only four days of training for their employees in both 1994 and 1995; that in-house training staffs in 31% of the organizations polled were downsized (reducing staff-to-employee ratios); and that budget cuts reduced internal training expenditures (Bassi & Cheney, 1996).

Recently, Zenger (1996) argued that training has become peripheral to organizations rather than integrated into the organization. Zenger reminds organizations that the purpose of training is to improve behavior on the job. "Training should be an expression by the organization about the worth of its people and its eagerness to invest in them" (Zenger, 1996, p. 51). Recent trends in training indicate that the size of many training departments is shrinking while the use of training sources outside the organization is increasing (Bassi et al., 1996). This trend could send a negative message to employees. For instance, as the reliance on outside consultants increases, trainers will have less information about the organizational factors that impact training effectiveness and the jobs for which they will be developing training. Outside consultants, who need to approach training from an entrepreneurial perspective, may react by developing training programs that are not a solid fit to the organization's needs. This could perpetuate employees' perceptions that training is not valued.

The circumstances under which many organizations provide training for their employees sends a powerful message that the organization does not value training. For instance, organizations often send employees to training after there has been an accident. Consequently, the message employees receive is that training is punishment for making a mistake rather than a preventative intervention that keeps them from making mistakes.

Additionally, many organizations rely heavily on the use of on-the-job training rather than organized training programs. This shifts the burden of responsibility of training new employees away from the organization (specifically, the human resources, training, or personnel department) and toward line managers and employees. The employee who conducts the training is often naïve to training design and learning theories and probably inattentive to the quality of training the new employee is receiving. Perhaps worse is the message to employees that training is considered an incidental activity, and not valued as a means to prepare employees.

Finally, the selection and notification of training participants can also be an indication of how much the organization values the training. Organizations are often reluctant to send their most valuable or productive employees to training because they do not wish to have their best employees away from work. Therefore, the least productive or most junior person is often sent to training as a "representative" who will bring the training back to the work group. In addition, employees are often ordered to training whether they believe they need it or not. Research indicates that trainee perceptions of their commitment to attend training, appropriateness and satisfaction with the training program, and motivation to learn are higher when the individual has the choice to attend a training program rather than being ordered to attend (Baldwin, Magjuka, & Loher, 1991; Hicks & Klimoski, 1987). However, mandatory training may be perceived as more valuable (Tannenbaum & Yukl, 1992). Further, many organizations do not monetarily support employees who seek additional (outside) training.

Research needs. Clearly, it is important to ensure that employee motivation to train is as high as possible upon entering training (see Tannenbaum et al., 1993). Research is needed to explicate organizational factors that affect trainee motivation to learn and apply the newly acquired skills. Along with selection and notification, such factors as trainee choice to attend training, training climate, and the way training is publicized in an organization all require further research. Perhaps more importantly we need to start focusing more on the design and delivery of on-the-job training (OJT). Since it is through how OJT is structured that organizations may communicate to employees the value of learning, it is necessary for us to focus more on how to deliver effective OJT. Therefore, research is also needed to develop techniques to prepare those employees who act as job coaches and deliver on-the-job training. These individuals require skills, such as coaching, feedback and assessment skills, in order to successfully deliver this type of training. However, there is little research available to guide preparation of such skills.

Myth 3: Training is Too Expensive

Organizations that spend millions of dollars on training are beginning to cut training budgets and training expenditures per employee (Bassi & Cheney, 1996). These cuts are indicative of an underlying belief in many organizations that *training is too expensive*. Organizations may have this misperception because their training efforts are misdirected. Training can be expensive for organizations if it is developed without thoughtful planning or purchased without sufficient scrutiny. For example, training wastes resources when it is purchased or developed without first adequately determining the training need. Moreover, in the absence of evidence that the training program is effective, money may be spent needlessly on ineffective training. The cost of the program and other hidden charges incurred with off-site training, such as travel, lodging, and per diem, compounded by the added expense of the loss of productivity due to the time away from work sums to a considerable expenditure, especially if the training fails to live up to the organization's expectations.

Rather than an expense, training should be viewed as an investment in the organization's employees (McSparran, 1993). McSparran (1993) suggests that wasted training and consequently wasted organizational resources, such as money, can be traced to four sources: failing to identify training deficiencies prior to training; not focusing enough attention on practice and feedback during training; failing to distinguish between a lack of skill and a lack of motivation; and failing to assess the trainability of employees. Like all good investments, the consumer or developer should thoroughly research the product before committing to the expense. For instance, being a good consumer involves assessing the organization's training needs and selecting (or developing) a validated training program that matches those needs. When developing training, it means investing in hiring training experts, rather than relying on task experts who may lack an understanding of the instructional process (more will be said about this later).

Conversely, there may be expense in *not* providing training. This is difficult to quantify but there need to be attempts to address the issue of how much better the organization could be if its employees were better trained. A more direct expense organizations may encounter for not training is litigation. For example, in *Canton v. Harris* (489 US 378), a police officer brought suit against the City of Canton in part because he was inadequately trained to perform duties that were required of him (Leazes, 1995). The *Canton* case is only one among a growing number. Leazes (1995) suggested that the best defense against the risk of law suits is a proactive decision not be vulnerable to risk by integrating training into daily operations. In this way, organizations are "minimizing harmful mistakes due to a lack of knowledge or understanding of required procedures, poor or outdated skills, or negligent attitudes that results in careless, poor quality service" (p. 176).

Research needs. In some respects, addressing and ameliorating the rest of the myths in this paper could dispel this myth. In other words, the better able the science of training is to provide sound principles to the practice of training, the more obvious training effectiveness will be. In addition, this myth could be addressed directly if more operational tests of training effectiveness were conducted and reported. Unfortunately, it is difficult to find field studies that demonstrate training effectiveness and value. Hence, organizations are left to select or develop training without empirical data to support the selected approach or to help select an approach from several. The development of models linking individual training outcomes to organizational outcomes and objectives may provide insight for increasing training utility and value (Kozlowski, Brown, Weissbein, Cannon-Bowers, & Salas, in press). More meta analyses of existing training effectiveness studies, such as one conducted by Burke and Day (1986), would also be useful.

*Myth 4: Applying the Instructional Systems Design Process
Ensures Training Success*

There is a simplistic assumption that *applying procedures such as the Instructional Systems Design (ISD) approach ensures that training will be successful.* The problem with this type of thinking is that the ISD process is often applied blindly, with little or no consideration for how well each of the steps fits the training situation. The consequence too often is a "check the box" mentality, where organizations believe that engaging in the steps in the process will guarantee success. Application of the ISD process tends to focus only on a small aspect of a training program rather than considering training as an ongoing system of experience (Kozlowski, 1998). Furthermore, it isolates training from the broader organizational context (Kozlowski & Salas, 1997).

This is not to say that the ISD process is without value. Quite the contrary, such a systematic approach to training is far better than a haphazard one. However, applying the process simplistically, without necessary thought, **and experience**, will not ensure success.

Myths Associated with Pre-Training Factors

As described previously, pre-training factors include individual, organizational, and motivational characteristics that exist prior to the conduct of training. Pre-training factors are often forgotten in the design of training programs, yet trainees' expectations, motivation, and self-efficacy have been shown to impact training effectiveness (Tannenbaum et al., 1991). The following myths relating to pre-training factors have been identified.

Myth 5: All Trainees Come Equally Ready for Training

Many training programs assume that *all trainees come equally ready for training.* Trainees often do not enter a training program equated on ability, prior knowledge, experience, or motivation (Tannenbaum et al., 1993). When an individual enters a training program, he/she brings his/her experiences or prior knowledge, ability, motivation, and expectations to the training session; collectively, these factors help define a person's "trainability" (Cannon-Bowers et al., 1995). Research on trainability tests indicates that they can predict short-term training success and job performance (Gordon, Cofer, & McCullough, 1986; Robertson & Downs, 1989; Tannenbaum et al., 1993; Wexley, 1984). The consequences of making the assumption that all trainees will be equally ready for training are that the training may be ineffective for the majority of the attendees. Advanced trainees may already know the material; while less advanced trainees may not have the requisite knowledge or experience to benefit from it.

On the bright side, research into several pre-training (or pre-practice) interventions indicates that several rather simple additions implemented prior to training can help to make training more effective (see Cannon-Bowers, Rhodenizer, Salas & Bowers, 1998). These include advance organizers, attentional advice, pre-training briefs, preparatory information, metacognition, and goal orientation. In all cases, these interventions can help to bring all trainees to a level where they will benefit maximally from training.

Myths Associated with Training Design and Delivery

As described by the training effectiveness model, training factors include variables directly related to the design and conduct of training. These factors are best described as corresponding to an instructional systems design model, which includes components such as a needs assessment, statement of training objectives, and the development and delivery of the instructional phase. These factors follow a rational approach to training, which begins with a needs assessment and determination of competencies.

Needs assessment phase. First, the needs assessment phase of training provides the foundation for a training program by analyzing training needs at the organizational, person, and task level. Essentially, a determination of where within the organization training is needed, who needs training, and what tasks they need to learn must be made. Providing training without first answering these questions can lead to the development of ineffective, or use of inappropriate, training programs. Most notably, trainees may not learn skills required to successfully perform their job. For example, team training may be given within a department supposedly to increase teamwork. However, if teams are not the primary work unit, then employees will not use the skills they acquired in training. Conversely, individuals who work in a team may receive training to perform job tasks, but never undergo team training. Unfortunately, it often appears that the latest training fad is often the highest training priority. Gordon (1994) reminds trainers to tend to organizational needs and goals rather than getting sidetracked by the latest trends in training.

It is often the case that training programs are designed based on a simplistic view of the task demands, which are gleaned from the task analysis. Training developers need to have a rich and detailed understanding of the trainee's job responsibilities and competencies. Training developers can begin to assemble this information through various forms of task analysis, including techniques such as team task analysis, cognitive task analysis, and organizational analysis. This understanding must drive training as well as measurement design. The following myths have been identified that relate to this phase of training design.

Myth 6: Performing a Task Analysis Ensures Training Success

Does *performing a task analysis ensure training success*? Clearly, not all task analyses are created equal. Task analysis data can be collected through at least nine different techniques, such as interviewing subject matter experts and job incumbents, questionnaire data, observation, and print media (see Goldstein, 1991 for others). However, each of the different techniques used to collect job and task analysis data has advantages as well as disadvantages that can influence the quality and accuracy of the information obtained.

The type of data that is collected is equally important. It is necessary to select the most appropriate type of task analytic procedure. For instance, a behavioral task analysis reveals the steps needed to perform a complex task, while a cognitive task analysis uncovers the strategies and cues experts use in performing a task (Klein, 1995). Similarly, a team task analysis examines the components of teamwork (Bowers, Baker, & Salas, 1994), but does not yield information about individual requirements. Consequently, the effectiveness of any task analysis is dependent upon the degree to which the appropriate analysis approach has been selected and employed.

A task analysis is crucial to the needs assessment phase of training development since this information then forms the basis upon which the training program is designed. However, a task analysis is not the only part of the needs assessment that must be performed. It is equally important to determine *who* and *where* in the organization training is needed. It is possible that training may not be the answer to the organization's problem. In addition, recently, organizations have begun to acknowledge that training needs should be integrated with and support the organization's goals and strategic directions (Martocchio & Baldwin, 1997). Therefore, providing training based solely on a task analysis will not yield the most effective training program.

Further, it is important to assess the information that is revealed during a task analysis and carefully select those skills that should be trained. Not all of the skills revealed during a task analysis should be trained. Training should focus on those skills which are used frequently and are necessary for successfully performing the task (Goldstein, 1993). Finally, attention must be given to how *best* to train targeted KSAs within organizations.

Research needs. Research to improve the task analysis process as it applies to training development is required. In particular, if the relationship between task analysis results and training effectiveness were made, it could highlight the importance of investing in the task analysis before developing training. For example, knowing that one task analysis process led to training that was more effective than another would help to highlight the importance of this process.

Two other areas of research related to task analysis are crucial. First, research is needed to better establish task analysis techniques for teams (Baker, Salas, & Cannon-Bowers, 1998). Given the prevalence of teams it is necessary to better understand team competencies and their relationship to training.

Second, research into cognitive task analysis is required. Cognitive task analysis (CTA) seeks to document the cognitive aspects of the task such as decision making, problem solving, planning and the like (Gordon & Gill, 1997). Efforts to extend and apply CTA in organizations are required as well as the development of easy-to-use tools for conducting one.

Myth 7: Subject Matter Experts Can Articulate Training Needs

It is common practice to consult subject matter experts during the needs analysis. Data from American Society for Training and Development's benchmarking forum indicates that 63% of the polled organizations reported that training was developed based on requests from field or line managers while only 28% either always or frequently relied on a training needs analysis report to initiate training (Bassi & Cheney, 1996). This practice stems from the belief that *subject matter experts can articulate training needs*. It is hard to deny that subject matter experts (SMEs) can provide an excellent source of information about their task domain, and the use of experts during the needs analysis phase is often recommended. However, the use of subject matter experts has limitations as well, particularly when they are the only source of data. The utility of subject matter experts is limited for two reasons. First, subject matter experts have domain knowledge, but not necessarily knowledge about the organization's goals. Second, eliciting task domain knowledge that is useful from subject matter experts can be difficult.

The first problem, that many organizations have the misconception that subject matter experts know enough about organizational goals to define training needs, is an unspoken assumption in many cases. Clearly, SMEs can provide insightful information about the knowledge, skills, and attitudes that are required to successfully perform on the job, and can rate the importance of these tasks (Cranny & Doherty, 1988; Meister, 1985). Therefore, subject matter experts are often relied upon to articulate training needs at the task level. However, subject matter experts may have little insight into the organizational and person analysis aspects of a needs analysis. For example, a subject matter expert may not know that an organization is planning to update equipment in the near future. New equipment may change the skills that need to be trained. Since training is provided in the context of an organization, organizational variables, such as climate, growth, and development, impact training and should not be neglected. An organizational analysis will reveal the ways in which these variables will influence training.

A second reason that SMEs are not always the best source of data is that, while they probably know much about their task domain, they will have a hard time articulating what they know. This is because expert knowledge seems to be encoded in a way that makes it difficult for experts to express (Cooke, 1994). In fact, SMEs often cannot explain how they perform their jobs; however, they are often relied upon for information concerning cognitive tasks, such as decision making, and problem solving. Consequently, a science has evolved investigating techniques to elicit knowledge from experts (Cooke, 1994). In a comprehensive review of knowledge elicitation techniques, Cooke (1994) outlines the many techniques that can be used to elicit knowledge from experts. Some of these techniques include interviews, observations, critical incident approaches, process tracing, and conceptual techniques. The demand placed on the expert changes as a function of the technique.

Even when they are used, SME input must be treated with caution. In the end, SME ratings are subjective judgments. Hughes and Prien (1989) investigated SME job analysis judgments and found that a panel of SMEs had lower interrater agreement than is usually found for panel judgments. Additionally, SMEs linked all of the tasks to a job skill thus resulting in uninterpretable data (Hughes & Prien, 1989). Landy and Vasey (1991) found that SME demographic characteristics, specifically sex and experience, influenced SME responses on a job questionnaire. Hence, when an organization chooses to use SME input in developing training, it must be done with care.

Research needs. We again note that research into knowledge elicitation and cognitive task analysis is needed. This is particularly true as tasks become more complex and require higher order skills. In addition, continuing research into the task rating process is needed. In particular it is important to determine how characteristics of the SME effects the judgements they make, and how such judgements may be made more accurate.

Training and Instructional Design

Once the needs assessment has been completed and instructional objectives have been derived, the training program is designed. During this phase, the training developer is faced with a number of decisions that greatly impact the learning environment. The training developer must determine the most appropriate medium to present information (e.g., lecture, on-the-job, computer based). Training developers also must remember principles of learning, select appropriate exercises, and consider whether individual or team training is needed (Salas & Cannon-Bowers, 1997). The following myths apply.

Myth 8: Individual Training is the Same as Team Training

Many organizations are beginning to rely on teams rather than individuals to perform work tasks. This shift has lead to a resurgence of research concerning teams and team training (Guzzo & Dickson, 1996). However, many organizations mistakenly believe that *individual training is the same as team training.* However, the principles used to train individuals do not always apply to teams. In fact, the literature is inundated with guidelines and principles for training individuals, but many analogous guidelines are absent from the team training literature (Swezey & Salas, 1992). In the absence of guidance, it is tempting to employ principles and guidelines used in individual training to teams.

Teams are more than just a collection of individuals (see Salas, Dickinson, Converse, & Tannenbaum, 1992). Further, contrary to popular belief, effective teamwork does not occur automatically. When individuals come together in a group, a team develops or matures over time (Morgan, Glickman, Woodard, Blaiwes, & Salas, 1986), but it does not necessarily perform most effectively. Instead, investigators have determined that the most efficient method for developing teamwork skills is to train teamwork behaviors directly (Nieva, Fleishman, & Reick, 1978; Salas, Cannon-Bowers, & Johnston, 1997).

For example, Morgan et al. (1986) emphasized that there are two distinct types of skills that team members need to acquire. Those skills include task-oriented skills as well as team-oriented skills. Task-oriented skills can be acquired individually; however, team-oriented skills are best acquired in team training. According to Cannon-Bowers et al. (1995), the nature of the training developed should rest on the nature of competencies (both team and individual) required for effective performance.

Research needs. While strides have been made in understanding teams and team training (see Salas & Cannon-Bowers, in press; Salas & Cannon-Bowers, 1997), there is still much to be done. For example, the teamwork competencies established by Cannon-Bowers et al. (1995) require further specification and empirical validation. Efforts to assess teamwork KSAs along the lines of Campion and associates (i.e., Campion, Papper, Medsker, 1996; Stevens & Campion, 1994) are also worthwhile.

In addition, further work regarding team training interventions is needed. Empirical evaluation studies that demonstrate the effectiveness and applicability of particular strategies, such as team debriefing skills and team self-correction, would be especially useful.

Myth 9: Lecture-Based Training is Superior to Other Training Methods

According to ASTD, lecture based training comprises approximately 70% of all training delivery time (Bassi & Cheney, 1996). This prevalence of training time spent in classroom (i.e., lecture-based) training logically leads to an assumption that many organizations feel that lecture-based training is superior to other training delivery mechanisms (Dulworth & Shea, 1995). Although lecture based training may be quite efficient as a method to train large numbers of people simultaneously, it is not always very effective. However, the scientific literature certainly provides guidance for designing other types of training that enhances learning, retention, and transfer. In fact, this literature **does not lead to the conclusion** that *lecture-based training is superior to other training methods.* Some of the methods which training developers have available to them include paper-and-pencil exercises, role play simulations, computer based training, and simulators. Clearly, the effectiveness of the method varies with the specific nature of the competency being trained and the size of the group.

For instance, paper-and-pencil exercises are often used in training declarative knowledge. Role play simulations are effective for training interpersonal skills, human relation skills (Bass & Vaughan, 1966), and in managerial training. Computer-based training (CBT) can be used to meet a variety of training objectives and to acquire a number of skills ranging from equipment maintenance to interpersonal skills. The advantages of CBT are in data collection and learner flexibility. Simulators are effective at training motor skills, perceptual-motor skills, and procedural skills.

In designing a training program, it is necessary to examine the skills that are being trained in order to determine the most appropriate method to use for practice (Salas & Cannon-Bowers, in press). It appears as though there is a tendency to use a training exercise that happens to be the current trend or *en vogue*. For instance, management games are widely used as a training method, but it is often unclear what trainees learn from the exercise. Moreover, they are often not evaluated on the skills they should be learning, such as interpersonal and problem solving skills (Tannenbaum & Yukl, 1992).

Myth 10: Practice Makes Perfect

One common misconception about training is that *practice equals training*. This assumption can be readily observed in many tasks in which trainees are provided with practice time without the benefit of instructor guidance. However, Sterns and Doverspike (1989) distinguish practice from training by stating that training attempts to "control the nature of experience acquisition," while simple practice lacks control and direction (p. 314). In other words, a training program identifies the skills that should be practiced, determines how these skills should be practiced, and then provides feedback in attempt to maximize the benefits from the time spent in practice.

Clearly, practice is a necessary means to acquire and hone skills. In fact, Anderson (1983) proposed that individuals learn through direct experience with the task. Consequently, the importance of practice sessions during and after training cannot be overstated. However, research concerning the nature of practice indicates that the *conditions* of practice influence the effectiveness of practice (Cannon-Bowers, Salas, Bowers, & Rhodenizer, 1996). Essentially, conditions before, during, and after practice can impact the effectiveness and utility of the practice session. The training system must take into consideration all of the conditions that can affect practice. Practice in the absence of training, therefore, may not produce desired effects. An individual may mislearn the task or develop inappropriate mental representations of the task (Cannon-Bowers, Salas, & Converse, 1993; Kieras & Bovair, 1984).

It should also be noted that training which provides simple *exposure to a task is* not sufficient for learning. Although task exposure sessions can be informative and those who attend may become familiar with the task, they are not training. The operational difference between task exposure and training is that after exposure to a task, an individual may have enough knowledge to know how and where to get the appropriate information about the topic. In contrast, after training an individual will have the skills to perform to task. Overall, very little information will be retained after simple exposure (Goldstein, 1993).

On-the-job training, in many cases, may actually be exposure to the task rather than training. Typically, on-the-job training is informal and is conducted by a job incumbent. On-the-job training is often provided in lieu of designing a formal training program and incurring the associated costs. When on-the-job training is conducted informally, with no clearly defined objectives and no practice and feedback, there is no assurance that the trainee will learn all of the skills needed to perform the job or trainees may learn more slowly via trial and error (Tannenbaum, 1997). Essentially, all of the elements that contribute to successful training are absent. Taking these elements into consideration, on-the-job training can be designed such that effective training results (Sullivan & Miklas, 1985). In fact, on-the-job training could have advantages over classroom training because transfer is nearly guaranteed since the practice environment is identical to task performance environment (Goldstein, 1993).

Research needs. In general, researchers need a better understanding of how and why practice leads to learning (see Cannon-Bowers et al., 1998). For instance, the context of practice, part vs. whole task practice, the timing of practice, and the type of practice can all impact the effectiveness of the practice session. There is a need to emphasize practice in the context of the training system. In other words, research must focus on realizing the most appropriate methods for practicing complex tasks, such as those found in many work settings and examining transfer issues as they relate to practice (Schmidt & Bjork, 1992). Additionally, practice is nearly meaningless in the absence of feedback and learning strategies.

Trainers/Instructors

When considering the learning environment, trainer or instructor characteristics provide an important source of variance. It has been suggested that the trainer's characteristics can make or break a training program (Goad, 1982). The trainer plays an important role in training and can have a profound impact on the trainee and his/her success. Consequently, the mechanisms for selecting and preparing trainers must be carefully planned and executed. The following myth applies.

Myth 11: Everyone is an Expert in Training

Training development and delivery does not require magical powers or special innate abilities; however, it is erroneous to believe that *everyone is an expert in training*. Training development and delivery does require training in the requisite knowledge, skills, and attitudes in order to be successful in designing training. According to the ASTD's 1995 HRD Executive Survey, approximately 73% of respondents believe that training managers and staff lack the requisite skills and should "expand their competencies" in order to be prepared for the future trends in training (Bassi et al., 1996). This research indicates that many training practitioners in the field may not be adequately prepared to design and deliver training.

Being a successful instructor requires an understanding of learning principles that can positively benefit a trainee and the learning environment. Unfortunately, making the most competent employees the trainers, without proper preparation, will not ensure competent instruction. For instance, a trainer must be skilled at two levels: attitudinal (emotive) as well as instructional. That is, part of the trainers' task is to manage the attitudinal and emotive aspects of training. Trainers must be able to assess trainees' interest and stimulate or maintain the level of motivation during training (Goad, 1982). Trainers must also be knowledgeable about techniques for maximizing the learning environment. They must be educated about learning principles, exercise selection, media selection, and providing practice and constructive feedback. Trainers, like all other professionals, must be trained in order to perform their jobs (Lewick-Wallace & Jask, 1988). Goldstein (1993) presents a list of characteristics of good trainers, which includes, but is not limited to, organizational skills, clarity in presentation of material, and motivational in terms of goal setting and encouraging learning.

Research needs. The entire question of instructor training requires attention. Moreover, where research exists, results must be documented and translated into guidelines that may be used to guide the preparation of instructors. In addition, as with other myths, empirical data, particularly from field studies, is needed to determine the relationship between instructor training and overall training effectiveness.

Equipment/Media

Trainers have a wide variety of training technologies from which to choose. Effective training requires that technology be applied thoughtfully, paying special attention to the knowledge and skills that are to be trained. For example, Patrick (1992) believes an important point to make with respect to the use of computers in training is that the "technology per se does not guarantee that the quality of training is good" (p. 435). Rather, the underlying principles used in the training development are more important than the technology.

Myth 12: Simulation of Any Kind is Training

Since simulators are being incorporated into many training programs (Tannenbaum & Yukl, 1992), there is a tendency to think that *simulation of any kind is training*. Simulations can vary according to complexity (e.g., in-basket exercise to full-mission combat simulation) and fidelity (e.g., role play and desk-top computer simulations to motion-based simulation). Training developers must be cognizant of the fact that simulation and simulators are tools that can be used to facilitate training (Flexman & Stark, 1987). They provide an opportunity to practice in an environment that replicates important task characteristics. In many cases, simulators provide a safe practice medium for skill acquisition in high risk environments, such as aviation. Simulators also provide a highly reproducible and configurable environment that can provide practice on skills for low frequency events. It is important to remember when designing training programs, that simulation is a tool for training. Learning does not automatically result from the experience. It is important to control the environment by defining learning objectives, providing appropriate instruction, practice, and feedback, as well as debriefing trainees. Tannenbaum and Yukl (1992) emphasized the importance of a well designed training program that has been carefully planned, taking into consideration developmental stages of learning, in order to successfully incorporate simulation into a training program.

While the potential benefits of simulation have been discussed, the issue of training effectiveness remains. Tannenbaum and Yukl (1992) pointed out potential problems in evaluating the effectiveness of simulation-based training. For instance, simulators are often used in conjunction with other training methods, such as lecture, discussions, and demonstrations. Additionally, simulators are often used to train multiple learning objectives, some of which cannot be readily measured. A meta-analysis of simulator studies concluded that they can be an effective method of training aviators (Hays, Jacobs, Prince, & Salas, 1992). However, the effectiveness of simulators for training managerial skills has not been as clearly established (Tannenbaum & Yukl, 1992).

Another issue with respect to simulation is fidelity. Fidelity is the "degree to which the simulator represents the system and its environment" (Flexman & Stark, 1987, p. 1031). There is an assumption in the training community that high fidelity simulation provides better training than low fidelity simulation. The traditional belief is that the training environment should mimic the work environment as closely as possible in order for transfer to occur. However, fidelity is not limited to the physical features of a simulator or simulation exercise. Cognitive tasks, which often do not require equipment (e.g., training games, role play, etc.), require cognitive or functional fidelity (Hays & Singer, 1989). In these cases, the cognitive processes, such as decision making, must remain consistent across the training simulation task and the job task.

It is important to determine the degree of simulation fidelity that is needed when developing effective simulation exercises (Jacobs & Dempsey, 1993). It is not necessary to assume that a high fidelity simulation is needed to meet the specified training objectives. Research exists which demonstrates that skills can be trained in low fidelity simulation and transferred favorably (Flexman & Stark, 1987; Johnson, 1981; Patrick, 1992). For example, aircraft cockpit procedures have been trained successfully using low fidelity training devices. According to Johnson (1981), two categories of information are associated with the fidelity of procedural training devices: cueing the trainee and providing feedback.

Research needs. In order to successfully employ tools such as simulation, research is needed to more fully take advantage of its unique characteristics. For example, in helping trainees to develop appropriate knowledge structures, it may be necessary to determine and replicate important cues in the environment that experts use to guide performance. Research into the nature of complex environments, and the manner in which experts interact with those environments, can help guide the development of simulations (Stout et al., 1997).

It would also be of value to investigate methods to formalize simulation based training (Cannon-Bowers et al., 1998; Oser, Cannon-Bowers, Dwyer, & Salas, 1997). Scenario-based training must link the targeted training objectives with the exercise events, which in turn must drive the performance measures and feedback developed to accomplish training. Research to generate principles of simulation-based training would be useful.

Myth 13: "Free Play" is Training

In many tasks, trainees are allowed task exposure without the benefit of instructor guidance or performance feedback. This practice is indicative of the belief that *free play is training* because the trainee is engaged in exploratory learning about the task and is developing skills. Although exploratory learning may be effective in some instances (Kamouri, Kamouri, & Smith, 1986), evidence suggests that exploratory learning is not optimal (Bayman & Mayer, 1984; Frederiksen & White, 1989; Njoo & de Jong, 1993). In fact, free play is *not* training for several reasons. First, training is based on learning objectives that are typically absent in free play. Second, training occurs in a controlled environment in which the trainee is acquiring specific skills through performance measurement and feedback. Last, free play does not insure that skills acquired will transfer to the working environment.

Exploratory learning or training "refers to the acquisition of new information through activities initiated and controlled by the learner, which may include hypothesis testing and generation of positive and negative examples" (Kamouri et al., 1986, p. 172). It is believed that exploration training can impact learning in some tasks because it relies on the trainees "intrinsic interest" in the subject. Then, perhaps exploratory learning may be beneficial for some trainees but not all. Much of exploratory learning is based on analogical reasoning and the ability to apply existing knowledge to new situations. There are two potential problems with exploration-based training. First, it requires a well developed knowledge base that can be used to form hypotheses for testing (Frederiksen & White, 1989). Second, in the absence of a well developed knowledge base, erroneous conclusions may be reached and the trainee is left with incomplete knowledge concerning the topic.

Free play should not necessarily be abandoned as a training exercise. There are certainly ways in which free play can be turned into an effective training technique. For instance, experts may benefit from setting their own objectives and goals, with the help of trainers, to be obtained during free play sessions (Charney, Reder, & Kusbit, 1990). Additionally, technology may also be a viable method to support discovery learning (Bennett, 1992; de Mul & van Oostendorp, 1996).

Research needs. Clearly, there is a place for free-play or discovery learning. However, in general, the principles of practice we alluded to earlier should be applied to ensure that task exposure is managed and leads to effective training. Research is needed to uncover when discovery learning is appropriate and for what types of tasks.

Myths Associated with Post-Training Factors

Once a training program has been designed and developed, it is necessary to evaluate the training in order to determine its effectiveness. Goldstein (1993) defined evaluation as "the systematic collection of descriptive and judgmental information necessary to make effective training decisions related to the selection, adoption, value, and modification of various instructional activities" (p. 147). Essentially, the organization must determine whether the trainee learned the training objectives. Kirkpatrick (1959) suggested that training should be evaluated on four levels of criteria: reaction, learning, behavior, and results. Further, Kraiger et al. (1993) have proposed a construct-oriented approach to training evaluation, which is based on the notion that learning outcomes are multidimensional. However, industry often neglects training evaluation. A recent survey indicated that two thirds of the training departments surveyed did not evaluate their programs to "determine the return on their training investment" (Bassi et al., 1996, p. 11). Grove and Ostroff (1990) have identified several obstacles to training evaluation including managerial apathy toward evaluation, lack of understanding of evaluation techniques, and fear that the program will not be evaluated favorably. These obstacles have lead to some myths concerning training evaluation. These myths are discussed below.

Myth 14: Assessing Training Effectiveness Involves a Pre-test and Post-test Only

The problem with this mentality is that it is an oversimplification of the training process because it treats training as a "black box" (Salas, Burgess, & Cannon-Bowers, 1995). In other words, it tends to ignore the process of learning or skill acquisition, emphasizing instead the outcome that is observed as a function of training. This way of thinking causes training personnel to overlook the underlying mechanisms of task accomplishment and does not support a diagnosis of what needs to be changed in order to improve performance. For example, if a trainee does not perform well in training, the chances of improving this performance in subsequent episodes is enhanced when it can be determined that the cause of the error was a deficiency in a particular aspect of task knowledge. Such information suggests a remedial strategy that can ameliorate the deficiency. Conversely, if it is determined that a trainee does not perform well, but not why this is the case, performance problems can persist. Hence, performance is not maximized because important assessment information is not used to its full advantage.

It is also the case that in a competitive, turbulent environment, we need information that will enable us to act quickly (e.g., make changes, flex the workforce up or down). This means that we need to have a better handle on how well people are prepared to do their jobs, and how well they actually perform. It also implies that we are collecting information that will enable us to understand whether or not performance is acceptable (with respect to some standard), and if not, how it must be changed. The important point here is that this can all be done "on-line" in the actual work environment (i.e., that adjustments are made quickly in response to changing situational or task demands).

All of this suggests that a dynamic assessment process be adopted where performance information is tracked and interpreted as the task is being performed (Cannon-Bowers & Salas, 1997). This strategy is believed to be useful because it provides a more detailed picture of what employees know, and their level of proficiency. It also better supports the feedback and remediation process. When measuring dynamically, feedback can be specific, timely and process-oriented, all of which have been shown to improve performance (Dempsey & Sales, 1993).

Research needs. As noted, the concept of dynamic assessment (i.e., measuring performance in real time) is worthy of future study. This requires that we investigate how best to capture and interpret the moment-to-moment changes that occur in performance. In addition, methods to record performance (e.g., with the help of automation) on-line are required.

Another area of research that requires further study concerns the nature of typical versus maximal performance (Sackett, Zedeck & Fogli, 1988). This line of work suggests that an employee's actual, day-to-day performance levels (i.e., typical performance) will be lower than performance levels established in tests or single performance episodes (maximal performance). Importantly, typical performance describes what can actually be expected in the organization; hence measuring maximal performance can be misleading. If it were possible to measure performance dynamically, it would be easier to assess typical performance.

Myth 15: A Positive Reaction to Training Means that Learning Has Occurred

The first level of evaluation suggested by Kirkpatrick (1959) is the trainee's reaction to training. According to ASTD's Benchmarking Forum companies, 94% of the training courses offered were evaluated at trainees' reaction level (Bassi et al., 1996). Therefore, it seems that trainers believe that *a positive reaction to training means the trainee is learning*. Unfortunately, there is not necessarily a connection between the trainee's reaction to the training program and the trainee's learning (Campion & Campion, 1987). In fact, a meta analysis of studies evaluating training at each of Kirkpatrick's four-level model found a weak relationship between reactions and learning (Alliger, Tannenbaum, Bennett, Traver, & Shotland, 1997). The interpretation of the trainee's reaction to training should not be used to conclude that the trainee learned anything.

This is not to say that trainee reactions have no place in training evaluation. Alliger et al. (1997) found that questions concerning the utility of the training were moderately related to transfer. They can also provide an organization with information concerning the programs receptivity in the organization. Asking trainees for their reaction to the training program may be a quick and inexpensive way to determine if the program should be offered again. They can also indicate a trainee's motivation to attend training. Measuring trainee reaction in a mandatory training program may also provide insight into faults in the training design. The point is that relying on reactions, as the only evaluation data will not provide a true picture of the value of the training.

It is also important to consider the trainee's characteristics when completing a reaction questionnaire. A trainee may react negatively because he/she is not learning or perhaps because the training program did not meet his/her expectations. Alternatively, the trainee could be reacting to the trainer or to other organizational factors. For example, the trainee's reaction may be more indicative of the organization's climate than of any specific training feature.

Research needs. Further efforts to understand trainee reactions and their relationships to trainee motivation are required. Such research should include investigations of factors that contribute to trainee reactions, how reactions are associated with other outcomes (e.g., learning), and how they are related to training effectiveness. For example, trainee reactions may not predict learning, but they may influence the likelihood that the trainee will apply newly learned skills, as suggested by Alliger et al. (1997).

Myth 16: An Attitudinal Change Means a Behavioral Change

Some training programs are designed to effect a change in attitude. This may be a valuable exercise unless the assumption underlying such a program is that *an attitudinal change means a behavioral change*. In actuality, the reverse appears to be true. Typically, an attitudinal change is seen to follow a behavioral change

(Gagne & Briggs, 1979). Hence, training programs designed to change behavior must not concentrate solely on changing attitudes.

One of the most notable attitudinal training programs is found in aviation. Crew Resource Management (CRM) concentrates on crew member attitudes and behaviors and their impact on safety (Wiener, Kanki, & Helmreich, 1993). CRM seeks to improve teamwork between captains and co-pilots by fostering positive attitudes toward teamwork (Helmreich, 1984). The primary method of evaluation in CRM training is self-report questionnaires concerning the trainees' attitudes after training. The problem with this approach is that even if attitudes change, they are not necessarily predictors of behavior change (Smith, 1991). The trainee may report an attitudinal change regardless of whether a behavioral change has or will occur. Moreover, the "demand characteristics" in such a situation may cause trainees to report changes that they think are desirable. For instance, a trainee may not want to jeopardize his/her job by appearing to not have benefited from the training or, perhaps does not wish to revisit the training.

It should be noted that some training may be designed specifically to change attitudes. Our point is that this type of training, by itself, cannot be expected to change behavior or affect skill levels. However, it can be useful as a means to prepare trainees for more behaviorally-oriented training.

Research need. The impact of attitude change and its relationship to behavior change requires further research. As noted, there may be occasions in which an attitudinal change is the target of training. However, the relationship between attitude change and behavior is complex. Research is needed to further understand this relationship in training.

Myth 17: A Cognitive Change Means a Behavioral Change

Learning, by definition, is a cognitive change. However, it is false to assume that *a cognitive change means a behavioral change*. Learning does not always result in a behavioral change. An examination of the stages of learning may clarify this point.

Learning progresses through stages (Anderson, 1987). The first stage is referred to as the declarative stage. During this stage, declarative information (i.e., facts) is acquired. For instance, a trainee may rehearse the steps involved in executing a task. The second stage is knowledge compilation. During this stage, declarative knowledge becomes proceduralized. That is, the trainee learns to execute the procedure by performing the sequence of events involved in the procedure. The third stage is proceduralization. During this stage, an individual uses the knowledge that has been learned and applies it in a particular context. As the individual practices these skills, automaticity in response develops so that the trainee will be able to use the skill when required (Anderson, 1992).

Given this framework, it may be more apparent how a cognitive change (i.e., to the knowledge base of the trainee) can occur in the absence of behavioral (skill) changes. Returning to the example above, a trainee who receives a written test, as an evaluation of skill acquisition, may appear to have acquired the skill. In fact, he/she may not be able to successfully *execute* the skill; instead, he/she may have acquired the underlying knowledge (declarative knowledge), but not have learned the skills necessary to perform effectively. In order to determine whether a behavioral change has occurred, the skill must be measured behaviorally, not cognitively.

Research needs. The relationship between learning and behavior is complex. While trainees obviously need to learn targeted knowledge, doing so does not guarantee that skills have been learned. Further research into this relationship is required. In addition, there may be factors such as self-efficacy (or individual differences variables) that decrease the probability that a behavior change will be evident. Research is needed to investigate these factors as well.

Myth 18: Paper-and-Pencil Measures are a Good Indicator of Training Success

Evaluation of training includes a measurement of the skills that have been learned. The belief that *paper-and-pencil measures are a good indicator of training success* is not completely accurate. Performance measurement is only meaningful if the appropriate measures are employed. The appropriate measures are those which link back to the objectives of the training program. Therefore, a paper-and-pencil measure is inappropriate for the measurement of behavioral tasks. Paper-and-pencil tests measure declarative knowledge at best. However, the goal of most training programs is teaching trainees to perform a set of skills that will be used on the job. Therefore, performance measurement should be behavioral in nature.

In general, performance measurement is an essential tool for training. According to Cannon-Bowers and Salas (1997), performance measures in training must serve multiple purposes (Cannon-Bowers & Salas, 1997), address multiple levels (Prince, Brannick, Prince, & Salas, 1997), and tap multiple components which include cognitive outcomes, behavior skill-based outcomes, and attitudinal outcomes (Kraiger et al., 1993). Performance measurements must also provide an assessment or evaluation of trainee's progress and contributes crucial performance information that can be used to provide the trainee with feedback or knowledge of results. Similarly, performance measurement also aids in remediation, by determining skills, which are targeted for additional training.

Taken together, this body of research suggests that performance measurements need to be designed with a purpose and be theoretically rooted. They need to capture the complexity of learning, and describe, evaluate, and diagnose performance. In order to be most effective, performance measures need to be event-based, that is, based on training objectives, take into consideration the task requirements, and be linked to feedback.

Research need. Clearly, further work designed to improve the performance measures typically employed in training is required. Also, according to Cannon-Bowers and Salas (1997), performance measures in training must be diagnostic. Research is needed to improve the diagnosticity of measures employed in training. This means that a deeper understanding of the task, how it is performed, and what indicates that it is being properly learned must be achieved. In particular, it is necessary to investigate the relationship between observed performance, that is, those aspects of behavior that can be readily tracked, with what the underlying knowledge and skill levels might be. Such technology is crucial if intelligent and instructorless training systems are to be developed (Steele-Johnson & Hyde, 1997).

Transfer of Training

Obviously, training is provided in order to enhance job performance. Therefore, it is essential to determine whether the skills learned in the training program actually increase performance on the job. In order for transfer to occur, the trainee must learn the skills, generalize the skills to the work setting, and retain the skills for subsequent use. Unfortunately, training often does not successfully transfer to the work environment (Baldwin & Ford, 1988). Despite the billions of dollars that are spent on training, few organizations invest the time and money in validating training by assessing transfer. This may occur for several reasons (e.g., organizations may not want to find out that their training is worthless after a large investment of time and money). Alternatively, organizations may assume that transfer will occur because evidence of skill acquisition is evident. This leads to the final myths that we discuss.

Myth 19: A Behavioral Change During Training Ensures Transfer of Training

According to ASTD's benchmarking forum, only 13% of training courses were subjected to Kirkpatrick's Level 3 Evaluation (i.e., that behavior change observed in training transferred to the job) (Bassi & Cheney, 1996). Because training programs rarely investigate transfer, many organizations erroneously conclude that *a behavioral change during training ensures transfer to the job.* Positive training transfer refers to the "degree to which trainees effectively apply the knowledge, skills, and attitudes gained in a training context to the job" (Baldwin & Ford, 1988, p. 63). While, it may seem intuitively obvious that material learned in training should automatically transfer to the work setting, research indicates that training transfer is a complex phenomena that is influenced by multiple factors.

In fact, Baldwin and Ford (1988) specify the two conditions of transfer: the generalization of the material acquired during training to the work setting and long-term retention of the skills. According to this model, trainee characteristics, training design, the work environment, and learning and retention directly and indirectly influence the conditions of transfer. Tracey et al. (1995) found that both the transfer climate and continuous-learning culture directly effected post-training behaviors. Additionally, Thayer and Teachout (1995) identified the climate for transfer and transfer enhancing training activities as key constructs in their training transfer model. According to their model, goal setting, relapse prevention, and self-efficacy can be used to enhance learning and transfer. In-training transfer enhancing activities include "overlearning, fidelity, varied practice, principles/ meaningfulness, self-monitoring cues, relapse prevention, goal setting, and top management support" (Thayer & Teachout, 1995, p. 8). Goldstein (1993) provided some general guidelines concerning training transfer. Essentially, it is important to keep the responses that trainees make consistent from the training environment to the job environment. The task stimuli, on the other hand, does not need to be the same. Negative transfer may occur if the task stimuli in the practice situation requires a different response than the transfer environment.

Training design can contribute to learning during training which fails to transfer in several ways. For instance, it may be the case that the trainee never actually learned the task, and that observed behavioral change may have been due to temporary performance effects (Druckman & Bjork, 1991; Schmidt, 1991). Essentially, the trainee may not have learned the skill well enough to retain it. Training programs must be designed to provide adequate practice and feedback in order to enhance long term retention and application of the skills.

Research needs. While quite a bit of recent research has been directed at understanding transfer, further research is needed. Specifically, it is necessary to better understand the factors that enhance and hinder transfer. In addition, further work to increase the diagnosticity of measures (as noted earlier) will contribute to the improvement of transfer. This is because diagnostic measures will help to highlight the causes for ineffective transfer. For example, if it is determined that the trainee acquired targeted skills, but does not transfer them, effort to examine the target environment would be in order. Conversely, if it is determined that transfer is low because trainees did not fully acquire skills, remedial training would be in order.

Myth 20: Training Lasts Forever

It sometimes appears as though organizations believe that *training lasts forever*; that is, that once training has been delivered and evaluated, the information that is learned will be remembered forever. This belief is evident if one examines the statistics provided on training. For instance, two common reasons that organization initiate training is (1) the introduction of new products and (2) changes in methods or procedures (Bassi & Cheney, 1996). Both of these suggest that the training will focus on new KSAs. Additionally, employees receive on average only four days of training; therefore, it is doubtful that this training involves refresher courses or remedial training.

Table 2. Research Needs

Myth	Explanation	Research Questions
Myth 1. Organizations use training to train people.	Organizations use training for many purposes: to communicate with employees, provide awareness, protect from litigation, achieve "political correctness", improve image, provide a perk or reward, provide a quick fix and to check off a box.	• How do the messages organizations send to employees impact training effectiveness? • What implicit organizational goals are achieved via training?
Myth 2. Organizations send a message that training is valued.	In fact, organizations take actions that indicate the contrary; for example, they allow few days for training, downsize training staff, cut training budgets, conduct informal OJT and do not pay attention to the notification process.	• How do organizational factors influence trainee motivation to enter training? • How does the notification process impact learning during training? • How do we best prepare on-the-job trainers?
Myth 6. Performing a task analysis ensures training success.	There are many types of task analysis techniques and these do not lead to equivalent data. Care must be given in selecting a task analysis method.	• What is the relationship between task analysis data and training effectiveness?
Myth 7. Subject matter experts can articulate training needs.	SMEs have excellent knowledge of their jobs, however they may not have access to other sources of data such as person and organizational analyses. In addition, in complex tasks, SMEs may have difficulty articulating their knowledge, and may require sophisticated knowledge elicitation techniques	• Which techniques are superior at knowledge elicitation for task and cognitive task analyses?
Myth 8. Individual training is the same as team training.	Team skills are needed in addition to task skills. Teams develop competitiveness that must be identified, trained and assessed in addition to individual KSAs.	• Which team competencies should be trained and how best are they trained?

(continued)

Table 2. (Continued)

Myth	Explanation	Research Questions
Myth 10. Practice makes perfect.	Practice is a complex phenomenon, and the conditions under which practice occurs can impact its effectiveness. Additionally, feedback is needed in addition to practice.	• How can practice be optimized during training?
Myth 11. Everyone is an expert in training.	Task experts are not necessarily training experts. Therefore, it is not the case that task experts can select effective training or serve as instructors without preparation.	• Which characteristics are most influential in training? • What is the relationship between instructor training and training effectiveness?
Myth 12. Simulation of any kind is training.	Simulation is a tool for training. It provides a practice environment, but requires defined training objectives, performance measures and feedback.	• Which environmental cues are essential to the development of knowledge structures? • How should trainers develop scenarios?
Myth 13. Free play is training.	There may be times when discovery learning is appropriate. However, free play will most often require training objectives, performance measures and feedback. A guided practice strategy may be more effective.	• How can discovery learning be structured to ensure learning? • Can technology best support discovery learning?
Myth 14. Assessing training effectiveness involves a pre-test and post-test only.	Simple measures of training effectiveness disregard the process of learning and do not facilitate error correction or diagnosis of of performance problems.	• How can dynamic assessment be achieved? • What is the relationship between typical performance and maximal performance?
Myth 15. A positive reaction to training means that learning has occurred.	In general, there is not a strong relationship between reactions and learning. Reactions may be an important outcome of training, however, as they may indicate motivation to transfer.	• What is the relationship between reactions and outcomes and training effectiveness?

(continued)

Table 2. (Continued)

Myth	Explanation	Research Questions
Myth 16. An attitudinal change means a behavioral change.	Attitudinal change may be an important outcome of training if it indicates a trainee's propensity to transfer newly learned skills. However, attitude change should not be equated with behavior change.	• How can an attitudinal change be brought about?
Myth 17. A cognitive change means a behavioral change.	When trainees demonstrate a cognitive change it cannot be assumed that a behavioral change will also occur. It is possible that trainees have experienced a change in declarative knowledge, but do not have the skills (or procedural knowledge) to affect a behavioral change.	• How do trainee characteristics (self-efficacy) influence behavioral changes? • What is the relationship between declarative and procedural knowledge?
Myth 18. Paper-and-pencil measures are a good indicator of training success.	Paper-and-pencil measures do not assess important aspects of learning such as procedural knowledge or skill. Measures must be tailored to the type of learning that is presumed to have occurred.	• How can the diagnosticity of performance measures be improved?
Myth 19. A behavioral change during training ensures transfer of training.	The relationship between training performance and training transfer is a complex one. Many factors affect the transfer environment, and may have an impact on the success of transfer. In addition, factors in the training environment (e.g., variety of practice) may also impact transfer.	• What factors enhance transfer? • How can the diagnosticity of performance measures be improved?
Myth 20. Training lasts forever.	Training must continually be updated to keep pace with changing technology. In addition, skills decay over time and must be continually refreshed and retrained.	• At what rate do skills decay? • How can relapse prevention be applied to training?

This is a problem because research indicates that skills, particularly procedural skills, decay when they are not used or practiced (Annett, 1979; Hagman & Rose, 1983; Schendel & Hagman, 1982). Refresher training and relapse prevention programs are necessary to maintain the skills that are learned during training (Druckman & Bjork, 1991). The rate at which skills decay depends on the initial learning (practice and feedback) and the type of task that was learned (declarative or procedural knowledge). Refresher training provided at frequent intervals is only one possible intervention to aid in the long-term retention of skills. Relapse prevention has recently emerged as a promising approach to long-term maintenance of skills learned during training (Marx, 1982).

Research needs. Two areas of research are suggested here. First, it is necessary to better understand how various knowledge and skills decay. Such information will be useful in developing refresher training schedules. Moreover, such information must be translated into a format that can be used by organizational trainers.

Second, research is needed to understand more completely how the relapse prevention process might be applied to training. Interventions that can be used as follow-up strategies to ensure that training gains do not decay must be investigated.

Many of these myths and research needs correspond to those myths presented in Table 2.

ADDRESSING THE MYTHS

There are a number of factors that have contributed to the creation and perpetuation of many of the myths that we have discussed. For example, the continued simplistic view of training (trainee → training program → trained worker), or the fact that our field is still "faddish" and dominated by practitioners, or the lack of a viable translation mechanism between scientists and practitioners or maybe each of these myths have a grain of truth; taken together all of these may have generated the myths.

Another factor is that many of the myths that have been identified and discussed in this review exist because of the chasm that separates research and practice. While further research is needed to refine training, more energy should be focused on disseminating what is already known about training to organizations and promoting reciprocity (Salas et al., 1997). Too often, research is published in academic journals that are not accessible to practitioners (Quiñones & Ehrenstein, 1997). The burden of bridging this gap also falls on researchers. Often times, academicians publish research that is difficult to apply in organizational settings; in fact, practitioners often suggest that more applied research is needed (Sind-Prunier, 1996). In addition, researchers could be more direct in explaining the potential application of their work within organizational settings.

Part of the burden also falls on practitioners. Practitioners who are interested in providing better training programs within their organizations have the opportunity to seek out the state-of- the-art in training research by attending national conferences and workshops of training development. Practitioners should examine training fads more critically when purchasing training. Rather than relying on the praises of a training program, they should demand to see evidence, such as evaluation studies of training success. Although there are often organizational constraints (cost and time) to consider in training development, adhering as closely as possible to instructional systems design and taking into consideration the pre-training and post-training factors that influence training effectiveness may advance training design in many organizations.

Both researchers and practitioners should be creative in their attempts to bridge the gap between research and practice. One approach which appears to be gaining popularity is the "Bridging the gap" sessions that are being held at national conferences. A second approach has been the inclusion of practitioners papers in referred journals, such as *Personnel Psychology, Training Research Journal* and *The International Journal of Aviation Psychology*.

The Internet may provide a forum in which researchers and practitioners can begin to bridge the gap. Misic and Hill (1994) encourage organizations to support Internet use in organizations. The prevalence and ease of use of the Internet makes it a candidate to use as a medium for increasing communication between researchers and practitioners. Several possibilities for Internet uses include: setting up sites which answer training design and implementation questions, a site that reviews training programs that can be purchased from companies that sell training, a chat room for researchers and practitioners to interact in real time, and training design tutorials for practitioners.

CONCLUDING REMARKS

It is remarkable to witness the advances that have been made in training. Significant progress has been made in a number of areas. We now know more about learning, training motivation, training evaluation, and training effectiveness than 10 years ago thanks to the resurgence of training research in the industrial/organizational psychology, human factors, military psychology and applied cognitive domains. To quote other colleagues in similar fields "the training field is alive and well" and growing. While much needs to be done, the future looks challenging and promising.

The future looks bright because a number of events keep us motivated and looking forward to new advances in the science and practice of training. First, considerable investments in training research and development are still available and there is no evidence of drying out. The military continues to provide funding for improving the training of its workforce. This investment could not only benefit our military readiness but the private sector as well. The private sector will benefit from all of the findings, lessons learned, tools, instructional strategies and technologies that continue to be designed, tested, and evaluated by government laboratories (Cannon-Bowers & Salas, 1998). Partnerships between government and the private sector will probably emerge in the next millennium and we hope that this will be a better way of conducting business. It is in the best interest of both to learn from each other and leverage resources as well as knowledge.

Second, technological advances will continue to re-shape how we view learning and training in the next millennium. As computer systems become faster and more capable, only our imagination on how to design and deliver training will be our limit. We could envision, in the near future, intelligent, multi-media systems capable of diagnosing skill deficiencies and providing remediation on-line at the employees' workstation. Research on intelligent tutoring and cognitive diagnosis will take us there.

Third, as stated earlier, we predict that our next revolution in training will come from understanding dynamic assessment. While an important piece of training is being worked on (i.e., performance measurement), and advances made (see Brannick, Salas, & Prince, 1997), this concept of dynamic assessment must be understood and researched. The implications of this concept for on-the-job training and guided practice are immeasurable.

Finally, the need for translating our science into practice (and vice versa) is imperative. If we really expect to impact organizations, we must continue to seek ways to communicate our findings. We must not forget, as scientist-practitioners, that we conduct science to solve organizational problems. It is our mandate to continue to provide answers (with the best science that we have) to human resource specialists, managers, and organizations on how to best design and deliver training.

So, the next millennium looks promising for the training field, as we dispel the myths and we embark into uncharted territory, we look forward to the next set of theoretically-driven principles of training, and to new cognitive-based instructional strategies that help organizations achieve training effectiveness.

ACKNOWLEDGMENT

The views expressed in this chapter are those of the authors and do not reflect the official position of the organizations with which the authors are affiliated. The authors would like to thank Steve W. J. Kozlowski and Scott I. Tannenbaum for their valuable and insightful comments on earlier versions of this chapter.

REFERENCES

Alliger, G.M., Tannenbaum, S.I., Bennett, W., Traver, H., & Shotland, A. (1997). A meta-analysis of the relations among training criteria. *Personnel Psychology, 50,* 341-358.

Anderson, J. (1992). Automaticity and the ACT* theory. *American Journal of Psychology, 105,* 165-180.

Anderson, J. (1987). Productions systems, learning, and tutoring. In D. Klahr, P. Langley, & R. Neches (Eds.), *Production system models of learning and development* (pp. 437-458). Cambridge, MA: MIT Press.

Anderson, J. (1983). *The architecture of cognition.* Cambridge, MA: Harvard University Press.

Annett, J. (1979). Memory for skill. In M.M. Gruneberg & P.E. Morris (Eds.), *Applied problems in memory* (pp. 215-247). London: Academic Press.

Baker, D.P., Salas, E., & Cannon-Bowers, J.A. (1998). Team task analysis: Lost but hopefully not forgotten. *The Industrial-Organizational Psychologist, 35*, 79-83.

Baldwin, T.T., & Ford, J.K. (1988). Transfer of training: A review and directions for future research. *Personnel Psychology, 41*, 63-105.

Baldwin, T.T., Magjuka, R.J., & Loher, B.T. (1991). The perils of participation: Effects of choice of training on trainee motivation and learning. *Personnel Psychology, 44*, 51-65.

Bass, B.M., & Vaughan, J.A. (1966). *Training in industry: The management of learning.* Belmont, CA: Wadsworth.

Bassi, L.J., Benson, G., & Cheney, S. (1996). *Trends: Position yourself for the future.* Alexandria, VA: American Society for Training and Development.

Bassi, L.J., & Cheney, S. (1996). *Results from the 1996 benchmarking forum.* Alexandria, VA: American Society for Training and Development.

Bayman, P., & Mayer, R.E. (1984). Instructional manipulation of users' mental models for electronic calculators. *International Journal of Man-Machine Studies, 20*, 189-199.

Bennett, K. (1992). The use of on-line guidance, representation aiding, and discovery learning to improve the effectiveness of simulation training. In J.W. Regian & V.J. Shute (Eds.), *Cognitive approaches to automated instruction* (pp. 217-241). Hillsdale, NJ: Lawrence Erlbaum.

Bowers, C.A., Baker, D.P., & Salas, E. (1994). The importance of teamwork in the cockpit: The utility of job/task analysis indices for training design. *Military Psychology, 6*, 205-214.

Brannick, M.T., Salas, E., & Prince, C. (1997). *Team performance assessment and measurement.* Mahwah, NJ: Lawrence Erlbaum.

Burke, M.J., & Day, R.R. (1986). A cumulative study of the effectiveness of managerial training. *Journal of Applied Psychology, 71*, 232-245.

Campion, M.A., & Campion, J.E. (1987). Evaluation of an interview skills training program in a natural field setting. *Personnel Psychology, 40*, 675-691.

Campion, M.A., Papper, E.M., & Medsker, G.J. (1996). Relations between work team characteristics and effectiveness: A replication and extension. *Personnel Psychology, 49*, 429-452.

Cannon-Bowers, J.A., Burns, J.J., Salas, E., & Pruitt, J.S. (1998). Advanced technology in decision-making training: The case of shipboard embedded training. In J.A. Cannon-Bowers & E. Salas (Eds.), *Making decisions under stress: Implications for individual and team training* (pp. 365-374). Washington, DC: American Psychological Association.

Cannon-Bowers, J.A., Rhodenizer, L., Salas, E., & Bowers, C.A. (1998). A framework for understanding pre-practice conditions and their impact on learning. *Personnel Psychology, 51*(2), 291-320.

Cannon-Bowers, J.A., & Salas, E. (1997). A framework for developing team performance measures in training. M.T. Brannick, E. Salas, & C. Prince (Eds.), *Team performance assessment and measurement: Theory, methods and applications* (pp. 45-62). Hillsdale, NJ: Lawrence Erlbaum.

Cannon-Bowers, J.A., & Salas, E. (1998). *Making decisions under stress: Implications for individual and team training.* Washington, DC: American Psychological Association.

Cannon-Bowers, J.A., Salas, E., Bowers, C.A., & Rhodenizer, L. (1996, April). *Conditions of practice: Optimizing learning and performance.* Paper presented at the annual meeting for the Society of Industrial Organizational Psychologists, San Diego, CA.

Cannon-Bowers, J.A., Salas, E., & Converse, S. (1993). Shared mental models in expert team decision making. In N.J. Castellan (Ed.), *Current issues in individual and group decision making* (pp. 221-246). Hillsdale, NJ: Lawrence Erlbaum.

Cannon-Bowers, J.A., Salas, E., Tannenbaum, S.I., & Mathieu, J.E. (1995). Toward theoretically based principles of training effectiveness: A model and initial empirical investigation. *Military Psychology, 7*, 141-164.

Carnevale, A.P., Gainer, L.J., & Villet, J. (1990). *Training in America: The organization and strategic role of training.* San Francisco: Jossey-Bass.

Charney, D., Reder, L., & Kusbit, G.W. (1990). Goal setting and procedure selection in acquiring computer skills: A comparison of tutorials, problem solving, and learner exploration. *Cognition & Instruction, 7,* 323-342.

Cooke, N. (1994). Varieties of knowledge elicitation techniques. *International Journal of Human-Computer Studies, 41,* 801-849.

Cranny, C.J., & Doherty, M.E. (1988). Importance rating in job analysis: Note on the misinterpretation of factor analyses. *Journal of Applied Psychology, 73,* 320-322.

Dempsey, J.V., & Sales, G.C. (Eds.). (1993). *Interactive instruction and feedback.* Englewood Cliffs, NJ: Educational Technology.

De Mul, S., & van Oostendorp, H. (1996). Learning user interfaces by exploration. *Acta Psychologica, 91,* 325-344.

Dipboye, R.L. (1997). Organizational barriers to implementing a rational model of training. In M.A. Quiñones & A. Ehrenstein (Eds.), *Training for a rapidly changing workplace: Applications of psychological research* (pp. 31-60). Washington, DC: American Psychological Association.

Druckman, D., & Bjork, R. (1991). *The mind's eye: Enhancing human performance.* Washington, DC: National Academy Press.

Dulworth, M., & Shea, R. (1995). Six ways technology improves training. *HR Magazine, 40*(5), 33-36.

Eliade, M. (1963). *Myth and reality* (W.R. Trask, trans.). New York: Harper & Row.

Flexman, R.E., & Stark, E.A. (1987). Training simulators. In G. Salvendy (Ed.), *Handbook of human factors* (pp. 1012-1038). New York: John Wiley & Sons.

Ford, J.K., Kozlowski, S.W.J., Kraiger, K., Salas, E., & Teachout, M.S. (Eds.). (1997). *Improving training effectiveness in work organizations.* Mahwah, NJ: Lawrence Erlbaum.

Ford, J.K., Quiñones, M.A., Sego, D.J., & Sorra, J.S. (1992). Factors affecting the opportunity to perform trained tasks on the job. *Personnel Psychology, 45* (3), 511-527.

Frederiksen, J., & White, B. (1989). An approach to training based upon principled task decomposition. *Acta Psychologica, 71,* 89-146.

Gagne, R.M., & Briggs, L.J. (1979). *Principles of instructional design.* New York: Holt, Rinehart and Winston.

Goad, T.W. (1982). *Delivering effective training.* San Diego, CA: University Associates.

Goldstein, I.J. (1989). Critical training issues: Past, present, and future. In I.L. Goldstein & Associates (Eds.), *Training and development in organizations* (pp. 1-21). San Francisco: Jossey-Bass.

Goldstein, I.L. (1991). Training in work organizations. In M.D. Dunnette & L.M. Hough (Eds.), *Handbook of industrial and organizational psychology* (2nd ed., Vol. 2, pp. 507-619). Palo Alto: Consulting Psychologists Press.

Goldstein, I.L. (1993). *Training in organizations: Needs assessment, development, and evaluation.* Pacific Grove, CA: Brooks/Cole.

Gordon, J. (1994). Madame Z. *Training, 31,* 8.

Gordon, M.E., Cofer, J.L., & McCullough, P.M. (1986). Relationship among seniority, past performance, interjob similarity, and trainability. *Journal of Applied Psychology, 71,* 518-521.

Gordon, S.E., & Gill, R.T. (1997). Cognitive task analysis. In C.E. Zsambok & G. Klein (Eds.), *Naturalistic decision making* (pp. 131-140). Mahwah, NJ: Lawrence Erlbaum.

Grove, D.A., & Ostroff, C. (1990). Program evaluation. In K. Wexley & J. Hinrichs (Eds.), *Developing human resources.* Washington, DC: BNA Books.

Guzzo, R.A., & Dickson, M.W. (1996). Teams in organizations: Recent research on performance and effectiveness. *Annual Review of Psychology, 47,* 307-338.

Hagman, J., & Rose, A. (1983). Retention of military tasks: A review. *Human Factors, 25*(2), 199-213.

Hays, R.T., Jacobs, J.W., Prince, C., & Salas, E. (1992). Flight simulator training effectiveness: A meta-analysis. *Military Psychology, 4,* 63-74.

Hays, R.T, & Singer, M.J. (1989). *Simulation fidelity in training system design: Bridging the gap between reality and training.* New York: Springer-Verlag.

Helmreich, R. (1984). Cockpit management attitudes. *Human Factors, 26,* 583-589.

Hicks, W.D., & Klimoski, R.J. (1987). Entry into training programs and its effects on training outcomes: A field experiment. *Academy of Management Journal, 30,* 542-552.

Hughes, G.L., & Prien, E.P. (1989). Evaluation of task and job skill linkage judgments used to develop test specifications. *Personnel Psychology, 42,* 283-292.

Jacobs, J., & Dempsey, J. (1993). Simulation and gaming: Fidelity, feedback, and motivation. In J. Dempsey & G. Sales (Eds.), *Interactive instruction and feedback* (pp. 197-228). Englewood Cliffs, NJ: Educational Technology Publications.

Johnson, S. (1981). Effect of training device on retention and transfer of a procedural task. *Human Factors, 23*(3), 257-272.

Kamouri, A.L., Kamouri, J., & Smith, K.H. (1986). Training by exploration: Facilitating the transfer of procedural analogical reasoning. *International Journal of Man-Machine Studies, 24,* 171-192.

Kieras, D., & Bovair, S. (1984). The role of a mental model in learning to operate a device. *Cognitive Science, 8,* 255-273.

Kirkpatrick, D.L. (1959). Techniques for evaluating training programs. *Journal of the American Society of Training Directors, 13*(3-9), 21-26.

Klein, G. (1995). The value added by cognitive task analysis. *Proceedings of the Human Factors and Ergonomics Society 39th Annual Meeting* (pp. 530-533). Santa Monica, CA: Human Factors and Ergonomics Society.

Kozlowski, S.W.J. (1998). Training and developing adaptive teams: Theory, principles, and research. To appear in J.A. Cannon-Bowers & E. Salas (Eds.), *Making decisions under stress: Implications for individual and team training* (pp. 115-153). Washington, DC: American Psychological Association.

Kozlowski, S.W.J., Brown, K.G., Weissbein, D., Cannon-Bowers, J.A., & Salas, E. (in press). A multilevel approach to training. In K.J. Klein & S.W.J. Kozlowski (Eds.), *Multilevel theory, research and methods in organizations.* San Francisco: Jossey Bass.

Kozlowski, S.W.J., & Salas, E. (1997). An organizational approach for the implementation and transfer of training. In J.K. Ford & Associates (Eds.), *Improving training effectiveness in workplace organizations* (pp. 247-290). Hillsdale, NJ: Lawrence Erlbaum.

Kraiger, K., Ford, J.K., & Salas, E. (1993). Application of cognitive, skill-based, and affective theories of learning outcomes to new methods of training evaluation. *Journal of Applied Psychology, 78,* 311-328.

Kraiger, K., & Jung, K.M. (1997). Linking training objectives to evaluation criteria. In M.A. Quiñones & A. Ehrenstein (Eds.), *Training for a rapidly changing workplace: Applications of psychological research* (pp. 151-175). Washington, DC: American Psychological Association.

Landy, F.J., & Vasey, J. (1991). Job analysis: The composition of SME samples. *Personnel Psychology, 44,* 27-50.

Leazes, F.J. (1995). Pay now or pay later: Training and torts in public sector human service. *Public Personnel Management, 24*(2), 167-180.

Lewick-Wallace, M., & Jask, R.C. (1988). Tips for effective on-the-job training. *Performance & Instruction, 27,* 17-18.

Martocchio, J.J., & Baldwin, T.T. (1997). The evolution of strategic organizational training: New objectives and research agenda. In G.R. Ferris (Ed.), *Research in personnel and human resources management* (Vol. 15, pp. 1-46). Greenwich, CT: JAI Press.

Marx, R.D. (1982). Relapse prevention for managerial training: A model for maintenance of behavior change. *Academy of Management Review, 7,* 433-441.

McSparran, M.K. (1993). Training: No automatic return on investment. *Beverage World, 112,* 122.

Meister, D. (1985). *Behavioral analysis and measurement methods.* New York: John Wiley & Sons.

Misic, M.M., & Hill, J.A. (1994). Keys to success with the Internet. *Journal of Systems Management, 45*, 6-10.
Morgan, B.B., Glickman, A.S., Woodard, E.A., Blaiwes, A.S., & Salas, E. (1986). *Measurement of team behaviors in a Navy environment* (NTSC Tech. Rep. No.TR-86-014). Orlando, FL: Naval Training Systems Center.
Nieva, V.F., Fleishman, E.A., & Reick, A. (1978). *Team dimensions: Their identity, their measurement and their relationships* (Fin. Tech Rep. No. DAHC19-78-C-0001). Washington, DC: Advanced Research Resources Organization.
Njoo, M., & de Jong, T. (1993). Exploratory learning with a computer simulation for control theory: Learning processes and instructional support. *Journal of Research in Science Teaching, 30*, 821-844.
Noe, R.A. (1986). Trainees' attributes and attitudes: Neglected influences on training effectiveness. *Academy of Management Review, 11*, 736-749.
Noe, R.A., & Schmitt, N. (1986). The influence of trainee attitudes on training effectiveness: Test of a model. *Personnel Psychology, 39*, 497-523.
Oser, R.L., Cannon-Bowers, J.A., Dwyer, D.J., & Salas, E. (1997). Establishing a learning environment for JSIMS: Challenges and considerations [CD-Rom]. *Proceedings of the 19th Annual Meeting of the Interservice/Industry Training, Simulation and Education Conference* (pp. 145-153), Orlando, FL.
Patrick, J. (1992). *Training: Research and practice.* New York: Academic Press.
Prince, A, Brannick, M., Prince, C., & Salas, E. (1997). The measurement of team process behaviors in the cockpit: Lessons learned. In M.T. Brannick, E. Salas, & C. Prince (Eds.), *Team performance assessment and measurement: Theory, methods, and applications* (pp. 289-310). Mahwah, NJ: Lawrence Erlbaum.
Quiñones, M.A. (1997). Contextual influences on training effectiveness. In M.A. Quiñones & A. Ehrenstein (Eds.), *Training for a rapidly changing workplace: Applications of psychological research* (pp. 177-200). Washington, DC: American Psychological Association.
Quiñones, M.A., & Ehrenstein, A. (1997). Introduction: Psychological perspectives on training in organizations. In
M.A. Quiñones & A. Ehrenstein (Eds.), *Training for a rapidly changing workplace: Applications of psychological research* (pp. 1-10). Washington, DC: American Psychological Association.
Robertson, I.T., & Downs, S. (1989). Work-sample tests of trainability: A meta-analysis. *Journal of Applied Psychology, 74*, 402-410.
Roullier, J.Z., & Goldstein, I.L. (1997). The relationship between organizational transfer climate and positive transfer of training. In D.F. Russ-Eft & H.S. Preskill (Eds.), *Human resource development review: Research and implications* (pp. 330-347). Thousand Oaks, CA: Sage.
Sackett, P.R., Zedeck, S., & Fogli, L. (1988). Relations between measures of typical and maximum job performance. *Journal of Applied Psychology, 73*, 482-486.
Salas, E., Burgess, K.A., & Cannon-Bowers, J.A. (1995). Training effectiveness techniques. In J. Weimer (Ed.), *Research techniques in human engineering* (pp. 439-475). Englewood Cliffs, NJ: Prentice-Hall.
Salas, E., & Cannon-Bowers, J.A. (1997). Methods, tools, and strategies for team training. In M.A. Quiñones & A. Ehrenstein (Eds.), *Training for a rapidly changing workplace: Applications of psychological research* (pp. 249-280). Washington, DC: American Psychological Association.
Salas, E., & Cannon-Bowers, J.A. (in press). The anatomy of team training. To appear in L. Tobias & D. Fletcher (Eds.), *Handbook on research in training.* New York: Macmillan.
Salas, E., Cannon-Bowers, J.A., & Blickensderfer, E.L. (1997). Enhancing reciprocity between training theory and practice: Principles, guidelines, and specifications. In J.K. Ford & Associates (Eds.), *Improving training effectiveness in workplace organizations* (pp. 291-322). Hillsdale, NJ: Lawrence Erlbaum.

Salas, E., Cannon-Bowers, J.A., & Johnston, J.H. (1997). How can you turn a team of experts into an expert team?: Emerging training strategies. In C. Zsambok & G. Klein (Eds.), *Naturalistic decision making* (pp. 359-370). Mahwah, NJ: Lawrence Erlbaum.

Salas, E., Dickinson, T.L., Converse, S.A., & Tannenbaum, S.I. (1992). Toward an understanding of team performance and training. In R.W. Swezey & E. Salas (Eds.), *Teams: Their training and performance* (pp. 3-29). Norwood, NJ: Ablex.

Schendel, J.D., & Hagman, J.D. (1982). On sustaining procedural skills over a prolonged retention interval. *Journal of Applied Psychology, 67,* 605-610.

Schmidt, R. (1991). Frequent augmented feedback can degrade learning: Evidence and interpretations. In J. Requin & G. Stelmach (Eds.), *Tutorials in motor neuroscience* (pp. 59-75). London: Kluwer Academic.

Schmidt, R., & Bjork, R. (1992). New conceptualizations of practice: Common principles in three paradigms suggest new concepts for training. *Psychological Science, 3*(4), 207-217.

Sind-Prunier, P. (1996, September). Bridging the gap research/practice gap: Practitioners' opportunity for input to define research for the rest of the decade. *Proceedings of the Human Factors and Ergonomics Society 40th Annual Meeting* (pp. 865-867). Santa Monica, CA: Human Factors and Ergonomics Society.

Smith, K.A. (1991). *Task-assertiveness in teams: Implications for training.* Master's Thesis.

Smith, E.M., Ford, J.K., & Kozlowski, S.W.J. (1997). Building adaptive expertise: Implications for training design strategies. In M.A. Quiñones & A. Ehrenstein (Eds.), *Training for a rapidly changing workplace: Applications of psychological research* (pp. 89-118). Washington, DC: American Psychological Association.

Smith-Jentsch, K.A., Jentsch, F.G., Payne, S.C., & Salas, E. (1996). Can pretraining experiences explain individual differences in learning? *Journal of Applied Psychology, 81,* 110-116.

Steele-Johnson, D., & Hyde, B.G. (1997). Advanced technologies in training: Intelligent tutoring systems and virtual reality. In M.A. Quiñones & A. Ehrenstein (Eds.), *Training for a rapidly changing workplace: Applications of psychological research* (pp. 225-248). Washington, DC: American Psychological Association.

Sterns, H.L., & Doverspike, D. (1989). Aging and the training and learning process. In I.L. Goldstein & Associates (Eds.), *Training and development in organizations* (pp. 299-332). San Francisco: Jossey-Bass.

Stevens, M.J., & Campion, M.A. (1994). The knowledge, skill, and ability requirements for teamwork: Implications for human resource management. *Journal of Management, 20,* 503-530.

Stout, R.J., Cannon-Bowers, J.A., & Salas, E. (1997). A team perspective on situational awareness (SA): Cueing training [CD-Rom]. *Proceedings of the 19th Annual Meeting of the Interservice/Industry Training, Simulation and Education Conference* (pp.174-182), Orlando, FL.

Sullivan, R.F., & Miklas, D.C. (1985). On-the-job-training that works. *Training and Development, 39,* 118-120.

Swezey, R.W., & Salas, E. (1992). Guidelines for use in team-training development. In R.W. Swezey & E. Salas (Eds.), *Teams: Their training and performance* (pp. 219-246). Norwood, NJ: Ablex.

Tannenbaum, S.I. (1997). Enhancing continuous learning: Diagnostic findings from multiple companies. *Human Resource Management Journal, 36*(4), 437-452.

Tannenbaum, S.I., Cannon-Bowers, J.A., Salas, E., & Mathieu, J.E. (1993). *Factors that influence training effectiveness: A conceptual model and longitudinal analysis* (Technical Report 93-011). Orlando, FL: Naval Training Systems Center.

Tannenbaum, S.I., Mathieu, J.E., Salas, E., & Cannon-Bowers, J.A. (1991). Meeting trainee's expectations: The influence of training fulfillment on the development of commitment, self-efficacy, and motivation. *Journal of Applied Psychology, 76,* 759-769.

Tannenbaum, S.I., & Yukl, G. (1992). Training and development in work organizations. *Annual Review of Psychology, 43,* 399-441.

Thayer, P.W., & Teachout, M.S. (1995). *A climate for transfer model* (Interim Technical Paper AL/HR-TP-1995-0035). Brooks Air Force Base, TX: Air Force Material Command.

Tracey, J.B., Tannenbaum, S.I., & Kavanagh, M.J. (1995). Applying trained skills on the job: The importance of the work environment. *Journal of Applied Psychology, 80*, 239-252.

Wexley, K.N. (1984). Personnel training. *Annual Review of Psychology, 35*, 519-551.

Wiener, E.L., Kanki, B.G., & Helmreich, R.L. (1993). *Cockpit resource management*. San Diego, CA: Academic Press.

Zenger, J.H. (1996). The painful turnabout in training. *Training and Development, 50*, 48-51.

MANAGERIAL DISCRETION, COMPENSATION STRATEGY, AND FIRM PERFORMANCE:
THE CASE FOR THE OWNERSHIP STRUCTURE

Henry L. Tosi, Luis Gomez-Mejia, Misty L. Loughry, Steve Werner, Kevin Banning, Jeffery Katz, Randall Harris, and Paula Silva

ABSTRACT

In this paper, we examine managerial discretion and the firm's compensation strategy. When managers have substantial discretion, they are able to manage the firm in ways that place their interests over those of stockholders. The bases for this argument rest in agency theory (Fama & Jensen, 1983) and managerial capitalism (Berle & Means, 1932; Marris, 1964). We report on a set of studies that examine how managerial discretion is acquired and institutionalized, how it is related to the firm's overarching compensation strategy, and how managerial discretion can be constrained to improve firm performance.

INTRODUCTION

Compensation strategy refers to choices a firm makes from the repertoire of options that are available to design, administer, and allocate pay to its members with the ostensible objective to improve the firm's competitive advantage and, ultimately, its financial performance (Gomez-Mejia & Balkin, 1992; Barkema & Gomez-Mejia, 1998). The main body of theory and research on compensation strategy is built around this premise. It retains the distinctive profit maximizing flavor from neoclassical economics in that the objective is to rationally select those compensation choices that best fit the requirements of the firm's contexts.

One line of research focuses on the overarching features of the compensation system at the corporate or business unit level. To this end, several strategic pay choices have been identified by various researchers, including pay mix (variable versus fixed), long- versus short-term orientation, egalitarian versus hierarchical pay structures, centralization versus decentralization of pay decisions, mechanistic versus organic procedures, and the like (for an extensive review, see Gomez-Mejia, 1992; Gomez-Mejia & Balkin, 1992). Most of this research has used a contingency perspective to determine the profile of strategic compensation choices that would lead to improved firm performance, given organizational characteristics (e.g., life cycle stage, size, and R&D intensity), business unit strategies, corporate strategies, and the environment (e.g., industry structure and instability).

A second stream of research on compensation strategies with this perspective focuses on top executives, mainly the CEO, and is based primarily on agency theory (Ross, 1973; Spence & Zeckhauser, 1971). It addresses the question of how CEO compensation strategies adopted by the firm can elicit executive decisions that enhance firm performance. For instance, paying the executive in stock options rather than bonuses linked to accounting based performance measures presumably leads to a long term vision and more investment in R&D for high technology firms which in turn fuels innovation, a requirement for successful performance. Following this tradition, studies have correlated CEO compensation strategy choices (e.g., variable pay, adoption of various stock based plans, pay versus market, target difficulty, and the like) to a variety of factors such as firm size, performance, diversification, and others. This research, which has generally led to mixed results, is reviewed elsewhere (see Gomez-Mejia, 1994; Wiseman & Gomez-Mejia, 1998).

There is a theoretical and empirical basis for questioning the validity of the maximizing assumptions underlying much compensation strategy research. It is based on the theory of managerial capitalism (Baumol, 1967; Berle & Means, 1932; Marris, 1964) (referred to later in this paper as managerialism) in which it is argued that this maximizing premise is not applicable in large publicly held firms in which equity holdings are widely dispersed. The reason is that atomistic ownership results in weak stockholder influence, and leaves the management with a substantial level of discretion. This managerial discretion permits the management, particularly the CEO, to design compensation strategies that permit top executives to pursue their own interests, which often do not coincide with those of shareholders and, therefore, will lead to suboptimal outcomes for equity holders.

WHAT IS MANAGERIAL DISCRETION?

Since the mid 1980s, the concept of managerial discretion has been the subject of much more discussion and research in the management and organization literature than in prior periods. One impetus was the paper "Managerial Discretion: A Bridge Between Polar Views of Organizational Outcomes" by Hambrick and Finkelstein (1987); Another was the renaissance of the theory of managerial capitalism (Baumol, 1967; Berle & Means, 1932; Marris, 1964) as a basis for studies of strategic issues such as diversification and compensation. While these two approaches seem to have much in common because both use the term discretion, there are some important, critical differences between them that should be noted.

The Hambrick-Finkelstein (1987) concept of managerial discretion is much broader than that articulated in the managerial capitalism model. For them it is decidedly multi-level, while the concept from managerial capitalism is not. Discretion for Hambrick and Finkelstein is defined to result from three sources: the amount of variation permitted by the **environment**, the amount of **leeway afforded the executive by the firm** to make a wide variety of strategic choices, and the **ability of the executive** to create and implement a variety of alternatives (Finkelstein & Hambrick, 1990). This means that it is a function of the environment, the organization, and the person. In several of their studies, it appears that the major indicator of managerial discretion is the degree of latitude provided by the industry environment (e.g., Finkelstein & Boyd, 1998; Haleblian & Finkelstein, 1993; Hambrick & Abrahamson, 1995; Hambrick, Geletkarycz, & Frederickson, 1993; Rajagopalan & Finkelstein, 1992). It is for this reason that we think the core of their concept is **strategic flexibility**, or the capacity to make choices that locate the firm in a specific sector of its task environment.

The approach to managerial discretion in the theory of managerial capitalism is narrower. It focuses on the decision latitude that is a result of the fact that when markets are unable to discipline the management of the firm to maximize equity holders' returns, the firm's executives may be able to manage it in ways that are in their interests, to the detriment of equity holders. This can happen when equity holdings are widely dispersed so that no single stockholder possesses an adequate level of stock to control the manager, specifically the CEO (Berle & Means, 1932). The degree of managerial discretion is captured by the concept of **ownership structure**, which refers to the distribution of equity holdings. Depending upon the distribution of equity, firms are designated as **owner-controlled**, **management-controlled**, or **owner-managed**.

There is another important distinction between these two approaches. The Hambrick/Finkelstein concept appears to be pervaded by the rationality assumption of neoclassical economics, implying that managerial decisions are made to achieve goals to maximize firm returns, or that are in the owners' interests. This is very different from the effects of discretion in managerial capitalism, in which it is assumed that managerial behavior is self-interested and, when managers have substantial discretion, it may lead to behaviors that do not maximize firm performance and shareholder wealth.

We believe that the theoretical approach of managerial capitalism can cast a very different light on overall firm compensation strategy from the conventional neoclassically-based orientation. But it can only be a point of departure because the vast majority of research focuses on the way that top management, primarily the CEO, is paid. It does not address, in any significant way, a number of more broad issues of interest to scholars in organizational behavior and human resource management. Two of the most central of these questions are "How is discretion acquired and exercised?" and "Does the structure of CEO compensation filter to lower levels in the firm?"

In this paper, we address these issues to show how managerial discretion can usefully inform scholars about these sorts of issues. The paper is organized in four parts:

- **Part I. Agency Theory/Managerial Capitalism: Managerial Discretion as Leading to Opportunistic Behavior.** In this part of the paper we describe the two theoretical models (i.e., agency theory and managerial capitalism) that explain why CEOs may not act in the interests of equity holders, and instead seek to set their own interests ahead of the owners.
- **Part II. The process of acquiring managerial discretion.** In this section, we show how CEOs acquire discretion that permits them to act in their own interests, and we specifically describe three studies. The first shows how the CEO in management-controlled firms is able to negotiate an agency contract that provides greater discretion than for the CEO in owner-controlled firms. The second study clarifies how the CEO in the management-controlled firm is able to decouple compensation from firm performance, reducing compensation risk. The third study illustrates the differences in evaluation criteria used by boards in management-controlled and owner-controlled firms.
- **Part III. The effects of managerial discretion on firm compensation strategy.** Part III extends the current body of research on CEO compensation strategies to the more broad structure of the overall pay strategy of the firm, examining how managers and others within the firm are paid. This discussion provides a clearer picture about how the ownership structure affects incentive alignment across a wide range of firm members.
- **Part IV. Controlling the CEO: An analysis of the effects of incentive alignment and monitoring on firm performance and CEO decision making.** The results in Part III are consistent with other research that shows how CEOs with discretion can shift risk to equity holders. In Part IV,

we report on two studies that demonstrate how the negative effects of managerial discretion on equity holders might be reduced.

PART I. AGENCY THEORY/MANAGERIAL CAPITALISM: MANAGERIAL DISCRETION AS LEADING TO OPPORTUNISTIC BEHAVIOR

The fundamental theoretical concepts that interest us are found in agency theory and the theory of managerial capitalism. In these theories, when managers have substantial freedom and latitude to act, unrestricted by market forces or internal governance mechanisms, they are likely to act in ways that further their own interests and not necessarily those of the equity holders.

Agency Theory

Agency theory views a firm as a "legal entity that serves as a nexus for a complex set of contracts (written and unwritten) among disparate individuals" (Jensen, 1983, p. 326; Ross, 1973; Spence & Zeckhauser, 1971). The agency relationship is defined as "a contract under which one or more persons (the principal(s)) engage another person (the agent) to perform some service on their behalf which involves delegating some decision making authority to the agent" (Jensen & Meckling, 1976, p. 308). In the agency literature that focuses on the internal management of the firm, the typical view is that shareholders or the board of directors are principals, and top managers, more specifically the CEOs, are agents.

The theory and derivative research are based on three key assumptions (Baiman, 1990; Hunt & Hogler, 1990). They are:

1. Both owners and managers are rational and self-utility maximizing with unlimited computation ability. Without rational self-utility maximizing managers, owners could easily establish convergence between their goals and those of the managers.
2. The manager has private information that the owner cannot learn without costs, creating information asymmetry. If perfect information were freely available, there would be no agency problem since owners could simply monitor their managers (McGuire, 1988).
3. The manager is work and risk averse, otherwise owners could just specify in the contract what managers need to do and assume that the managers will do it. If managers were not risk-averse, owners could just shift the risk to the managers, making payment completely dependent on the outcome.

Owners, through board of directors, face two problems with the CEO: moral hazard and adverse selection. *Moral hazard* is the lack of effort put forth by the

CEO or the misuse of firm resources to cater to the CEO's interests. It can occur when the contract is based on imperfect measures of behavior (Baiman, 1982), such as departmental output. It can be solved by an information system that eliminates information asymmetry or with monitoring and contracts that appeal to a manager's self-seeking nature (Baiman, 1982; Jensen & Meckling, 1976; Simon, 1991). *Adverse selection* is the misrepresentation of ability by the CEO. This can occur if CEOs are motivated to misrepresent their private information to achieve their own goals. Adverse selection can be resolved through information that reveals the CEOs' private information or by risk sharing which minimizes returns for CEOs who misrepresent themselves.

Solving these problems results in agency costs for the owner. These are any costs incurred to assure that the CEO executes decisions that are in the best interest of the owner. Owners minimize agency costs by balancing the cost of monitoring, the cost of risk shifting, and the cost of unresolved agency problems (Jensen & Meckling, 1976). Monitoring costs are those that arise from the need to institute incentive schemes, monitoring procedures, supervision, added hierarchical levels to the organization, information systems, budgeting systems, reporting procedures and boards of directors (Eisenhardt, 1989; McGuire, 1988). Bonding expenditures are costs implicit in sub-optimal risk sharing. Residual losses are the costs arising from unresolved agency problems.

Monitoring As A Way to Minimize the Agency Problems

Monitoring is the direct or indirect observation of the manager's action or behavior (Jensen & Meckling, 1976). Holding costs constant, it is always beneficial to the owner unless the CEO's actions can only have a positive effect on the owner (Holmstrom, 1979; Shavell, 1979). When outcomes are measurable, managers' goals can be aligned with owners' goals by making pay contingent on outcomes, but this transfers undesired risk to the manager since the manager's behaviors are not the only factor that determines the outcome. If risk sharing were optimal, monitoring would be unnecessary since the contract is self-enforcing through manager and owner goal alignment (McGuire, 1988; Singh, 1985).

Monitoring can solve both adverse selection and moral hazard, but it may be costly. Generally, monitoring involves gathering information on (1) the manager's effort, (2) exogenous factors that might affect firm performance, and (3) the outcomes (McGuire, 1988). If information about all three is available, no agency problem exists since a contract can be based on the appropriateness of the manager's actions. Without knowledge of the manager's actions, owners are exposed to moral hazard or adverse selection. Without knowledge of the random external variables which may affect the success of the manager's efforts, owners can not judge the appropriateness of the manager's efforts, and may be rewarding (or punishing) the manager for circumstances beyond the control of the manager. Without knowledge of the outcome, the owner cannot evaluate the relationship between the manager's actions and other external factors affecting the outcome.

Research has supported the theoretical assertions about the role of monitoring. Anderson (1985) found that the greater the difficulty of evaluating a salesperson's performance (outcome), the more likely the firm was to substitute monitoring for commission. Similarly, Eisenhardt (1988) found that the more programmable (monitorable) jobs were, the less likely sales people would be paid in risk-sharing commissions. She also found that as monitoring increased (i.e., as measured by span of control), the use of risk-sharing commissions decreased and that the use of monitoring was a better predictor of low risk-sharing when monitoring was easier (i.e., when jobs were programmable) than when monitoring was more difficult. Conlon and Parks (1990) found that when devising a new pay agreement, performance contingent compensation (risk sharing) was more likely when owners could not monitor a manager's efforts.

Incentive Alignment and the Control of Managers

The general assumption is that owners as principals are risk neutral because, as stockholders, they can diversify risk through a portfolio of investments, while agents (managers) are risk averse because they have no such option of diversification other than within the firm. Empirical studies demonstrate that managers are risk averse. In studying mergers, Hand, Lloyd, and Modani (1983) concluded that mergers occurred to "enable managers to diversify the risk of their employment income" (p. 9). Amihud and Lev (1979) found greater diversification and more conglomerate acquisitions in firms where managers (agents) were relatively less controlled by owners. They concluded that risk reduction through conglomerate mergers as a merger motive was "plausible for managers who are striving to decrease their 'employment risk'. The empirical findings…are consistent with this managerial motive" (p. 615).

There is substantial theoretical demonstration of the effects of risk sharing. Androkovich (1990) showed that risk-sharing attributes and inherent incentive effects of a payment scheme are very closely inter-linked. Shavell (1979) showed that if the agent were risk averse, compensation would always depend to some extent on the outcome, but never leave the agent bearing all the risk. Also, the fee would always depend to some extent on information which the owner has about the agent's effort. Basu, Lal, Srinivasan, and Staelin (1985) showed that the weaker the link between an agent's effort and outcome, the less the principal should share risk with the agent. Thus, the greater the variance in the outcome caused by exogenous factors, the less risk-sharing is desired since risk averse agents will require excessive compensation.

Risk sharing is accomplished through incentive alignment; that is, rewarding agents for measurable results that are determined to be in the best interest of the owner. Thus, owners seek to devise compensation systems to minimize agency problems by linking CEO pay to performance. By basing a portion of an agent's reward on his or her behavior, it is possible to resolve moral hazard and adverse selection problems. Straight base salary is risk free, but any compensation tied to variable performance measures entails greater risk for CEOs and reduces risk for owners. Complete risk shifting could occur if CEOs were not risk averse, since 100% of the CEO's compensation could be tied to outcomes, completely aligning owner and CEO goals and essentially eliminating the agency problem.

Managerial risk aversion is costly because premiums are required to get managers to accept risk. For example, an experiment by Conlon and Parks (1990) showed that agents required risk premiums in return for accepting contingent pay. Eisenhardt (1988) found that greater outcome uncertainty was significantly and negatively related to risk sharing. Self-utility maximizing managers have no incentive to create moral hazard or adverse selection problems because it would inhibit the achievement of their desired rewards.

Theory of Managerial Capitalism

Like agency theory, managerial capitalism also posits that managers will maximize their self-interest at the expense of owners: there is no "justification for assuming that those in control of a modern corporation will also choose to operate it in the interest of the owners" (Berle & Means, 1932, p. 121). The core of the managerialism argument is that managers have discretion and control of the firm because of the dispersion of stock ownership. Many of the possible constraints on managers are eliminated when ownership is so widely dispersed that the gain to any individual stockholder (through an increase in share value) is greatly offset by the cost of that action (Hindley, 1970). This allows managers the autonomy to pursue maximization of their own utility through increased personal financial gain, increased power, working conditions above those required by competitive conditions, and job security or labor-leisure trade-off, goals that diverge from the profitability goals of owners (Berle & Means, 1932; McEachern, 1975).

Executives in charge of the firm will prefer organizational growth as a primary firm goal, though they recognize the need to maintain a satisfactory level of profits to obtain capital to further increase sales and protect against unwanted acquisitions (Marris, 1964; Baumol, 1967). Growth serves several functions for managers who control the firm. First, salaries are far more closely correlated with firm size than firm profits, thereby reducing income risk (Baumol, 1967). Second, firm growth leads to increased power, salary, and status for managers (Marris, 1964). Lastly, growth helps ensure survival because larger firms are more insulated against environmental threats, thus providing increased job security for managers (Galbraith, 1967).

The ownership structure. Discretion, or the control of a firm, "lies in the hand of the individual or group who have the actual power to select the board of directors, (or its majority), either by mobilizing the legal right to choose them—'controlling' a majority of the votes directly or through some legal device—or by exerting pressure which influences their choice" (Berle & Means, 1932, p. 69). The degree of managerial discretion is defined in terms of the ownership structure, which refers to the distribution of equity holdings. The most common approach is to differentiate between manager-controlled, owner-managed, and owner-controlled firms using an equity concentration measure, or the minimum percentage of stock held by the largest single non-management shareholder (Hunt, 1986). A firm is classified as **owner-controlled** when there is a single non-management equity holder with at least a 5% stock holding. Firms that are not designated as owner-controlled are **manager-controlled** unless a member of management holds more than 5% of the outstanding equity. When a member of management holds 5% or more of the outstanding stock, firms are classified as **owner-managed**.

Support for managerialism. There are numerous studies from accounting (e.g., Hunt, 1985), economics (e.g., McEachern, 1975), management (e.g., Tosi & Gomez-Mejia, 1989), and sociology (e.g., Allen, 1981) that have shown self-serving managerial behaviors in management-controlled firms. For example, Grabowski and Mueller (1972) pitted a managerialism model against a pure stockholder welfare maximization model in dividend and investment decisions. They concluded that the managerial model was conceptually and statistically superior. In a study of merger decisions, Lewellen, Loderer, and Rosenfeld (1985) found that senior managers with large personal stockholdings in their firms were less likely to engage in acquisitions that reduce shareholder wealth than were managers with small stockholdings.

Other research shows how different, specific classes of top management decisions may transfer greater risks and higher agency costs to owners when managers control the firm:

1. Managers may choose accounting methods which state results in ways more favorable to them than to stockholders (Biddle, 1980; Biddle & Lindahl, 1982; Bowen, Noreen, & Lacey, 1981; Dyl, 1989; Groff & Wright, 1989; Hagerman & Zmijewski, 1979; Holthausen & Leftwich, 1983; Morse & Richardson, 1983; Sunder, 1973, 1975).
2. Managers may make investment decisions that are less optimal for owners, but which minimize managerial downside risk (Coffee, 1988; Hill & Hansen, 1989; Hill, Hitt, & Hoskinsson, 1988; Morck, Schleifer, & Vishny, 1989; Walsh & Seward, 1990).
3. Managers may undertake acquisitions and mergers that transfer higher agency costs to owners (Amihud & Lev, 1981; Halpern, 1973, 1983; Herman & Lowenstein, 1988; Jarrell, Brickley, & Netter, 1988; Kroll, Simmons, & Wright, 1990; Magenheim & Mueller, 1988; Ramanujan & Varadarajan, 1989; Ravenscraft & Scherer, 1987; Roll, 1987).

4. Managers may use internal political strategies to block organizational control mechanisms intended to provide checks on managerial discretion (Alderfer, 1986; Coughlan & Schmidt, 1985; Fierman, 1990; Meyer & Rowan, 1977; Salancik & Pfeffer, 1980; Schleifer & Vishny, 1988; Williams, 1985).
5. Managers may use organization resources to insulate themselves from the disciplining effects of external markets (Angelo & Rice, 1983; Bradley & Wakeman, 1983; Dann & DeAngelo, 1983; Malatesta & Walking, 1988; Ryngaert, 1988).

However, the results of research that examine the differences in performance of management-controlled and owner-controlled firms are mixed (Hunt, 1986). Several studies have found no support for the managerialism view of the relationship between the ownership structure and performance (Elliott, 1972; Holl, 1975; Kamerschen, 1968; Madden, 1982; Radice, 1971). Other studies, though, support the hypothesized relationship between the ownership structure and firm performance (Boudreaux, 1973; Krause, 1988; McEachern, 1975, 1976; Monsen, Chiu, & Cooley, 1968; Palmer, 1973; Shelton, 1967; Stano, 1976).

There are a number of explanations for these inconsistent findings about firm performance. One is that there may be no actual differences in manager-controlled and owner-controlled firms, and when differences are found in studies, they are due to sampling error. Another possible reason is that ownership distribution is used as a distant inferential proxy for actual monitoring. When controlling for differences in actual monitoring (i.e., measured by a behavioral scale), ownership distribution by itself shows little difference in firm performance (Tosi & Gomez-Mejia, 1994). A third reason that differences may not appear in some of the studies is because the nature of accounting methods used in owner-controlled and management-controlled firms may lead to an understatement of the reported differences between them (Hunt, 1985; Krause, 1988; Salamon & Smith, 1979). There is evidence that management-controlled firms overstate earnings, attempting "to control the information in annual accounting reports in a manner which causes firm performance to be misrepresented" (Salamon & Smith, 1979, p. 327). This suggests that if management-controlled firms overstate their earnings, the reported differences between them and owner-controlled firms would be underestimated, implying that actual differences may be in fact greater than actually reported in financial sources.

PART II. THE PROCESS OF ACQUIRING MANAGERIAL DISCRETION

Most of the research that focuses on managerial discretion examines its effects on CEO compensation and strategic choices at the firm level. In this part, we were interested in another question, "How is it that a CEO can acquire and institutionalize discretion?" The typical explanations of CEO selection and succession (i.e., human capital theory, tournament theory, and organizational politics) are not helpful, as they only explain how one obtains a position. We are more interested in how one obtains power. We undertook three studies to examine this process. In the first we assessed how the agency contract was negotiated and what were the differences between owner-controlled and management-controlled firms (Banning & Tosi, 1998). The second study focused on ways that CEOs of management-controlled firms reduced their compensation risk (Tosi & Gomez-Mejia, 1989). Finally, in the third study, we demonstrated how different criteria are used by boards of directors in the process of setting CEO pay (Silva & Tosi, 1998).

Negotiating the Agency Contract: Who Sets the Rules of the Game

Because CEOs are in a more vulnerable and tenuous position during their first ten years (Ocasio, 1994), it is perfectly rational that they attempt to solidify their position as early as possible, beginning with the process of negotiating their contract and continuing after being appointed. There is both the incentive and an opportunity to negotiate a favorable agency contract prior to the time of succession. At the same time, it is not possible to have a perfectly fashioned complete contract because neither the CEO nor the board of directors knows all of the contingencies, including the problems of adverse selection and moral hazard, that a contract should cover. The result is a more general and ambiguous "relational contract" that has both explicit and implicit elements (Milgrom & Roberts, 1992). This is useful for the incoming CEO because it makes it easier to develop a defensive post-succession strategy. For instance, new CEOs might find it easy to engage in opportunistic behavior during the contracting process because of private information, adverse selection, and moral hazard. A new CEO could, for example, negotiate very favorable terms by virtue of information known only to him or her. Another possibility is post-contractual opportunism, which permits CEOs to take advantage of the loopholes in the relational contract because all contingencies cannot be known before hand, and this is particularly true when the CEO is not subject to monitoring by a strong board of directors.

Because there are no perfectly fashioned contracts available for us to study, we thought that one way the acquisition of discretion can be understood would be to analyze how elements of the agency relationship (i.e., the incentive structure, monitoring and strategic choices of the firm) differ after succession in management-controlled and owner-controlled firms. This would provide clues about how the CEO could strengthen his position in management-controlled firms relative to the CEO in owner-controlled firms. For example, it is likely that the contract in owner-controlled firms will provide less latitude for the CEO to construct a strong self-defense strategy because there should be (1) stronger incentive alignment, (2) a board of directors that is likely to monitor the CEO more actively, (3) strategies that are more in the interests of owners and, ultimately, higher firm performance. We can therefore state, as a general proposition, that the contract for CEOs in management-controlled firms will be negotiated with the owners in a relatively weak bargaining position, providing CEOs with opportunities to strengthen their position after appointment. This should result in a contract for CEOs with (1) the power to defeat incentive-alignment agreements, (2) the ability to alter monitoring arrangements, and (3) discretion to choose firm strategies that enhance their own position.

Negotiating Incentive Alignment

We expect that the new CEO will seek a contract with higher pay and lower compensation risk than the previous incumbent (Harris & Raviv, 1979), and this is most likely to be the case in management-controlled firms. The result is that after succession:

1. The level of CEO compensation will be greater in management-controlled firms than in owner-controlled firms.
2. The CEO's compensation risk will be lower in management-controlled firms than in owner-controlled firms.

Negotiating the Monitoring Arrangements

Because the board of directors is the conventional mechanism for monitoring the CEO, we expect that new CEOs will seek authority to modify the board's composition in ways that consolidate their own influence. First, they may actively encourage existing directors to leave the board, replacing them with more sympathetic directors who are more similar to themselves. Directors appointed by a CEO are likely to act to protect the CEO's interest because, in many ways, it is also in their own interest (Main, O'Reilly, & Wade, 1995; Westphal & Zajac, 1995).

Second, after succession, CEOs are likely to increase the proportion of outside directors on the board, especially when the firm has weak equity holders. Outside directors must rely on the firm's managers (especially the CEO) for information about the firm's performance, and as a result, they may be subject to manipulation by the CEO (Baysinger & Hoskisson, 1990). They are also likely to be more sympathetic than insiders to the agendas of CEOs, in that executive pay is higher in firms with a greater proportion of outside board members, and this relationship holds when controlling for firm size and performance (Main et al., 1995; O'Reilly, Main, & Crystal, 1988; Westphal & Zajac, 1995). Further, inside directors may have more to gain by ousting an incumbent CEO. The reason, according to tournament theory (Lazear & Rosen, 1981; Ocasio, 1994), is that, as senior managers, inside directors may actively compete for the chief executive's position. Finally, Fizel and Louie (1990) and Baysinger, Kosnik, and Turk (1991) found that greater concentrations of outside directors resulted in lower CEO accountability. Thus,

Managerial Discretion

although outside directors may ostensibly be appointed to represent owners and limit CEO power, it appears that these board members may actually augment the chief executive's influence (Ocasio, 1994). This would suggest that following succession:

3. The ratio of outside directors to inside directors will be greater in management-controlled firms than in owner-controlled firms.
4. The turnover of directors will be greater in management-controlled firms than in owner-controlled firms.

New CEOs can also seek to protect their positions by appointing outside directors who are highly compensated by their own firms. Several studies show that CEO pay is higher when their boards are composed of outside directors who are highly compensated in their own firm (Kosnik, 1990; Main et al., 1995; O'Reilly et al., 1988). O'Reilly and his colleagues (1988) attribute this to norms of reciprocity that would lead these directors to be more likely to approve of high levels of compensation for the new CEO. Therefore, not only might powerful CEOs attempt to stack their boards with outsiders, they would search for highly paid outsiders. So we expected that after succession:

5. The level of external compensation received by outside directors will be greater in management-controlled firms than in owner-controlled firms.

In addition to reducing monitoring by the board by (1) stacking it with outside members who are dependent on the CEO for information, (2) removing some board members or failing to renew their terms, and (3) seeking highly compensated outside board members who may be more inclined to approve of higher executive compensation, another way to be a more powerful CEO is to hold the CEO and board chair positions simultaneously. The additional influence that results from holding both the CEO and board chair positions may result in less effective monitoring of the CEO's activities (Dalton & Kesner, 1985). As a result, we expected that after succession:

6. The appointment of new CEOs to the position of chair of the board of directors will be more prevalent in management-controlled firms than in owner-controlled firms.

Negotiating Strategic Decision Influence

The contract for powerful CEOs is likely to provide them with discretion that will permit them to seek sales growth or diversification, both strategies that are likely to reduce their own employment risk (Amihud & Lev, 1981; Hill & Snell, 1988, 1989). Diversification involves expansion of the firm, and this increased size serves the interests of the CEOs. Diversification that is unrelated to the firm's core business, however, leads to lower performance, whether in comparison to single-business firms or related diversifiers (Bettis, 1981; Dubofsky & Varadarajan, 1987; Rumelt, 1982). In addition, because firm size (sales volume) is the single most significant determinant of the level of executive pay (Ciscel & Carroll, 1980; Gomez-Mejia, Tosi, & Hinkin, 1987), a firm's level of unrelated diversification and CEO compensation are closely tied. Further, previous research shows that management-controlled firms engage in significantly more unrelated diversification than owner-controlled firms (Amihud & Lev, 1981; Baysinger et al., 1991; Hill & Snell, 1988, 1989). This led us to suspect that after succession:

7. Unrelated product-market diversification will be greater in management-controlled firms than in owner-controlled firms.
8. Related diversification will be lower in management-controlled firms than in owner-controlled firms.

Succession and Firm Performance

Perhaps the most effective defensive strategy for new CEOs is to improve firm performance. We nevertheless expect that new CEOs in management-controlled firms will be able to negotiate contracts that provide strong defensive mechanisms, and this will have a detrimental effect on firm performance. This is suggested by those studies that demonstrated that several strategies more common in management-controlled firms, such as increasing unrelated diversification and decoupling CEO pay from firm performance, are associated with lower financial performance (Amit & Livnat, 1988; Hill & Snell, 1988; Rumelt, 1982; Tosi & Gomez-Mejia, 1994). So, after succession:

9. Levels of financial performance will be higher in owner-controlled firms than in management-controlled firms.

We studied these propositions in a sample of publicly held firms in which the CEO had been replaced during the period 1988 to 1991. The main question of this study was how the dependent variables, reflecting changes in the agency contract, are affected by whether the firm is owner-controlled or management-controlled. The dependent variables were (1) CEO compensation, (2) CEO compensation risk, (3) firm performance, (4) corporate strategy, (5) CEO/Chair duality (whether the CEO also holds the board chair position) (Dalton & Kesner, 1985), (5) outsider ratio (i.e., the number of outside directors on the firm's board divided by the total number of board members), (6) director turnover (i.e., the percentage of directors who leave the board in each year), and (7) outside director compensation (i.e., the cash pay received by outside directors in their positions as executives at their own

Managerial Discretion 177

firms) (O'Reilly et al., 1988). Control variables were firm size, the origin of the new CEO, and the reason that the previous CEO left the firm.

We found that the negotiated incentive alignment was more favorable for CEOs in management-controlled firms. The level of CEO pay was higher and the compensation risk lower in firms controlled by managers.

New CEOs in management-controlled firms also had negotiated monitoring arrangements that were more favorable. For new CEOs in management-controlled firms, the ratio of outside directors was higher after succession than in owner-controlled firms. However, the rate of director turnover was similar for both management-controlled and owner-controlled firms. There was no significant difference in director pay as a function of the ownership structure. However, there was a stronger relationship between director pay and CEO pay in management-controlled firms and, because management-controlled CEOs are more highly paid than owner-controlled CEOs, we expect that there are pay differentials in the direction suggested. The data failed to support our intuition that new CEOs in management-controlled firms would also be appointed board chair in their early years in the firm. We do show later, in another study, that there is a difference in the duality of CEOs between management-controlled and owner-controlled firms (Silva & Tosi, 1998), suggesting that accession to the board chair probably occurs later for management-controlled CEOs, but is less likely to happen for owner-controlled CEOs.

The strategic decisions following succession were related to the ownership structure. While there was no difference in unrelated diversification between management-controlled and owner-controlled firms following succession, CEOs in management-controlled firms tended to decrease the degree of related diversification in the years following their appointment.

Finally, firm performance after succession was affected by the ownership structure: Management-controlled firms did not perform as well as owner-controlled firms.

Decoupling CEO Pay from Performance: Processes Inside the Board

To the extent that CEOs in management-controlled firms are able to institutionalize the power and control, as we have just shown above, it should be reflected in the relationship between the CEOs and the boards of directors, as well as other key actors involved with the management of the firm. Starting with this premise, we conducted a study of various organizational actors and stakeholders on the CEO compensation processes (Tosi & Gomez-Mejia, 1989). There were three main questions addressed in this study. The first was "Is there greater reported incentive alignment and monitoring in owner-controlled firms than in management-controlled firms?" The second question was, "Does compensation risk differ between management-controlled and owner-controlled firms?" Finally, we assessed the different levels of influence that the various actors had in the CEO compensation setting processes.

We surveyed the chief compensation officers (CCOs) of manufacturing companies randomly selected from the American Compensation Association membership directory. The compensation officers reported on the following factors: (1) CEO incentive alignment and monitoring of the compensation setting process, (2) CEO compensation risk, (3) influence of various stakeholders in the CEO compensation process (the CEO, major stockholders, boards of directors, the compensation committee, and outside consultants), (4) CEO tenure, (5) CEO origin, (6) firm performance, (7) firm size, and (8) the ownership structure.

The results, consistent with agency theory and managerialism, showed that monitoring of the CEO compensation process is higher in owner-controlled firms than in management-controlled firms. In addition, there is more pay risk in owner-controlled firms than in management-controlled companies.

There were interesting results for the different levels of influence of the various actors involved in fixing the CEO compensation structure. In management-controlled firms, the CEOs themselves and management consultants have significantly more influence than their counterparts at owner-controlled firms. On the other hand, in owner-controlled firms, the influence levels were higher for major stockholders, the board of directors, the compensation committee, and the chief compensation officer. These results were expected because in management-controlled companies, board appointments are usually controlled by management, whereas in owner-controlled firms the board is more likely to be chosen to represent the owners interests (Herman, 1981).

We also showed that the level of CEO compensation monitoring is related to higher compensation risk in bonus and long-term income for both management-controlled and owner-controlled firms, but that the relationship is stronger at owner-controlled firms. Furthermore, the level of monitoring is negatively related to CEOs' influence over their own pay packages in both management-controlled and owner-controlled firms, but is much stronger in owner-controlled firms. Monitoring is negatively related to the use of compensation consultants in management-controlled firms, and positively related to the influence of chief compensation officers in both types of firms. While in management-controlled firms the level of monitoring was not related to the influence of major stockholders or the board of directors, these relationships were strong and positive at owner-controlled firms.

On the Nature of CEO Performance Evaluation by the Board of Directors

One would expect that if the CEO is more influential in the compensation setting process that the influence will be exerted in ways that present the appearance of rationality to the board of directors that must ratify the pay decisions. This would be necessary, particularly, because as we show later, the pay of CEOs in management-controlled firms is less sensitive to changes in firm performance and more sensitive to changes in firm size, whereas pay/performance sensitivities are stronger in owner-controlled firms.

This led us to consider that different CEO performance evaluation criteria might be used in owner-controlled and management-controlled firms (Silva & Tosi, 1998). Our fundamental intuition was that the criteria for CEOs in owner-controlled firms will more likely be objective measures of firm performance, while in management-controlled firms these measures will be less critical and will be supplemented by criteria that reflect the CEOs leadership and management skills. Of course, these latter factors are more subjectively evaluated and weighted and will, therefore, accrue to the benefit of the management-controlled CEO.

We surveyed chief compensation officers who were members of the American Compensation Association to assess these issues and found that the CEOs of management-controlled firms:

1. Are more likely to hold the position of board chair simultaneously.
2. Are less likely to have their performance reviewed anonymously than CEOs in owner-controlled firms.
3. Are less likely to have their performance evaluated by financial performance measures such as earnings per share, economic value added (EVA), and profitability.
4. Have greater weight placed on qualitative criteria as opposed to quantitative criteria in their performance evaluation, and these qualitative criteria have become increasingly important in recent years.
5. Have more emphasis placed on their managerial skills (e.g., planning, handling problems, representing the firm) than CEOs in owner-controlled firms.

Summary

These studies show how new CEOs acquire and institutionalize discretion. The consistent theme is that new CEOs who are in more favorable positions, relative to those charged with monitoring managerial decision making, are able to negotiate agency contracts with more generous terms. The contractual mechanisms through which the institutionalization of power occurs in management-controlled firms differ from those in owner-controlled firms. In management-controlled firms, the contracts of CEOs who are appointed board chair permit them more actively to affect strategies that minimize their risk. However, this does not improve firm performance as in the owner-controlled firms in our study (Hunt, 1986; McEachern, 1975; Ware, 1975).

In addition, CEOs in management-controlled firms, especially when they also hold the position of board chair, negotiate compensation contracts with higher pay levels than their predecessors and less pay risk, while pay risk is higher for CEO successors in owner-controlled firms. Thus, compensation that is already high because CEOs are in management-controlled firms (Gomez-Mejia, Tosi, & Hinkin, 1987; McEachern, 1975) is not likely to suffer much from the poorer performance of the firm.

Another favorable aspect of the negotiated agency contract in management-controlled firms is that after succession, the proportion of outside board members is higher than in owner-controlled firms. As we have already noted, these sorts of boards are more likely to act in support of the CEO, both with respect to compensation and strategic choices. Indeed, not only were outside ratios higher in manager-controlled firms, higher outside ratios were associated with higher levels of total CEO pay and lower levels of pay risk. The results are consistent with previous findings that CEOs in management-controlled firms have more influence over board members with respect to their pay levels and pay risk than those in owner-controlled firms (Tosi & Gomez-Mejia, 1989), and that this influence is also associated with lower performance (Tosi & Gomez-Mejia, 1994).

The CEOs in management-controlled firms also manage early in their tenure to reduce their risk by managing the strategic choices of the firm. After succession, performance-enhancing related diversification is lower in management-controlled firms while it increases in owner-controlled firms, probably because related diversification has few short-term benefits for the CEO (Amihud & Lev, 1981; Harris & Tosi, 1998; Hill & Snell, 1988, 1989). Ultimately, the negotiated agency contract appears to have negative effects on the firm's financial performance. In manager-controlled firms, firm performance is lower than in the owner-controlled firms.

The other two studies provide insight about the nature of the CEO relationship with the board of directors and other influential actors within the firm responsible in one way or another for monitoring the CEO. Both show that CEO monitoring and incentive alignment are lower in management-controlled firms. For example, in management-controlled firms, the CEO is the most influential actor in the determination of his own compensation structure. The board, major stockholders, and the compensation committee are much weaker.

One of the strongest actors is the compensation consultant. The compensation consultant works at the pleasure of the board, stacked with outsiders for whom there exists reciprocal dependence with the CEO. The consultant, working to the requirements of the board, can easily formulate a compensation package very favorable to the CEO, both in terms of the total level of pay and compensation risk.

We have some ideas about how this happens in management-controlled firms when firm performance is low. In this case, primary reliance on the financial indicators would lead to recommendations of lower pay and higher pay risk. Because that is clearly not in the interest of the CEO and the top management, other justification must be found, a task that is relatively simple for a competent consultant. First, the consultant can argue that the job of the CEO is indeed complicated and simple financial measures will not capture its totality. Therefore, the board of directors should expand the criteria used to determine compensation beyond the objective financial indicators (ROE, ROA, ROI and EVA) that are of primary importance in owner-controlled firms. These qualitative, subjective criteria should, the argument continues, be used because they reflect the CEOs leadership and managerial ability, in addition to those more conventional economic indicators.

If this scenario occurs as our intuition implies, then it is quite easy to justify higher CEO pay on the basis of these subjective criteria in the face of lower economic performance. One reason is that, as attribution theory suggests, the board favorable to the CEO will be positively disposed toward the CEO and overestimate the importance of the subjective criteria and underestimate the relevance of the financial measures when they do not reflect high performance. In addition, because CEO evaluations are less likely to be anonymous in management-controlled firms, one who does negatively assess the CEO will be known, permitting the CEO to act in ways to minimize his opposition. Thus, relying on these more imperfect measures when there are good outcome measures that could be used results in transferring both higher agency costs and more risk to equity holders.

The conclusion from these studies is that CEO monitoring and incentive alignment are used less frequently in management-controlled firms where they are needed most to combat opportunistic behavior on the part of CEOs with high levels of discretion because of the nature of the negotiated agency contract and what the CEO can do to institutionalize his discretion. The complicated task requirements of the CEO position as well as the nature of the CEO compensation package, particularly the long-term incentives and bonuses portions, allow the designer of the compensation package to mask the true amount of compensation and the nature of the relationship between the CEO's pay and firm performance.

PART III. THE EFFECTS OF MANAGERIAL DISCRETION ON FIRM COMPENSATION STRATEGY

We now know how the CEO in the management-controlled firm can affect the structure of the compensation by negotiating a favorable agency contract, dominating the process of setting his/her pay and affecting the nature of the evaluation criteria. We also know from a substantial body of other research that in management-controlled firms, CEOs are paid more and have less compensation risk than those in owner-controlled firms. Up to now, the implicit assumption has been that aligning the interests of the owners and the CEO by making pay contingent on outcomes of interest to the owners is enough to motivate others at lower levels of the firm to act similarly to maximize owner returns, either because the CEO is powerful enough to influence them or because incentives at lower levels are similarly aligned. We examine this assumption in three studies that examine how the compensation strategy of the firm is related to the ownership structure. These studies show that in management-controlled firms relative to owner-controlled firms (1)

CEOs use their discretion to extract more compensation with less compensation risk, (2) that similar compensation discrepancies occur also at upper management-levels and, (3) that the average employee does not have the same wage premium as managers but bears less compensation risk than his or her counterpart in owner-controlled firms.

CEO Compensation

The research on CEO pay tends to focus on how it relates to firm performance and firm size. The reason for understanding the performance effect on pay is obvious; compensation is the mechanism through with managerial incentives can be aligned with the interests of equity holders.

Firm size should affect pay for several reasons. For one thing, larger firms have more organizational levels and there tends to be an average 30% to 40% pay difference between levels. Also, according to economic theory (Roberts, 1959), CEO pay should increase with firm size because the CEO's marginal revenue product is greater in larger firms. Good management can substantially increase total profits at a large firm by increasing efficiency, and the CEO's salary will be absorbed into the expenses of a large number of units. On the other hand, at a small firm, even substantial increases in efficiency over fewer units will not justify a high CEO salary. Furthermore, firms usually hire CEOs from similarly sized companies, so there are different competitive wage markets for firms of various sizes (Roberts, 1959). Human capital theory (Agarwal, 1981; Becker, 1964) also supports an association between firm size and CEO pay because as CEOs' jobs become more complex and carry greater responsibility with growth, so the CEO might be compensated independently of performance. However, human capital variables have not accounted for significant variance in CEO pay in prior empirical studies (Agarwal, 1981; Dyl, 1988).

While these are the standard, rational arguments that are based on the assumption of the profit-maximizing manager, the managerialist position, as we noted earlier, is that CEOs in management-controlled firms will be able to reduce their compensation risk and extract higher salaries than those in owner-controlled firms. This leads to the proposition that the strongest determinants of pay will differ as a function of ownership structure.

We tested this proposition in a study of the compensation of CEOs in manufacturing firms in the United States whose compensation package was reported in *Business Week* for the four year period between 1979 and 1982 (Gomez-Mejia et al., 1987). The dependent variables were: (1) total compensation, (2) salary, bonuses and long-term income, and (3) changes in total compensation and salary, bonus and long-term income. The independent and control variables were (1) firm performance, (2) firm size, (3) CEO origin and (4) the ownership structure.

We found that the size of the firm was a significant predictor of absolute amount of total compensation and all of its components in management-controlled firms.

Changes in CEO bonuses and total compensation were strongly and positively related to firm size. CEOs hired from outside the company received much larger bonuses and increases in bonuses and in total compensation during the study period than did incumbents or those promoted from within the organization. Firm performance, however, was related to the absolute amount of bonuses, the long-term portion of CEOs' compensation, and total compensation, not salary.

On the other hand, in owner-controlled firms, firm performance was the best predictor of CEO pay. It was significantly related to the total amount of CEO compensation, base salary, bonuses, and long-term income. Furthermore, changes in total compensation and all of its components were positively related to the company's financial performance. Changes in the size of the firm and whether or not the CEO was hired from outside the company had no effect on any aspect of CEO pay at owner-controlled companies. An important point is that in owner-controlled companies performance explained four times more variance in total compensation than in management-controlled firms. A later study by Hambrick and Finkelstein (1995) examined CEO pay in management-controlled and owner-controlled firms from a similar perspective and their results were consistent with ours.

Managerial Compensation Strategy:
Managerial Incentives Within the Firm

We have already said that the unspoken assumption in the literature that deals with incentive alignment and monitoring is that all that is necessary is that both control mechanisms can be applied to CEOs, and that is enough. We were interested in how incentive alignment was structured for managers within the firm, and how it is related to firm performance. The reason is that lower-level managers are the agents of owners as well as the upper-level managers, but they also act as principals who must monitor those lower in the hierarchy (Fama & Jensen, 1983; Stiglitz, 1975). This creates potential agency problems because the agent-monitors must bear the cost of monitoring, but they receive little benefit from it unless they are themselves monitored or have shared ownership risks/incentives (Baker, Jensen, & Murphy, 1988). Therefore, firms that do not have performance-compensation alignment at the top layers of management would not be expected to have it at lower levels.

Furthermore, even in firms with incentive alignment at the top layers, control loss would be expected as monitoring is decreased at each hierarchical level (Williamson, 1967). Because CEO incentive alignment and performance monitoring are weaker in management-controlled than in owner-controlled firms, it is expected that there will be weaker incentive alignment throughout the management layers in management-controlled firms. It is also expected that excessive compensation will be a greater agency cost at lower managerial layers because paying above-market wages will enable higher-level managers to attract and retain quality employees, thus reducing their effort while increasing the cost not to themselves, but to owners. In addition, because pay levels increase with each hierarchical level, increasing subordinates' pay would tend to increase the top-manager's pay.

We studied the compensation structure of upper level managers in a sample of management-controlled firms, owner-controlled firms, and owner-managed firms (Werner & Tosi, 1995). Specifically, these questions were:

1. Do management-controlled firms have higher management pay levels than owner-controlled firms?
2. Is managerial pay sensitive to changes in performance in owner-controlled firms and to changes in size in management-controlled firms in the same way as CEO pay?
3. Are a greater proportion of managers eligible for long-term performance incentives in owner-controlled firms than in management-controlled firms?
4. Are a greater proportion of managers eligible for bonuses in owner-controlled firms than in management-controlled firms?
5. Does bonus compensation constitute a higher proportion of the managerial compensation package in owner-controlled firms than in management-controlled firms.

The dependent variables were: (1) base pay (the average salary of managers at a particular level in the firm), (2) total pay (average base pay plus average bonus for managers at a specific organizational level), (3) changes in total pay (the average change in this total for the study period for each manager), (4) the percentage of long-term performance eligible managers in a firm, (5) the percentage of bonus eligible managers in a firm, and (6) compensation risk. The independent and control variables were (1) the ownership structure, (2) years of education, (3) years of experience, (4) job level, (5) firm size, (6) financial performance, and (7) industry control dummy variables.

We found that for the study period, the average salary and bonuses of the top six executive levels in management-controlled companies were over $13,000 more per year than those at owner-controlled companies, and over $15,000 more than those at owner-managed firms. The gaps were the widest at the CEO level, where the chief executives at management-controlled companies averaged almost $115,000 more than those at owner-controlled companies, and over $219,000 more than those at owner-managed companies.

Pay was higher for all of the top six levels of management, with differences of about $45,000 at the second layer of management and decreasing to about $7,000 for the sixth layer. Below these upper levels, the differences between management-controlled and both owner-controlled and owner-managed companies were not statistically significant.

There were some important differences in pay sensitivity. Changes in base and bonus pay were related to both firm performance and the change in performance at the top levels of owner-controlled firms, but not in management-controlled firms.

There were some surprises in the findings for bonus and long-term incentives. More managers were eligible for long-term incentives and short-term bonuses in management-controlled firms than in owner-controlled firms. Because long-term incentives and short-term bonuses are more sensitive to performance in owner-controlled firms, one interpretation of these unexpected results is that the offering of long-term incentives and bonuses in management controlled firms serves to justify greater managerial pay under the masquerade that they represent pay at risk.

The Overall Compensation Strategy: Effects on the Average Employee

After examining the effects of the ownership structure on managerial pay, the question that remains is whether these managerial pay premia and lower levels of compensation risk in management-controlled organizations are passed to the average employee in the organization. There are some good reasons to expect that they might be. For example, paying above-market wages to all employees would make it easier for CEOs to accomplish their goals and could also be associated with prestige or bring CEOs' gratitude and loyalty from employees that could advance the CEOs' interests.

Two studies examined how two key facets of the overall compensation strategies of management controlled firms differed from those of owner-controlled firms (Tosi, Werner, & Gomez-Mejia, 1997). The first facet was the firms' strategies of paying above, at, or below market wages (Weber & Rynes, 1991). The second facet was the different approaches to the use of incentive based compensation, and how much of employees' pay should be at risk.

Our intuition was that management-controlled firms choose an above market wage strategy for three reasons. First, wage premiums reduce the managerial effort in selecting, training, and retaining quality employees. Second, a high-wage strategy would be expected to elevate managers' pay because pay increases linearly with hierarchical levels (Hills, Bergmann, & Scarpello, 1994; Milkovich & Newman, 1996; Simon, 1959). Third, the prestige associated with high compensation levels may cater to management's ego needs (Marris, 1964).

We also thought that management-controlled firms would make less use of performance incentives. First, it is not likely that managers who do not link their own pay to performance would do so for employees elsewhere in the firm. Second, while incentive pay increases performance, it reduces employee satisfaction (Schwab, 1974) and disrupts the social climate in companies (Whyte, 1949).

The first study was a survey of compensation practices survey sent to chief compensation officers of manufacturing firms with over 500 employees that examined how pay level strategies, pay for performance strategies, and pay at risk strategies are articulated as policies in management-controlled and owner-controlled firms. The dependent variables were the (1) degree to which the company's strategy was to pay wages and/or benefits above, at, or below the market level for their industry, (2) degree to which pay is tied to performance, (3) degree to which the firm's policy was to put employee pay at risk, and (4) percentage of variable compensation (bonuses and other pay incentives) for members at different organizational levels. The independent and control variables were firm size, firm performance, industry type, and the ownership structure.

The second study analyzed compensation practices using archival data from owner-controlled and management-controlled firms in 24 industries between 1989 and 1992. The dependent variables were (1) the firm's average pay per employee, adjusted by the type of industry, and (2) changes in average pay per employee. The independent variables were size, changes in size, firm performance, changes in firm performance and the ownership structure.

We found no differences in either the stated policies or the pay practices for pay levels in owner-controlled and management-controlled firms. The compensation premia paid to managers in management-controlled firms do not cascade down to the majority of employees.

However, we did find that management-controlled firms pursued compensation strategies with lower pay/performance sensitivity than owner-controlled firms in both studies. Owner-controlled firms reported a stronger pay/performance linkage than did management-controlled firms, as well as placing a larger proportion of employees' pay at risk for all employee levels. The level of pay risk decreases with each organizational level, but remains higher than that of management-controlled firms at all levels of owner-controlled firms. In addition, actual changes in wage levels were related to changes in the firm's financial performance in owner-controlled firms, but were related to change in size of the firm, not performance, at management-controlled firms.

Summary

The results of these studies show that the managerial discretion affects the way compensation is structured throughout the firm, not just for the CEO and the top levels of management, but also for other employees. CEOs in owner-controlled firms are paid more for performance and less for the scale of their operations than CEOs in management-controlled firms. The CEO compensation strategy in management-controlled firms avoids the vagaries of fluctuating performance, permitting the CEO to take advantage of the more stable factor of size, thus reducing compensation risk substantially. At the same time, they benefit from improvements in performance through the bonus system, taking additional compensation when performance is higher. They have the best of both worlds: the base salary is more strongly related to size of the firm, a relatively stable factor, but their bonuses are related to firm performance. On the other hand, CEOs in owner-controlled firms are in riskier positions; that is, they are primarily rewarded for performance, a more risky factor.

The nature of CEO compensation strategy is mimicked for top-level managers in both management-controlled and owner-controlled firms. For example, pay premiums were also found for upper-level managers in management-controlled firms, but not in owner-controlled firms. These upper-level managers in management-controlled firms were paid 8.2% more than managers in owner-controlled firms and the pay differences are exaggerated as a function of organizational level. For example, the average base and bonus of CEOs in management-controlled firms was $219,000 higher than that of a counterpart in an owner-managed firm. At the sixth hierarchical level, the average base and bonus differential was $7,000. These results are consistent with other studies of internal pay levels which found negative relationships between the pay of top-level managers in firms and equity concentration (Lambert, Larcker, & Weigelt, 1993; O'Reilly et al., 1988). However, there is no apparent economic justification for such premiums for the management-controlled firms in this study because they did not perform better than owner-controlled firms and they were significantly worse performers than owner-managed firms.

In addition to the substantial pay premiums, the firm's compensation strategy provides lower compensation risk for the upper level managers in management-controlled firms, reflecting the lower pay risk for CEOs in management-controlled firms (e.g., Gomez-Mejia et al., 1987; McEachern, 1975). In these firms, change in pay for the upper management group is decoupled from firm performance, and is sensitive to changes in the size of the firm. This differed from owner-controlled firms, in which changes in pay were more sensitive to changes in firm performance for the top-level managers.

However, the design of the managerial compensation strategy in management-controlled firms does not make this lower compensation risk transparent to the observer. The reason is that an examination of the strategy shows that in these firms there is a larger proportion of the upper management actually eligible for bonus, an element of compensation that is usually linked to firm performance. Furthermore, these managers actually receive a larger percentage of their pay in the form of bonuses even though these firms performed no better than owner-controlled firms.

So, to the observer, the compensation strategy of the management-controlled firm appears to have an economically rational basis, that it ostensibly provides incentives for maximizing firm performance. To us, this suggests one way in which the management-controlled firms use the compensation system to benefit managers. We believe that there are differences in the manner in which bonuses are determined, as a function of managerial discretion. We have already demonstrated that the criteria used for the evaluation of CEO pay in management-controlled firms is heavily influenced by subjective factors that extend beyond the expected financial performance indicators. Because compensation premia and the lack of pay risk cascade to lower managerial levels, it makes sense to us that these more subjective criteria will also. Think about how bonuses are determined: the typical approach is to allocate a percentage of profits to a bonus pool, which is then in turn allocated to individual managers by a "bonus committee" of higher level managers (Hills, 1987).

In owner-controlled firms, the triggering mechanism for the size of the bonus pool and the basis for its distribution to lower-level managers appears to be improvements in firm performance. In management-controlled firms, the triggering mechanism is not so simple and, further, the relevance and weight of these subjective criteria are under the control of the dominant management coalition (Tosi & Gomez-Mejia, 1989). It seems likely that the triggering mechanisms (i.e., those measures of performance used to determine bonuses) change from year to year, reflecting indicators that are more advantageous to the internal managers at the particular time that the bonus is being determined (Crystal, 1991, p. 15). Thus, bonuses do not reflect compensation risk in management-controlled firms, but rather serve as a discretionary mechanism that permits higher levels of pay to managers.

When we studied the overarching compensation strategy of the firm, the results were even more revealing about wage premia and pay risk in the types of firms we studied. For one thing, the compensation premia of the upper management levels in management-controlled firms were not shared by the average employee. The stated wage level policies of both types of firms were similar, but, and this is a key point, there was no difference in the wages actually paid.

The reason that this is a key point is that it identifies the average employee in the firm as another group to which the managers in management-controlled firms transfer costs. We think that this is so for two reasons. First, we have no doubt, given the data here and elsewhere, of the pay premia for the CEO and upper management in management-controlled firms. Second, if it is the case, as we have demonstrated, that the average compensation is the same between the two types of firms, then it can only be that way if there is greater disparity between the pay of the top management groups and those at lower levels in management-controlled firms than in owner-controlled firms. Thus, not only does the management not share the pay premium, but they appear to exploit the average employee, relative to owner-controlled firms.

The analysis of the overarching compensation strategy also provided evidence about how deeply embedded pay risk is in owner-controlled firms, and its absence in management-controlled firms; the average employee in owner-controlled firms bears compensation risk, those in management-controlled firms bear none. Thus, the nature of the incentive structure for non-managerial employees as a function of ownership structure is quite similar to that of the top-management.

This suggests a number of things to us about how discretion affects the firm's compensation strategy. First, CEOs in management-controlled firms are able to transfer risk to equity holders and receive higher pay after they have successfully negotiated a favorable contract by institutionalizing lower pay risk and higher compensation. It appears to us that this is executed in part by passing these same risk reducing strategies to upper-level managers, creating a relatively large group that has similar self-interests to the dominant coalition.

CEOs in owner-controlled firms work under a different compensation strategy, obviously designed with the interests of equity holders in mind. In owner-controlled firms, the intent is to place a greater amount of pay at risk in their compensation packages than is the case in management-controlled firms. The proportion of pay at risk decreases by organizational level, yet it remains higher for owner-controlled firms at all organizational levels.

Second, we believe that these differences are not readily discernable by simple observation of the compensation structure. We have shown that management-controlled firms create an illusion of risk sharing by distributing total pay in different ways from owner-controlled firms. This happens, we believe, because design of the compensation strategy permits them to "hide" the pay premiums that they are able to extract while at the same time appearing to remain competitive in terms of overall labor costs. For example, the management-controlled pay structure provides for more managers to be eligible for bonuses and, in fact, pays a larger proportion of total pay in the form of bonus. We believe that this sort of compensation structure is accomplished through the expertise of consultants who, in management-controlled firms, are dominated by the CEO and a CEO friendly board. These consultants are not likely to impose a compensation system that provides for lower pay and higher pay risk. This means that the CEOs probably actually manage their own pay, and perhaps that of the upper management group, to ensure higher pay and lower risk.

Third, we conclude that one reason for the lower performance of management-controlled firms is not only that managerial pay is not sensitive to firm performance, but also that these firms may take advantage of the average employee, relative to those in owner-controlled firms. The net effect is probably a culture in management-controlled firms that is not performance oriented, but is directed toward matters of top management survival and gain through political processes, not successes in product markets.

Finally, these studies raise some questions about the utility of models relying on managerial labor markets as disciplining mechanisms. Our results, and other studies that we have already cited, consistently show that the CEOs in management-controlled firms receive higher pay with lower pay risk and at the same time are able to retain their positions for longer periods, even under conditions of lower performance (e.g., Salancik & Pfeffer, 1980; McEachern, 1975).

PART IV. CONTROLLING THE CEO: AN ANALYSIS OF THE EFFECTS OF INCENTIVE ALIGNMENT AND MONITORING ON FIRM PERFORMANCE AND CEO DECISION MAKING

In Part I, we set out the basic theoretical underpinnings of our thesis that managerial discretion affects the compensation strategy of the firm. Because markets do not function effectively as disciplining mechanisms, the agency contract becomes the device through which principals seek to control agents. In the case of the CEO, this control is thwarted when the equity is so dispersed that owners, through boards, are not able to influence the management to act in their interests. In Parts II and III, our evidence shows how managers negotiate a contract that leads to higher pay and lower pay risk in management-controlled firms, and how this is manifested throughout the firm's compensation strategy. In this last section, we shift the focus toward the question of how to control the CEO through monitoring and incentive alignment.

We address this through the use of a laboratory study and with survey research methods more commonly used in the organizational sciences. These differ from the typical archival based approach in which the control of the CEO is studied. The primary reason is that in the conventional empirical work, there has been little research, let alone concern, on the construct validity of the commonly used archival measures. As one example, it is common to assess monitoring by using the proportion of outside to inside members of the board of directors. However, studies have shown that, different from the expected effects, higher proportions of outside directors are not likely to be good proxies for monitoring (Banning & Tosi, 1998; Ocasio, 1994, O'Reilly et al., 1988). If this is so, then the theoretical effects of monitoring have not been adequately assessed, making inferences about them from research results somewhat suspect.

The Relative Effects of Incentive Alignment and Monitoring on Managerial Decisions

This raises some important questions about the nature of control mechanisms to reduce agency costs. First, monitoring is conventionally assessed by the proportion of outside directors on the board, but we believe that this is, in fact, an indicator that is actually related to greater, not less, managerial discretion. Thus, it seems necessary to study monitoring in other ways in order to assess its effects relative to the effects of incentive alignment to determine how these two control mechanisms interact to curb managerial self-serving behaviors. Second, if agents are in a position to influence the contract, what are the consequences of subsequent managerial decisions governed by the contract?

A study by Zajac and Westphal (1994) examined the relationship between monitoring and incentive alignment as used by firms. Using the common measure of monitoring (i.e., the proportion of outside directors), they concluded that incentive alignment and monitoring act as substitutes for each other, and that the more monitoring is present, the lower the reliance on incentive alignment. In other words, firms with strong monitoring have low incentive alignment and vice versa. However, given our interpretation of the monitoring measure, it seems to us that just the reverse occurred. That is, over time, both incentive alignment and board monitoring decreased. Thus, the question of what happens to managerial decision making when incentive alignment is used as a substitute for monitoring or monitoring replaces incentive alignment is still open.

We addressed this in a laboratory experiment designed to disaggregate the two major components of the agency contract (i.e., monitoring and incentive alignment), and then examine their simple and interactive effects on managerial decisions that benefit or hurt the principal, considering the history of the CEO within the firm (Tosi, Katz, & Gomez-Mejia, 1997). We believe that the history of the CEO in the firm is important because several studies have shown that the nature of the agency contract appears to change over time (i.e., the sensitivity of pay to performance decreases as the CEO's tenure within the firm increases Gibbons & Murphy, 1992; Hill & Phan, 1991; Tosi & Gomez-Mejia, 1989). This suggests that social influence processes rather than market considerations become more important in the determination of CEO pay.

Simple Effects of Incentive Alignment and Monitoring

Both incentive alignment and monitoring should induce managerial decisions that benefit principals (i.e., reduce agency costs). In the case of incentive alignment, there are several reasons for this expectation. First, by tying agent's pay to performance desired by principals, the utility of principals and agents becomes more congruent so that managers are more likely to make decisions that maximize their total pay, yet also benefit principals (Groff & Wright, 1989; Kroll et al., 1990). Second, because managerial tasks are not programmable, financial incentives provide an efficient form of self-regulation (Eisenhardt, 1989). Thus, we expected that:

1. Incentive alignment will lead to agent decisions that promote the principal's welfare.

To the extent that unsupervised managers behave opportunistically (Fama, 1980; Hoskisson & Hitt, 1990), close supervision of managers should prevent them from making decisions that have a negative impact on principals. Theoretical papers by Shavell (1979) and Holmstrom (1979) also showed that any monitoring will be beneficial to principals, and will be increasingly so the more the actions of the agent may have negative consequences for the principal's outcomes. Thus:

2. Monitoring should be positively related with agents' decisions that are consistent with the welfare of principals.

Interaction of Monitoring and Incentive Alignment

Incentive alignment and monitoring should have a complementary effect on agents' decisions that benefit principals. First, monitoring is never perfect because of bounded rationality, low task programmability, and subjective assessments of the agents' actions. Likewise, incentive alignment mechanisms are not fool proof because agents may manipulate the payoff by taking advantage of unavoidable information asymmetries. For example, CEOs can manipulate accounting methods (e.g., Groff & Wright, 1989; Hunt, 1986) and reduce capital expenditures and research and development to improve profitability figures (Hill & Snell, 1989). Monitoring can reduce these problems by examining and interpreting the actions and motives of agents. Thus:

3. The presence of both monitoring and incentive alignment should have an additive positive effect on managerial actions that benefit principal.

The Impact of CEO Tenure

There is strong evidence that there is an increased decoupling of pay and performance as CEO tenure increases (e.g., Gibbons & Murphy, 1992; Hill & Phan, 1991; Murphy, 1986; Tosi & Gomez-Mejia, 1989). However, this has been interpreted differently. Some have argued that as information increases about the CEO (i.e., longer tenure within the firm), the owners and board do not need to rely on incentive alignment to control the agent (Murphy, 1986). Others argue that CEOs become more entrenched the longer they are in the firm, and this allows them to decouple pay and performance (see review by Walsh & Seward, 1990).

The escalation of commitment literature can help unravel the effect of tenure on managerial decisions, and how monitoring and incentive alignment may mediate that relationship (Staw, 1976). According to the escalation of commitment paradigm, CEOs are more likely to continue investments in bad projects when there are costs or losses from previous actions, these have occurred over an extensive period of time, and the decision maker cannot withdraw from the situation. Because short-term agents have had fewer opportunities to make decisions to which they might be strongly committed, we expect that:

4. CEOs with shorter tenures are more likely to make decisions that are consistent with the interests of principals than those with longer tenures.

Because long-term CEOs are more likely to escalate their commitment to previously poor decisions, tighter control by principals through monitoring and incentive alignment should lead to a reversal of escalation. If this is the case, this would be consistent with an entrenchment explanation for the gradual decoupling of CEO pay and performance as a function of tenure. Thus:

5. CEOs with longer tenure should make decisions that are more consistent with the interest of principals under conditions of high monitoring and high incentive alignment than under conditions of low monitoring and low incentive alignment.

We designed a variation of the escalation of commitment manipulation of the type used by Staw (1976). In this design, subjects make a series of investment decisions, then learn that the early decisions led to poor results. They are then asked to make a second decision similar to the first, but with the evaluative information from the first. Typically, regardless of the fact that their earlier decision was bad, they tend to continue to invest even more in the same failing strategy. Our modification of the design presented the subjects (i.e., undergraduate students at the University of Florida who took the role of CEO) with alternatives that reflected the sorts of outcomes predicted by agency theory and managerial capitalism. Subjects could choose between alternatives that either (1) resulted in greater sales growth and lower returns to equity holders or (2) lower sales growth and greater returns to equity holders.

The dependent variable was their choice of investing in a profit maximizing strategy or in a previously poor strategic choice made in an earlier period (i.e., the escalation level decision). The independent variables were (1) incentive alignment, (2) monitoring, and (3) CEO tenure.

A high incentive alignment condition linked CEO pay with a profit maximizing strategy, whereas a low incentive alignment condition linked CEO pay to a sales growth strategy, which was less advantageous to the principal. To create realism, participants received bonus points which were traded for cash-prize drawing tickets if the subjects' decisions were consistent with the compensation scheme described in their specific scenario. A high monitoring condition involved detailed justification by the subject of decisions made and close scrutiny by the administrator who reviewed and approved the decisions. In the low monitoring condition, the decisions of the subjects were not scrutinized by the experiment administrator. The tenure treatment was created by assigning subjects to either a long-term CEO or a short-term CEO condition, with the former made responsible for prior decisions as part of the simulation scenario.

We found that high incentive alignment was associated with profit maximizing decisions by the subjects, while high monitoring did not result in more profit-maximizing investments than low monitoring. Also, the combination of high incentive alignment and high monitoring did not lead to more profit maximizing decisions than high incentive alignment by itself.

CEO tenure, as expected, was an important variable. Subjects in the long-term CEO condition made lower average investments in the profit maximizing alternative than the new CEOs, and invested at higher levels in the previous strategy that produced poor results (an escalation of commitment effect). Most importantly, results showed that high monitoring/ high incentive alignment conditions reversed the escalation tendency for long-term CEOs because these subjects made significantly greater investments in the profit maximizing strategy than long-term subjects in the low monitoring/low incentive-alignment conditions.

These findings show that incentive alignment is more effective than monitoring in inducing agents to make decisions that are consistent with the interest of principals. This suggests that if firms substitute monitoring of the CEO for incentive alignment, it is likely to produce outcomes that are less beneficial to equity holders. The results also show that boards of directors are better off utilizing control mechanisms on long-term CEOs rather than leaving them unconstrained.

Incentive Alignment and Firm Performance

Earlier in this paper, we showed that CEOs with high discretion are able to insert themselves into the compensation process in ways that diminish the influence of the board of directors and major stockholders (Tosi & Gomez-Mejia, 1989) and, immediately above, how strong incentive alignment affects the decisions in ways that contribute to maximizing owner welfare. In neither of these studies did we demonstrate direct effects on firm performance. Here, we report a study that directly assessed the relationship between firm performance and incentive alignment.

The Nature of the Relationship Between Incentive Alignment and Firm Performance

While incentive alignment should have a positive impact on firm performance by preventing managers from pursuing self-serving objectives, it is possible that there are diminishing returns to incentive alignment, so that the relationship between this control mechanism and firm performance should be asymptotic rather than monotonic in nature (Tosi & Gomez-Mejia, 1994). There are two reasons for this prediction. First, when incentive alignment is very weak, increasing it should promote higher performance because the agent is more motivated to make decisions that improve the joint welfare of principal and agent. In other words, a win/win condition should prevent self-serving management behaviors because the CEO's interests are increasingly intertwined with those of the principals. But to the extent that CEOs have a limited influence on firm performance outcomes because of exogenous factors, the beneficial effect of incentive alignment on performance should decrease as the upper limits of performance that can reasonably be achieved through their deliberate actions is approached. In other words, once CEOs have done all they can, increases in CEO controls (i.e., incentive alignment) will do little to enhance firm performance (Tosi & Gomez-Mejia, 1994).

Second, if incentive alignment is exceedingly strong, managers may make decisions with negative consequences that counteract the potential benefits of increased levels of incentive alignment. For example, they may become overly conservative in their decisions for fear of jeopardizing their pay and losing their jobs (Baysinger & Hoskisson, 1990; Hoskisson & Hitt, 1988; Wiseman & Gomez-Mejia, 1998). Thus, we expect that for these reasons:

1. The relationship between CEO performance and incentive alignment should be asymptotic.

The Mediating Role of Ownership Structure

Assuming that ownership concentration and incentive alignment are closely correlated, as discussed in much of the literature reviewed earlier, one would expect to find that increments in incentive alignment should have more beneficial effects for the principal in management-controlled firms than in owner-controlled firms. This is because the level of incentive alignment in owner controlled firms is likely to be higher, and given diminishing returns to this control mechanism, approaching the upper limits of incentive would make it difficult to detect its unique effects, if any, on performance in owner controlled firms. On the other hand, if incentive alignment is lower for CEOs of management-controlled firms, observed firm performance should be more responsive to increases in incentive alignment. Thus:

2. The relationship between incentive alignment and firm performance is stronger in management-controlled firms than it is in owner-controlled firms.

We tested these questions in two separate samples (Tosi & Gomez-Mejia, 1994). In both samples, the chief compensation officer (CCO) responded to a survey that assessed the extent to which incentive alignment for the CEO was present. In addition, firm performance, ownership structure, and firm size were measured in both samples. The major difference between the two studies was that in Study 1 firm performance was a self-reported subjective assessment by the CCO, but in Study 2 archival performance indicators were used.

The results of the analysis of both samples supported the asymptotic effects of incentive alignment on firm performance. An analysis that separated the firms into low and high incentive alignment cohorts showed that the effect of incentive alignment on firm performance is much weaker in the high incentive alignment firms than in the low incentive alignment firms in both studies.

This research shows that higher incentive alignment is positively related to firm performance, but the effects are weaker at higher levels of incentive alignment. It also shows that incentive alignment was higher in owner-controlled firms. Because the relationship between incentive alignment and firm performance is asymptotic, and given that incentive alignment is higher in owner controlled firms, the findings are consistent with the expectation that when ownership is highly concentrated, additional incentive alignment has a weaker effect on performance than when ownership is highly dispersed.

Summary

The results of these studies show that control mechanisms do have an impact on executive decisions and subsequent firm performance, but that the effects are far more complicated than those postulated by agency theory. Traditional agency writers argue that incentive alignment and monitoring can diminish the moral hazard problem, and therefore improve the welfare of the principal by reducing agency costs. However, it is left unclear (1) which of the two control mechanisms seems to be most effective; (2) whether or not monitoring can serve as a substitute for incentive alignment and vice versa; (3) the extent to which monitoring and incentive alignment can reinforce each other so that their combined or complimentary effect is stronger than their singular effect; and (4) the degree to which there are diminishing returns to incentive alignment in terms of benefits or utility to principals.

Our results show that incentive alignment has a major effect on agent's decisions while monitoring does not. Furthermore, disaggregating these two control mechanisms indicates that their effects on agent's decisions are not complementary, nor can they be effectively used as substitutes for one another. In addition, contrary to those who argue that incentive alignment is less necessary for long-term CEOs, we show that it can help reverse the escalation of commitment tendency, leading to managerial decisions that are more favorable to owners.

The policy implications of these results are clear: incentive alignment exercises a singular positive effect on agent's decisions from the point of view of principal's objectives, and the beneficial effects become stronger as agent tenure increases. This casts doubts as to how beneficial it is for board of directors to subjectively assess CEO performance, and the practice followed by many firms of decoupling pay and performance as seniority of the CEO increases. This means that boards should establish strong incentive alignment mechanisms for the CEO, but be realistic as to how much the CEO can do to alter firm performance outcomes in response to increasingly higher levels of incentive alignment.

CONCLUSION

Like the vast body of other literature on firm strategic choices using the managerial capitalism paradigm, the set of studies here illuminates how managerial discretion affects the nature of the agency contract, its translation into the firm's compensation policies, and the ensuing positive consequences for the management and negative consequences for equity holders.

First, the firm performance criteria itself is more subject to manipulation in management-controlled firms. Because these firms tend to overstate earnings and are more likely to use subjective criteria for evaluation of CEO performance, any pay-for-performance plans for CEOs in management-controlled firms would tend to favor executives rather than shareholders.

Second, new CEOs in management-controlled firms can negotiate more favorable compensation contracts and are more likely to take strategic initiatives that augment their overall pay package.

Third, CEOs of management-controlled firms can use their discretion to neutralize boards of director power. They can obtain favorable treatment from the board as shown by lower pay risk, decoupling of pay and performance, greater use of subjective appraisals, less anonymity in the evaluation process, and the like. These practices are often legitimized through a network of consultants that indirectly or directly depend on the CEO's graces for their continued engagement with the firm.

Fourth, compensation strategies at the top of the hierarchy tend to cascade to lower levels in the firm. There is greater pay inequality and more managers are eligible for bonuses and long-term income that are *not* performance sensitive in management-controlled firms. Thus, these managers have more opportunities to secure greater pay in a manner that "looks good" to dispersed shareholders and other outsiders.

Fifth, incentive alignment systems can reduce agency costs among management controlled-firms. The shareholder welfare returns to investments in incentive alignment systems are greater if they can be implemented when the CEO has more entrenched power (those with high tenure), and when the level of incentive alignment is rather low (among management-controlled firms). This solution, however, is not easily implemented because of the embedded nature of CEO power in management-controlled firms. The reason is that should an active board threaten the CEO's compensation status, the CEO can take action to replace them with more sympathetic members. Other solutions that involve modification of the corporate governance laws to reduce the role of incumbent management in the selection of board members would be politically difficult to implement (Tosi, Gomez, & Moody, 1991). The most effective control probably emanates from market forces that lead to corporate takeover threats for those management-controlled firms that are not able to maintain some reasonable level of market performance.

Beyond these findings, there is a more general way that the research can be informative. We believe that the theoretical and empirical literatures of managerial capitalism and agency theory are fruitful sources of ideas for management scholars, casting a different light on the way that we address many issues in the field. For one thing, we think that there are a number of questions that could be examined with concepts and tools that are less familiar to those in other fields who do this sort of work. For example, the empirical work in these other fields relies almost exclusively on financial data from electronic archival sources. The convention is to create a proxy variable for a concept (e.g., discretion) by using one of these archival indicators (e.g., the percentage of stock held by a single equity holder), and then to draw behavioral inferences about it. This assumes that these financial proxies have isomorphic behavioral counterparts, an assumption that could easily be tested through the use of psychometric assessment of these variables, seeking to validate these financial proxies. This might lead to other methods of measurement or revisions of the theory as we learn more about the construct validity of these archival measures. For example, we believe that our laboratory study of incentive alignment and monitoring has several advantages over the use of firm-level econometric analyses of proxy variables created from databases such as Compustat and Compact Disclosure. First, monitoring and incentive alignment can be assessed directly instead of relying on distant inferential proxies for these variables such as the presence of more outside than inside board members. Second, a laboratory study allows for the manipulation of control mechanisms so that their simple and interactive effects on agent decisions can be isolated, allowing for stronger causal inferences.

For another thing, there are any number of questions that are more appropriate to our own discipline that can and should be assessed within conventional management frameworks instead of simply importing the theory and methods from economics. For example, it could be of interest to know how the cohesiveness of board members affects a range of strategic choices. The typical way that this is assessed in the current literature is to use demographic proxies for cohesiveness such as differences in age, levels of education, income, or other indicators of socioeconomic status. Another way to approach this same question would be to analyze the communication patterns of board members using network analysis (Krackhardt, 1990). This has proved to be a formidable tool for assessing individual power, an element that can have an important effect on board decisions. This could, we believe, provide greater insight into how power is distributed within the top management team and the board of directors, the effects on strategic choices, and how managers might use this discretion to further their own interests.

In a related vein, much of the agency literature in economics and finance is concerned with "optimal contracting" to align executive and shareholder interests. A more behavioral approach to agency questions would move away from a fixation on contracting issues and focus instead on the relationship that evolves between principal and agent under different ownership configurations. This would extend some earlier work into this domain that focuses on political and interpersonal aspects of principal/agent relations (Perrow, 1986), political issues (Zajac & Westphal, 1994), questions of social dynamics such as interpersonal trust (Zaheer & Venkatramen, 1995) or the psychological contract, and perceptions of the decision-making context (Wiseman & Gomez-Mejia, 1998). This approach could analyze the human element involved in principal-agent controls from a power perspective, recognizing the variability inherent in the nature of this relationship as a function of managerial discretion.

Finally, our involvement with this stream of research revealed to us an interesting paradox in the field of strategic management and organization studies. It is that there is often heavy criticism launched at the economics perspective because its approach is hyperrational, and not in touch with the realities of real life decision making. Yet, many who level this charge take precisely the same approach, unconsciously perhaps, in their own teaching and research. This occurs, as we note in the introduction to this paper, in most of the work in compensation strategy, selection, performance evaluation, and motivation approaches. It is most apparent however in the field of strategic management in the way that cases are taught, generally focusing on the ways that CEOs make strategic choices to maximize firm performance. We do not have empirical evidence of this last point on which we can rely. However, it is our intuition from our own experiences and those that have been reported to us by others who are extensively involved in the teaching of strategy that the focus of most courses, and especially the analysis of cases, that the solutions sought are those that maximize returns—and that means shareholder wealth. However, the literature reviewed here demonstrates that what is rational for top decision makers who enjoy a high level of discretion may or may not be rational from the point of view of firm owners.

APPENDIX

Measures of the Key Variables

We attempted to use similar measures in the different studies to maintain comparability. Except in the cases of the laboratory study (Tosi et al., 1997) and the survey studies (Silva & Tosi, 1998; Tosi & Gomez-Mejia, 1989, 1994), all data were gathered from conventional archival sources, including the COMPUSTAT and Compact Disclosure electronic data bases, the *Wall Street Journal*, Dun and Bradstreet's *Reference Book of Corporate Managements*, and proxy statements. Where necessary because the studies covered multiple years, all financial data were adjusted to constant dollars, using the first study year as the base.

1. **The Ownership Structure.** The ownership structure was measured with an equity concentration measure. It was operationalized by the presence of an external owner holding at least 5% of the outstanding equity (Hunt, 1986; McEachern, 1975). Firms in which at least one external owner held a 5% or greater equity stake were classified as owner-controlled. If no single individual, institution, or group held at least 5% of the outstanding shares, the firm was classified as management-controlled. When internal managers held at least a 5% equity stake, the firm was classified as owner-managed. The fact is that there were so few of these owner-managed firms in each study that these were grouped with the owner-controlled firms since it has been shown that these firms act like owner-controlled firms (McEachern, 1975).

There are two reasons why we believe that this variable is a valid proxy for managerial discretion. The first is that the measure has been meaningfully used in a large number of previous studies to classify firms as owner-controlled or management-controlled (e.g., Hunt, 1986; McEachern, 1975; O'Reilly et al., 1998). And, "...while it is true that the amount of stock necessary to control a firm effectively is unknown, as is the precise functional relationship between ownership concentration and control...if theory predicts differences depending on ownership classification, and a classification is found that is constant with the theory, then there is a reasonable assurance that the categorization has some validity" (Salancik & Pfeffer, 1980, p. 658). In addition, a study by Dyl (1988, p. 22) on the effect of ownership distribution on pay found that different measures of concentration are highly correlated and "yield very similar results."

The second reason that the concentration measure is correlated with other indicators of managerial discretion. We conducted a construct validity study, assessing the equity concentration measure against Mueller and Reardon's (1993) D-score (Banning & Tosi, 1998). Mueller and Reardon argued that equity concentration measures the potential for the use of managerial discretion, whereas the D score measures the actual exercise of managerial discretion.

The D score is calculated by taking the ratio of a firm's return on investments to its cost of capital. D scores equal zero when this ratio is equal to or greater than one, indicating that management has not invested beyond the point where the marginal return on investment equals the firm's cost of capital, which is the level that would maximize shareholder wealth. D scores of zero indicate that managerial discretion has not been exercised, either because managers lacked discretion to overinvest, or they elected to pursue other goals such as shareholder welfare. When the ratio of return on investment to cost of capital is less than one, management has invested beyond the level which maximizes shareholder value, and the D score is assigned a value of 1 minus the ratio, which provides a linear measure of the amount of overinvestment.

Managers are likely to overinvest when owners are weak and are not able to discipline them, so they will take advantage of their discretion to pursue investment goals that benefit themselves but reduce shareholder value. Mueller and Reardon (1993) have argued that managers in firms with concentrated equity will be forced to make decisions that maximize shareholder value; therefore, their D score is most useful in companies where the cost of capital exceeds the firm's return on investment because managers of these firms appear to pursue goals other than maximizing shareholder wealth.

We studied the relationship between the D-score and the equity concentration measure in two samples (Tosi & Gomez-Mejia, 1994). There were 850 firms in one sample and 810 firms in the second sample. These were publicly held manufacturing firms with at least 500 employees and $100 million in annual revenues. In addition to calculating the D score for all of the sample firms, data were collected on firm size, firm performance, and equity concentration. Industry dummy variables were used to control for possible industry effects, but none of the industry variables tested significant. The correlations among the variables were consistent across the two studies, which increases confidence in the validity of the construct. The two discretion measures were significantly correlated (sample 1, $r = .60$; sample 2, $r = .47$), and no other variable in the study correlated as strongly with the equity concentration measure as did the D score. Thus, the two measures appear to tap the same theoretical construct.

Further, the relationship between the D score and firm performance measures were relatively similar to the relationships between the equity concentration measure and the performance indicators. Both the D measure and the level of equity dispersion were strongly and negatively correlated with performance. Since both the equity concentration and D scores predict the same performance outcomes, it suggests criterion related validity for the equity concentration measure.

2. **Size**. Firm size was another measure used in most studies. Whereas there are a number of size indicators, our research strategy, wherever possible, was to factor analyze these various size measures and then construct a composite index constructed from the standardized number of employees, assets, and sales.
3. **Firm Performance**. There are several indicators of firm performance, each with its own strengths and weaknesses. Further, there is the standard argument over whether it is wiser to use accounting measures of performance (i.e., return on investment) or market measures of performance (i.e., change in the market value of the firm), because market measures are less subject to reporting biases by the management.

Our approach in all studies using archival data was to use a performance index. This was constructed the same way in each study. We first factor analyzed the conventional measures, then combined them into an index composed of several indicators, including return on equity, return on assets, return on investment, and the market-to-book ratio. These were industry-adjusted, as follows. First, we created an industry-average performance index similar to the firm-performance from the industry population averages as reported in Compustat for each of the industries represented in the sample. We then computed an annual firm-minus-industry deviation score using the difference of the firm-specific and industry-average performance indices (Hoskisson, Johnson, & Moesel, 1994). These deviation scores control for industry effects (Hoskisson et al., 1994).

4. **Industry controls**. Where necessary we controlled for industry through the use of dummy variables created from SIC codes.
5. **CEO Origin**. It has been shown that the compensation of CEOs who come from outside the firm is higher than for those who were promoted

from inside. Thus, we identified the CEO as coming from the internal or external labor market.
6. **CEO tenure.** The number of years the CEO held the position was a control variable in those studies that examined CEO compensation.
7. **Total Compensation.** One measure was the absolute level of cash pay for CEOs and other managers. In a study of the overall compensation strategy, we also used the average reported compensation per employee. This was determined from archival data bases of proxy statements. We consider this a suitable proxy for total compensation (Gomez-Mejia et al., 1987; O'Reilly et al., 1988), which often includes deferred pay as well as stock options and grants.
8. **Compensation risk.** Compensation risk was calculated by computing the proportion of total cash pay that was contingent on performance (i.e., a bonus) (Werner & Tosi, 1995).
9. **Related diversification** was measured according to a method developed by Wood (1971). Narrow spectrum diversification is defined as expansion, other than vertical integration, outside of the firm's 4-digit primary SIC code, but within a 2-digit SIC code industry. Narrow spectrum diversification is viewed as diversification closely related to a firm's core business (Varadarajan & Ramanujam, 1987).
10. **Unrelated diversification** was assessed according to a method developed by Wood (1971): broad spectrum diversification is defined as expansion, other than vertical integration, into a different 2-digit SIC code industry. Broad spectrum diversification is viewed as diversification less closely related to the firm's core business (Varadarajan & Ramanujam, 1987).

REFERENCES

Agarwal, N. (1981). Determinants of executive compensation. *Industrial Relations*, 20(1), 36-45.
Alderfer, C.P. (1986). The invisible director on corporate boards. *Harvard Business Review*, 64, 38-52.
Allen, M.P. (1981). Power and privilege in large corporations: Corporate control and managerial compensation. *American Journal of Sociology*, 86(5), 1112-1123.
Amihud, Y., & Lev, B. (1979). Risk reduction as a managerial motive for conglomerate mergers. *The Bell Journal of Economics*, 10, 605-617.
Amihud, Y., & Lev, B. (1981). Risk reduction as a managerial motive for conglomerate mergers. *The Bell Journal of Economics*, 12(2), 605-617.
Amit, R., & Livnat, J. (1988). Diversification and the risk-return trade-off. *Academy of Management Journal*, 31,154-166.
Anderson, E. (1985). The salesperson as outside agent or employee: A transaction cost analysis. *Marketing Science*, 4(3), 234-254.
Androkovich, R.A. (1990). Relative risk aversion, incentive effects, and risk sharing. *Atlantic Economic Journal*, 18(4), 38-41.
Angelo, H., & Rice, E.M. (1983). Antitakeover amendments and stockholder wealth. *Journal of Financial Economics*, 11, 329-359.

Baiman, S. (1982). Agency research in managerial accounting: A survey. *Journal of Accounting Literature, 1,* 154-210.

Baiman, S. (1990). Agency research in managerial accounting: A second look. *Accounting, Organizations and Society, 15*(4), 341-371.

Baker, G.P., Jensen, M.C., & Murphy, K.J. (1988). Compensation and incentives: Practice vs. theory. *The Journal of Finance, 43,* 593-616.

Banning, K.C., & Tosi, H. (1998). *Power and corporate control: CEO succession in owner and manager controlled firms.* Working paper, University of Florida, Gainesville, FL.

Barkema, H.G., & Gomez-Mejia, L.R. (1998). Managerial compensation and firm performance: A general research framework. *Academy of Management Journal, 41,* 135-145.

Basu, A., Lal, R., Srinivasan, V., & Staelin, R. (1985). Sales force compensation plans: An agency theoretic perspective. *Marketing Science, 4,* 267-291.

Baumol, W. (1967). *Business, behavior, value and growth.* New York: Harcourt, Brace, & World.

Baumol, W.J. (1959). *Business behavior, value and growth.* New York: The Macmillan Company.

Baysinger, B.D., & Hoskisson, R.E. (1990). The composition of boards of directors and strategic control: Effects on corporate strategy. *Academy of Management Review, 15,* 72-87.

Baysinger, B.D., Kosnik, R.D., & Turk, T.A. (1991). Effects of board and ownership structure on corporate R&D strategy. *Academy of Management Journal, 34,* 205-214.

Becker, G. (1964). *Human capital.* New York: Columbia University Press.

Berle, A., & Means, G. (1932). *The modern corporation and private property.* New York: Macmillan.

Bettis, R. (1981). Performance differences in related and unrelated diversified firms. *Strategic Management Journal, 2,* 379-394.

Biddle, G.C. (1980). Accounting methods and management decisions: The case of inventory costing and inventory policy. *Journal of Accounting Research, 18*(Supplement), 235-280.

Biddle, G.C., & Lindahl, F.W. (1982) Stock price reactions to LIFO adoptions: The association between excess returns and LIFO tax savings. *Journal of Accounting Research, 20,* 551-588.

Boudreaux, K. (1973). Managerialism and risk-return performance. *Southern Economic Journal, 39*(3), 366-372.

Bowen, R., Noreen, E.W., & Lacey, J.M. (1981). Determinants of the corporate decision to capitalize interest. *Journal of Accounting and Economics, 3,* 151-179.

Bradley, M., & Wakeman, L.M. (1983). The wealth effects of targeted share repurchases. *Journal of Financial Economics, 11,* 301-328.

Ciscel, D., & Carroll, T. (1980). The determinants of executive salaries: An econometric survey. *Review of Economics and Statistics, 62,* 7-13.

Coffee, J.C. (1988). Shareholders versus managers: The strain in the corporate web. In J.C. Coffee, L. Lowenstein, & S. Rose-Ackerman (Eds.), *Knights, raiders, and targets (pp. 77-134).* New York: Oxford University Press.

Conlon, E.J., & Parks, J.M. (1990). Effects of monitoring and tradition on compensation arrangements: An experiment with principal-agent dyads. *Academy of Management Journal, 33,* 603-622.

Coughlan, A., & Schmidt, R. (1985). Executive compensation, management turnover, and firm performance. *Journal of Accounting and Economics, 7,* 43-66.

Crystal, G.S. (1991). Why CEO compensation is so high. *California Management Review, 34,* 9-29.

Dalton, D.R., & Kesner, I.F. (1985). Organizational performance as an antecedent of inside/outside chief executive succession: An empirical assessment. *Academy of Management Journal, 28,* 749-762.

Dann, L.Y., & DeAngelo, H. (1983). Corporate financial policy and corporate control: A study of defensive adjustments in asset and ownership structure. *Journal of Financial Economics, 20,* 87-127.

Dubofsky, P., & Varadarajan, P. (1987). Diversification and measures of performance: Additional empirical evidence. *Academy of Management Journal, 30,* 597-608.

Dyl, A. (1988). Corporate control and management compensation: Evidence on the agency problem. *Managerial and Decision Economics, 9*, 21-25.
Dyl, E.A. (1989). Agency, corporate control and accounting methods-the LIFO-FIFO choice. *Managerial and Decision Economics, 10*, 141-145.
Eisenhardt, K.M. (1988). Agency and institutional theory explanations: The case of retail sales compensation. *Academy of Management Journal, 31*, 488-511.
Eisenhardt, K.M. (1989). Agency theory: An assessment and review. *Academy of Management Review, 14*, 57-74.
Elliot, J.W. (1972). Control, size, growth, and financial performance in the firm. *Journal of Financial and Quantitative Analysis, 7*, 1309-1320.
Fama, E., & Jensen, M. (1983). Agency problems and residual claims. *Journal of Law and Economics, 26*, 327-349.
Fama, E.F. (1980). Agency Problems and the theory of the firm. *Journal of Political Economy, 88*, 288-307.
Fierman, J. (1990). The people who set the CEO's pay. *Fortune, 121*(6), 58-66.
Finkelstein, S., & Boyd, B.K. (1998). How much does the CEO matter? The role of managerial discretion in the setting of CEO compensation. *Academy of Management Journal, 41*, 179-199.
Finkelstein, S., & Hambrick, D.C. (1990) Top-management team tenure and organizational outcomes: The moderating role of managerial discretion. *Administrative Science Quarterly, 35*, 484-503.
Fizel, J.L., & Louie, K.K.T. (1990). CEO retention, firm performance and corporate governance. *Managerial and Decision Economics, 11*, 167-176.
Galbraith, J.K. (1967). *The new industrial state*. Boston: Houghton-Mifflin.
Gibbons, R., & Murphy, K. (1992). Optimal incentive contracts in the presence of career concerns: Theory and evidence. *Journal of Political Economy, 100*, 468-506.
Gomez-Mejia, L. (1992). Structure and process of diversification, compensation strategy, and firm performance. *Strategic Management Journal, 13*, 381-397.
Gomez-Mejia, L. (1994). Executive compensation: A reassessment and a future research agenda. In G.R. Ferris (Ed.), *Research in personnel and human resources management*, (Vol. 12, pp. 161-222). Greenwich, CT: JAI Press.
Gomez-Mejia, L., & Balkin, D. (1992). *Compensation, organizational strategy, and firm performance*. Cincinnati: South-Western.
Gomez-Mejia, L., Tosi, H., & Hinkin, T. (1987). Managerial control, performance, and executive compensation. *Academy of Management Journal, 30*(1), 51-70.
Grabowski, H.G., & Mueller, D.C. (1972). Managerial and stockholder welfare models of firm expenditures. *The Review of Economics and Statistics, 54*, 9-24.
Groff, J.E., & Wright, C.J. (1989). The market for corporate control and its implications for accounting policy choice. In B.N. Schwartz & Associats (Eds.), *Advances in accounting* (Vol. 7, pp. 13-21). Greenwich, CT: JAI Press.
Hagerman, R.L., & Zmijewski, M.E. (1979). Some economic determinants of accounting policy choice. *Journal of Accounting and Economics, 1*, 141-161.
Haleblian, J., & Finkelstein, S. (1993). Top management team size, CEO dominance, and firm performance: The moderating roles of environmental turbulence and discretion. *Academy of Management Journal, 36*, 844-863.
Halpern, P. (1973). Empirical estimates of the amount and distribution of gains to companies in mergers. *Journal of Business, 46*, 554-575.
Halpern, P. (1983). Corporate acquisitions: A theory of special cases? A review of event studies applied to acquisitions. *Journal of Finance, 38*, 297-317.
Hambrick, D.C., & Abrahamson, E. (1995). Assessing managerial discretion across industries: A multimethod approach. *Academy of Management Journal, 38*, 1427-1441.
Hambrick, D.C., & Finkelstein (1995). The effects of ownership structure on conditions at the top: The case of CEO pay raises. *Strategic Management Journal, 16*, 175-193.

Hambrick, D.C., & Finkelstein, S. (1987). Managerial discretion: A bridge between polar views of organizational outcomes. In L.L. Cummings & B.M Staw (Eds.), *Research in organizational behavior* (Vol. 9, pp. 369-406). Greenwich, CT: JAI Press.

Hambrick, D.C., Geletkanycz, M.A., & Fredrickson, J.W. (1993). Top executive commitment to the status quo: Some tests of its determinants. *Strategic Management Journal, 14*, 401-418.

Hand, J.H. Lloyd, W.P., & Modani, N. (1983). Diversification of manager salary risk as a merger motive. *Akron Business and Economic Review, 14*, 7-9.

Harris, M., & Raviv, A. (1979). Optimal incentive contracts with imperfect information. *Journal of Economic Theory, 20*, 231-259.

Harris, R.D., & Tosi, H.L. (1998). *Strategy and structure in owner-controlled and management-controlled firms* (Working paper). Turlock: California State University.

Herman, E.S. (1981). *Corporate control, corporate power*. New York: Cambridge University Press.

Herman, E.S., & Lowenstein, L. (1988). The efficiency effects of hostile takeovers. In J.C. Coffee, L. Lowenstein, & S. Rose-Ackerman (Eds.), *Knights, raiders, and targets* (pp. 211-240). New York: Oxford University Press.

Hill, C., & Phan, P. (1991). CEO tenure as a determinant of CEO pay. *Academy of Management Journal, 34*, 712-717.

Hill, C., & Snell, S. (1988). External control, corporate strategy and firm performance in research-intensive industries. *Strategic Management Journal, 9*, 577-590.

Hill, C., & Snell, S. (1989). Effects of ownership structure and control on corporate productivity. *Academy of Management Journal, 32*(1), 25-46.

Hill, C.W., & Hansen, G.S. (1989). Institutional holdings and corporate R&D intensity in research intensive industries. *Academy of Management Best Papers Proceedings* (pp. 17-21). Washington, DC, 49th Annual Meeting of the Academy of Management.

Hill, C.W., Hitt, M., & Hoskinsson, R.E. (1988). Declining U.S. competitiveness: Reflections on a crisis. *Academy of Management Journal, 32*, 25-47.

Hills, F.S. (1987). *Compensation decision making*. Chicago: Dryden Press.

Hills, F.S., Bergmann, T.J., & Scarpello, V.G. (1994). *Compensation decision making*. Fort Worth, TX: Dryden Press.

Hindley, B. (1970). Separation of ownership and control in the modern corporation. *The Journal of Law and Economics, 13*, 185-222.

Holl, P. (1975). Effect of control type on the performance of the firm in the U.K. *The Journal of Industrial Economics, 23*(4), 257-271.

Holmstrom, B. (1979). Moral hazard and observability. *Bell Journal of Economics, 10*, 74-91.

Holthausen, R.W., &. Leftwich, R.W. (1983). The economic consequences of accounting choice. *Journal of Accounting and Economics, 5*, 77-117.

Hoskisson, R., & Hitt, M. (1990). Antecedents and performance outcomes of diversification: A review and critique of theoretical perspectives. *Journal of Management, 16*, 461-509.

Hoskisson, R.E., & Hitt, M.A. (1988). Strategic control systems and relative R&D intensity in large multiproduct firms. *Strategic Management Journal, 9*, 605-622.

Hoskisson, R.E., Johnson, R.A., & Moesel, D.D. (1994). Corporate divestiture intensity in restructuring firms: Effects of governance, strategy, and performance. *Academy of Management Journal, 37*, 1207-1251.

Hunt, H., III. (1986). The separation of corporate ownership and control: Theory, evidence, and implications. *Journal of Accounting Literature, 5*, 85-124.

Hunt, H.G. III (1985). Potential determinants of corporate inventory accounting decisions. *Journal of Accounting Research, 23*, 448-467.

Hunt, H.G. III, & Hogler, R.L. (1990). Agency-theory as ideology: A comparative analysis based on critical legal theory and radical accounting. *Accounting, Organizations, and Society, 15*, 437-454.

Jarrell, G.A., Brickley, J.A., & Netter, J.M. (1988). The market for corporate control: The empirical evidence since 1980. *Journal of Economic Perspectives, 2*, 49-68.

Jensen, M.C. (1983). Organization theory and methodology. *The Accounting Review, 58*, 319-339.

Jensen, M., & Meckling, W. (1976). Theory of the firm: Managerial behavior, agency costs and ownership structure. *Journal of Financial Economics, 3*, 305-360.

Kamerschen, D.R. (1968). The influence of ownership and control on profit rates. *The American Economic Review, 58*, 432-447.

Kosnik, R.D. (1990). Effects of board demography and directors' incentives on corporate greenmail decisions. *Academy of Management Journal, 33*, 129-150.

Krackhardt, D. (1990). Assessing the political landscape: Structure, cognition, and power in organizations. *Administrative Science Quarterly, 35*, 342-369.

Krause, D.S. (1988). Corporate ownership structure: Does it impact firm performance? *Akron Business and Economic Review, 19*(2), 30-38.

Kroll, M., Simmons, S., & Wright, P. (1990). Determinants of CEO compensation following major acquisitions. *Journal of Business Research, 20*, 349-366.

Lambert, R.A., Larcker, D.F., & Weigelt, K. (1993). The structure of organization incentives. *Administrative Science Quarterly, 38*, 438-461.

Lazear, E., & Rosen S. (1981). Rank-order tournaments as optimum labor contracts. *Journal of Political Economy, 89*, 841-864.

Lewellen, W., Loderer, C., & Rosenfeld, A. (1985). Merger decisions and executive stock ownership in acquiring firms. *Journal of Accounting and Economics, 7*, 209-231.

Madden, G.P. (1982). The separation of ownership from control and investment performance. *Journal of Economics and Business, 34*, 149-152.

Magenheim, E.B., & Mueller, D.C. (1988). On measuring the effects of acquisitions on acquiring firm shareholders. In J.C. Coffee, L. Lowenstein, & S. Rose-Ackerman (Eds.), *Knights, raiders, and targets* (pp. 211-240). New York: Oxford University Press.

Main, B., O'Reilly, C., & Wade, J. (1995). The CEO, the board of directors and executive compensation: Economic and psychological perspectives. *Industrial and Corporate Change, 4*, 293-332.

Malatesta, P.H., & R.A. Walking (1988). Poison pill securities; stockholder wealth, profitability, and ownership structure. *Journal of Financial Economics, 20*, 347-376.

Marris, R. (1964). *The economic theory of managerial capitalism*. New York: The Free Press.

McEachern, W.A. (1975). *Managerial control and performance*. Lexington, MA: D.C. Heath & Co.

McEachern, W.A. (1976). Corporate control and risk. *Economic Inquiry, 14*, 270-278.

McGuire, J. (1988). Agency theory and organizational analysis. *Managerial Finance, 14*(4), 6-9.

Meyer, J.W., & Rowan, B. (1977). Institutionalized organizations: Formal structure as myth and ceremony. *American Journal of Sociology, 83*, 340-363.

Milgrom, P., & Roberts, J. (1992). *Economics, organization, and management*. Englewood Cliffs, NJ: Prentice-Hall.

Milkovich, G.T., & Newman, J.M. (1996). *Compensation* (5th ed.). Chicago: Richard D. Irwin Publishers.

Monsen, R.J., Chiu, J.S., & Cooley, D.E. (1968). The effect of separation of ownership and control on the performance of the large firm. *Quarterly Journal of Economics, 82*, 435-451.

Morck, R., Schleifer, A., & Vishney R.W. (1989). Alternative mechanisms for corporate control. *American Economic Review, 79*, 842-853.

Morse, D., & Richardson, G. (1983). The LIFO/FIFO decision. *Journal of Accounting Research, 21*(1), 106-127.

Mueller, D.C., & Reardon, E. (1993). Rates of return on investment. *Southern Economic Journal, 60*, 430-453.

Murphy, K. (1986). Incentives, learning and compensation: A theoretical and empirical investigation of managerial labor contracts. *Rand Journal of Economics, 17*, 59-76.

O'Reilly, C.A., Main, B.G., & Crystal, G.S. (1988). CEO compensation as tournament and social comparison: A tale of two theories. *Administrative Science Quarterly, 33,* 257-274.

Ocasio, W. (1994). Political dynamics and the circulation of power: CEO succession in U.S. industrial corporations. *Administrative Science Quarterly, 39,* 285-312.

Palmer, J. (1973). The profit-performance effects of the separation of ownership from control in large U.S. corporations. *Bell Journal of Economics and Management Science, 4,* 293-303.

Perrow, C. (1986). *Complex organizations: A critical view* (3rd ed.). New York: Random House.

Radice, H. (1971). Control type, profitability and growth in large firms: An empirical study. *The Economic Journal, 81,* 547-562.

Rajagopalan, N., & Finkelstein, S. (1992). Effects of strategic orientation and environmental change on senior management reward systems. *Strategic Management Journal, 13,* 127-141.

Ramanujam, V., & Varadarajan, P. (1989). Research on corporate diversification: A synthesis. *Strategic Management Journal, 10*(6), 523-551.

Ravenscraft, D.J., & Scherer, F.M. (1987). *Mergers, sell-offs, and economic efficiency.* Washington, DC: The Brookings Institution.

Roberts, D. (1959). *Executive compensation.* Glencoe, IL: The Free Press.

Roll, R. (1987). Empirical evidence on takeover activity and shareholder wealth. In T.E. Copeland (Ed.), *Modern finance and industrial economics* (pp. 287-325). New York: Basil Blackwell.

Ross, S. (1973). The economic theory of agency: The principals problem. *American Economic Review, 52,* 134-139.

Rumelt, R. (1982). Diversification, strategy and profitability. *Strategic Management Journal, 3,* 359-369.

Ryngaert, M. (1988). The effect of poison pill securities on shareholder wealth. *Journal of Financial Economics, 20,* 377-417.

Salamon, G.L., & Smith, E.D. (1979). Corporate control and managerial misrepresentation of firm performance. *Bell Journal of Economics, 10,* 319-328.

Salancik, G.R., & J. Pfeffer (1980). Effects of ownership and performance on executive tenure in U.S. corporations. *Academy of Management Journal, 23,* 653-664.

Schleifer, A., & Vishny, R.W. (1988). *Managerial entrenchment.* Unpublished manuscript. Chicago: University of Chicago.

Schwab, D.P. (1974). Controlling impacts of pay on employee motivation and satisfaction. *Personnel Journal, 53,* 190-206.

Shavell, S. (1979). Risk sharing and incentives in the principal and agent relationship. *The Bell Journal of Economics, 10*(1), 55-73.

Shelton, J.P. (1967). Allocative efficiency vs. "X-efficiency": Comment. *The American Economic Review, 57,* 1252-1258.

Silva, P., & Tosi, H. (1998). *CEO evaluation processes* (Working paper). Gainesville: University of Florida.

Simon, H.A. (1959). Theories of decision making in economics and behavioral science. *American Economic Review, 49*(3), 253-283.

Simon, H.A. (1991). Organizations and markets. *Journal of Economic Perspectives, 3*(2), 25-44.

Singh, N. (1985). Monitoring and hierarchies: The marginal value of information in the principal-agent model. *Journal of Political Economy, 93,* 599-609.

Spence, M., & Zeckhauser, R. (1971). Insurance, information and individual action. *American Economic Review, 61*(2), 380-387.

Stano, M. (1976). Monopoly power, ownership control, and corporate performance. *The Bell Journal of Economics, 7,* 672-679.

Staw, B.M. (1976). Knee-deep in the big muddy: A study of escalating commitment to a chosen course of action. *Organizational Behavior and Human Performance, 16,* 27-44.

Stiglitz, J.E. (1975). Incentives, risk, and information: Notes towards a theory of hierarchy. *Bell Journal of Economics, 6,* 552-579.

Sunder, S. (1973). Relationship between accounting changes and stock prices: Problems of measurement and some empirical evidence. *Journal of Accounting Research, 11*(Supplement), 1-45.
Sunder, S. (1975). Stock prices and risk related accounting changes in inventory valuation. *The Accounting Review, 50,* 305-315.
Tosi, H., & Gomez-Mejia, L. (1989). The decoupling of CEO pay and performance: An agency theory perspective. *Administrative Science Quarterly, 34,* 169-189.
Tosi, H.L., & Gomez-Mejia, L.R. (1994). Compensation monitoring and firm performance. *Academy of Management Journal, 37,* 1002-1016.
Tosi, H.L., Gomez-Mejia, L.R., & Moody, D.L. (1991). The separation of ownership and control: Increasing the responsiveness of boards of directors to shareholders' interests? *University of Florida Journal of Law and Public Policy, 4,* 39-58.
Tosi, H.L. Werner, S. & Gomez-Mejia, L.R. (1997). *Sharing the wealth, sharing the risk: How the ownership structure affects the firm's compensation strategy* (Working paper). Gainesville: University of Florida.
Tosi, H.L., Katz, J.P., & Gomez-Mejia, L.R. (1997). Disaggregating the agency contract: The effects of monitoring, incentive alignment, and term in office on agent decision making. *Academy of Management Journal, 40,* 584-602.
Varadarajan, P., & Ramanujam, V. (1987). Diversification and performance: A reexamination using a new two-dimensional conceptualization of diversity in firms. *Academy of Management Journal, 30,* 380-397.
Walsh, J.P., & Seward, J.K. (1990). On the efficiency of internal and external corporate control mechanisms. *Academy of Management Review, 15,* 421-458.
Ware, R.F. (1975). Performance of manager versus owner controlled firms in the food and beverage industry. *Quarterly Review of Economics and Business, 15,* 81-92.
Weber, C.L., & Rynes, S.L. (1991). Effects of compensation strategy on pay decisions. *Academy of Management Journal, 34,* 86-109.
Werner, S., & Tosi, H.L. (1995). Other people's money: The effects of ownership on compensation strategy and managerial pay. *Academy of Management Journal, 38,* 1672-1691.
Westphal, J.D., & Zajac, E.J. (1995). Accounting for explanations of CEO compensation: Substance and symbolism. *Administrative Science Quarterly, 40,* 283-308.
Whyte, W.F. (1949). The social structure of a restaurant. *American Sociological Review, 54,* 302-310.
Williams, M.J. (1985). Why chief executives pay keeps rising. *Fortune, 111*(7), 66-76.
Williamson, O.E. (1967) Hierarchical control and optimum firm size. *Journal of Political Economy, 75,* 123-138.
Wiseman, R.M., & Gomez-Mejia, L.R. (1998). A behavioral agency model of managerial risk taking. *Academy of Management Review, 25,* 133-152.
Wood, A. (1971). Diversification, merger and research expenditures: A review of empirical studies. In R. Morris & A. Wood (Eds.), *The corporate economy: Growth, competition, and innovation potential* (pp. 428-453). Cambridge, MA: Harvard University Press.
Zaheer, A., & Venkatramen, N. (1995). Relational governance as an interorganizational strategy: An empirical test of the role of trust in economic exchange. *Strategic Management Journal, 16,* 373-392.
Zajac, E.J., & Westphal, J.D. (1994). The costs and benefits of managerial incentives and monitoring in large U.S. corporations: When is more not better. *Strategic Management Journal, 15,* 121-142.

PERSON-ENVIRONMENT FIT IN THE SELECTION PROCESS

James D. Werbel and Stephen W. Gilliland

ABSTRACT

Selection theory typically considers person-job fit as the basis for selecting job applicants. This paper suggests that selection theory should consider making fit assessments based on person-job fit, person-organization fit, and person-workgroup fit. Knowledge, skills, and abilities should be used to evaluate person-job fit. Values and needs should be used to assess person-organization fit. Interpersonal attributes and broad-based proficiencies should be used to assess person-workgroup fit. A facet model of selection decisions is developed that outlines the predictor and criterion variables in the selection process associated with each type of fit assessment. Predictor variables include work experience, educational experience, values, needs, broad based proficiencies, and interpersonal attributes. Criterion measures include performance, motivation, extra-role behaviors, work attitudes, retention, group cooperation, and group performance. Research related to the predictor and criterion domains is examined. Propositions are developed to suggest when a type of fit assessment is most appropriate to use with different organization conditions. In spite of potential legal problems with person-organization and person-workgroup fit, if biases can be reduced through structured processes to assess these different types of fit, organization effectiveness should improve if multiple types of fit are used to select new employees as opposed to using only person-job fit.

INTRODUCTION

Assessing applicant person-job fit is the traditional foundation for employment selection. Person-job fit occurs when applicants' proficiencies are congruent with the job requirements. From its simple inception evolving out of scientific management, the process of determining person-job fit increasingly gained sophistication with identification of both statistically reliable and valid processes that can be used to determine person-job fit. Person-job fit eventually achieved legal support with the development of the Uniform Guidelines on Employee Selection Procedures (1978). If employers use selection criteria based on person-job fit, they have commonly been able to successfully defend employment selection lawsuits.

In spite of significant improvements in the selection process through the use of person-job fit as opposed to unstructured selection processes (McDaniel, Whetzel, Schmidt, & Maurer, 1994), both practitioners (Montgomery, 1996) and academicians (Behling, 1998; Borman & Motwidlo, 1993; Kristoff, 1996) have suggested that person-job fit is less important than other types of fit. Carson and Stewart (1996) more radically proposed that person-job fit is no longer appropriate due to changes in the nature of work, organizations, and employer-employee relationships. These challenges to the traditional model of person-job fit call for a closer examination of the selection process to determine the extent to which person-job is a relevant approach to selection.

Changes in economic and social conditions create a need for organizations to be adaptive and responsive. Dynamic organizations and job demands may require different notions of selection than are offered through the relatively static assumptions of person-job fit. To discern the relevancy of person job-fit as an employment selection tool, we need to understand the historical employment conditions that lead to the dominant use of person-job fit in the selection process. We also need to consider historical changes in employer and employee relationships and then assess the relevancy of person-job fit for such relationships.

This paper examines the use of fit in the selection process. In particular, we first examine the historical reasons for the development and use of person-job fit for employment selection. Based on employers' current expectations of employees, we present specific criticisms of the exclusive reliance on person-job fit. To address these criticisms, we propose that employee selection should be expanded to include additional components of person-environment fit. We present a model for selecting employees that revolves around three types of fit: Person-job fit, person-organization fit, and person-workgroup fit. This model assumes that different types of fit may be relevant in different situations. For each type of fit, we discuss methods for identifying predictors of fit as well as consequences of fit. Assessing validity with regard to each type of fit is also considered. Finally, testable propositions and suggestions for practitioners to develop valid and unbiased selection processes are presented and discussed.

THE BASIS FOR PERSON-JOB FIT IN SELECTION

Person-job fit is viewed as a congruence between the demands of the job and the needed skills, knowledge, and abilities of a job candidate (Edwards, 1991). Traditionally, this fit is attained by first determining the demands of a job through a job analysis, which is used to identify the essential job tasks that an incumbent performs and the requisite skills, knowledge, and abilities to perform the job tasks. Recruiters and selection specialists then develop strategies to identify the extent to which candidates match the job requisites. These strategies may include resumes, tests, interviews, reference checks, and a variety of other selection tools. Finally, data are collected on the applicants and evaluated against the job requisites. If employee selection is based on a pre-determined structured process, it is likely to produce reliable and unbiased evaluations on which to select the most qualified job applicants.

The foundations for establishing person-job fit for employment selection emerged from the second principle of scientific management that called for the scientific selection of machines, material, and workers (Taylor, 1911). While Taylor offered little tangible direction about ways to scientifically select workers other than through testing, job analyses proved to be the foundation for the scientific selection of workers. Job analysis was also used to address problems with productivity and compensation. Ghorpade and Atchinson (1980) reported that enthusiasm for job analysis remained high throughout the thirties. As interest for scientific management declined, enthusiasm for job analysis as a panacea for employment issues also declined. Nonetheless, Ghorpade and Atchinson reported that interest in the use of job analysis in selection has remained persistent.

The initial enthusiasm for job analysis as a tool to aid in assessing person-job fit emerged from pre-industrial conditions that were incompatible with employer and employee relationships in the industrial age. Management at the turn of the century delegated significant and autonomous powers to supervisors who ran their shop according to their own rules (Gilson, 1924). There were few efforts to standardize processes and policies within the organization. Under this system, supervisors had two ways to recruit employees (American School of Correspondence, 1919). One method was for a supervisor to go to the company gate each day and select from people standing outside. Often, familial networks were used to influence a supervisor to pick a family member. At best, physical attributes were used to select applicants. The other method of recruitment was to use employment agencies that "were little better, often worse, than the foreign padrones" (American School of Correspondence, 1919, p. 19). This system was rife with ethical and employment problems. Efforts at establishing person-job fit were an attempt to provide some basis for standardizing employment practices and ensuring that systematic processes were used to select a qualified person to perform a job (Gilson, 1924).

The employment problems that motivated employers to adopt person-job fit, are no longer salient employment issues today. Many corporations have employment policies that formally or informally restrict nepotism (Ford & McLauglin, 1986). While employment agencies still exist, they usually have a professional reputation.

Nonetheless, there are currently two significant reasons for using person-job fit. First, validated and structured procedures for determining person-job fit have led to more effective selection of employees in comparison to unstructured techniques (Buckley & Russell, 1997; McDaniel et al., 1994). Second, the importance of person-job fit has been heightened by the passing of the Civil Rights Act, the results of subsequent court cases (e.g., *Griggs* vs *Duke Power*, 1971), and the development of the Uniform Guidelines. Employment selection procedures need to demonstrate validity and therefore, need to be based on unbiased job analysis procedures (Uniform Guidelines on Employee Selection Procedures, 1978). It has become important for employers to rely on sophisticated and legally defensible person-job fit criterion for employment selection.

CHALLENGES TO THE USE OF PERSON-JOB FIT

Challenges to the predominant use of person-job fit in the selection process stem from two perspectives. One calls for an expanded criterion domain, while the other calls for an expanded predictor domain.

Borman and Motwidlo (1993) called for an expanded criterion domain and suggested that selection practices should be based on factors associated with organizational effectiveness. Technical job performance, which is the primary basis for determining person-job fit, is one dimension of organizational effectiveness. However, Borman and Motwidlo suggested that individual behaviors other than technical performance are also determinants of organizational effectiveness. In particular, they suggested that prosocial behavior, organizational citizenship behavior, and organizational commitment contribute to organizational effectiveness. These determinants of organizational effectiveness are associated with support for organizational objectives, supportive interpersonal relationships, following organizational procedures and rules even when they are inconvenient, persisting with extra enthusiasm, effort and time commitments, and volunteering to perform tasks that are not normally required to perform the job. Traditional efforts at assessing person-job fit focus exclusively on the task orientation and ignore the support orientation.

There are three separate arguments for an expanded predictor domain beyond person-job fit. One argument suggests that universal selection criterion are more important than job specific selection criterion (Montgomery, 1996). Based on evidence collected from diverse work places and educational institutions, the U.S. Department of Labor (1991, 1992) reports that global competition and increased use of information technology mandates flexible production systems, employee empowerment, development of work teams, and increased use of technology. Furthermore, pressures to make these changes are increasing as global competition increases.

Thus, employers need to consider that employees will hold multiple jobs over the course of their employment with a company. Thinking and basic skills will help an employee attain high wages and be more valued by employers over a longer period of time than employees with narrow specific technical skills (U.S. Department of Labor, 1992). The Department of Labor reports called for improved educational practices in personal qualities, thinking skills, and basic skills, such as math and writing skills, to compensate for the changing workplace. This is relevant for selection because it may be easier to train technical skills and knowledge than to train personal qualities, thinking skills, and basic skills. Thus, selection may need to be based on broad-based proficiencies as opposed to specific job skills.

None of these issues associated with the changing demands of the workplace were prevalent when job analyses was created. At the turn of the century, efforts were needed to support a move from craft production to large-scale industrial production. Currently, large-scale industrial production is increasingly converting to flexible manufacturing practices (Bratton, 1993).

To some extent, the universalist argument is based on empirical evidence. Behling (1998), Ree and Earles (1992), and Hunter and Hunter (1984) reported that the single best predictor of successful job performance is a measurement of general intelligence. Montgomery (1996) reported that people low on dimensions associated with being a team player and being likable but high in proficiency were five times more likely to *not* be rehired than those who were high on both dimensions. He also suggested that failure on the job is more likely to be due to poor communication and interpersonal skills than to proficiency requirements. Therefore, interpersonal skills may be more critical for organizational productivity than technical skills.

A second argument for an expanded predictor domain calls for using person-organization fit in addition to person-job fit (Bowen, Ledford, & Nathan, 1991; Kristof, 1996). Advocates of this perspective observe that organizations are made up of political constituencies (Judge & Ferris, 1992). Chatman (1989) and Meglino, Ravlin, and Adkins (1989) suggested that political themes that run through organizations evolve from a set of corporate values. Corporate strategists might call these corporate visions. Regardless, it may be important to select people who share the values and visions of the organization (Bowen et al., 1991).

The concept of person-organization fit has had practitioner appeal for a long time. During the height of the scientific management movement, Gilson (1924) stated that "Some plants operating under scientific management consider that the *most* (italics added) essential qualification of an applicant is his *fitness for the organization* (original italics)" (p. 44). She then lists the qualities that are required in organizations that use scientific management and concludes that organizational analysis may be needed before conducting job analysis.

Values become important for selection when organizations are trying to promote major changes. For example, Gilson's (1924) concerns with using organizational fit stem from the political concerns in minimizing resistance from scientific management. Values may also become important to maintain a distinct corporate culture at organizations (Deal & Kennedy, 1982). However, most advocates of person-organization fit also acknowledge the need to select for technical competencies.

The third argument against a reliance on person-job fit stems from the total quality (TQ) literature. Job analysis, which provides the foundation for assessing person-job fit, is described as being "based on outdated ideas about jobs themselves" (Carson & Stewart, 1996, p. 56). The social construction of organizations is being modified as the organizational hierarchy changes to greater use of self-managed teams, and as we move toward increasing competition that requires an increasingly flexible work force (Bridges, 1994). The traditional way of organizing work by bundling tasks is being replaced with fluid and rapidly changing role responsibilities (Cascio, 1995).

Based on these assumptions, Cardy and Dobbins (1996) offer contrasting images between traditional human resource management (HRM) and total quality human resource management (TQHRM). The two perspectives have very different images about the nature of work. HRM is monothetic. Through typical job analysis, one can identify the most time efficient means to perform a job. In contrast, TQHRM is pluralistic. It assumes that there is equifinality and that there may be many effective and efficient ways to produce some product or provide some service. HRM is reductionistic and compartmentalized. It assumes that efficiency is increased through increased task specialization. In contrast, TQHRM is holistic. It takes a systems perspective of work situations. Performance can best increase by increasing system effectiveness. Thus, working as a team becomes critical for improving system effectiveness. Teamwork can not be readily determined by conducting a task-abilities job analysis. Finally, reflecting Borman and Motwidlo (1993), they also call for an expanded criterion domain. Traditional HRM focuses only on job performance. TQHRM calls for utilizing job satisfaction and organizational citizenship behaviors as valid organizational outcomes in addition to performance.

Conclusions

In summarizing the arguments surrounding person-job fit and job analysis, one wonders if there is a paradigm shift. The most radical criticism stemming from the

TQ literature suggests that the concept of job is inherently dysfunctional as we move to a post-modern society. Rather than a paradigm shift, we view the arguments as a swinging pendulum between the scientific management school of thought and the human relations school of thought. The scientific management school emphasizes labor productivity through increasing specialization. The human relations school emphasizes group dynamics and the importance of job satisfaction in creating effective organizations. Person-job fit and job analysis clearly stem from the scientific management school (Cardy & Dobbins, 1996). In contrast, the TQ movement with its emphasis on teams and job satisfaction reflects the concerns of human relations pioneers such as Roethisberger and Dickson (1933). Similarly, the call for an expanded criterion domain using person-organization fit is concerned about employee socialization into a group and the subsequent work related attitudes.

In our opinion, the pendulum in regards to employee selection has swung too far to the exclusive emphasis on the scientific management perspective. It needs to move to the middle between the scientific management and human relations perspectives.

The criticisms of the scientific management perspective are valid in many organizations, and they may be even more relevant in the presence of increased global competition. When jobs/tasks are constantly changing, the process of matching a person to some fixed job requirements becomes increasingly irrelevant. While the strong supporter of person-job fit may advocate constant updates of job analyses, this becomes problematic if one wishes to retain an employee who may no longer fit a job as the job changes. Ultimately, the changing job environment may lead to more outsourcing of labor, but relationships and continuity also have a role in promoting organizational effectiveness.

This thought parallels the development of the career concept. It has moved from a pre-industrial perspective that a career is a given trade that one performs for one's life, to a modern perspective of careers as a sequence of jobs, to a contemporary holistic boundaryless career which is viewed as a series of life experiences (Bird, 1994).

In spite of these criticisms of person-job fit, there are two enduring arguments that the proponents of changes have failed to successfully address. First, person-job fit has a long history of providing a structure that produces valid and reliable selection results. The proponents of change call for more flexibility in the selection process. If this creates increasingly unstructured selection processes, then these selection processes may become less reliable and valid.

Second, finding someone who fits the organization may be viewed as subtle ways to discriminate against protected categories of employees. Relatively recent class action suits against Texaco and Shoney's suggest that discrimination still exists in spite of 30 years of EEOC efforts to reduce it. The use of valid and reliable selection processes appears to reduce personal biases in the selection process and increase legal defensibility. A challenge will be to develop valid, reliable, and legally defensible selection processes for the less structured and more flexible aspects of fit.

To provide some balance to the exclusive use of person-job fit in the selection, we argue there is a need to develop pluralistic models to the selection process to accommodate the needs and the circumstances of many employment decisions. In particular, we suggest that there is a need to focus on finding fit at the job level, the group level, and the organization level. Furthermore there is a need to develop structured, valid, and reliable selection processes that address these different levels of fit. To facilitate this development, we propose a new model of the different types of fit in the employment selection process. We also discuss research regarding both predictor and criterion aspects of the different types of fit, and consider implications for the validation process.

AN EXPANDED MODEL OF FIT IN SELECTION

Previous literature (Muchinsky & Monahan, 1987) has suggested two ways in which fit can be conceptualized: *supplementary* fit and *complementary* fit. Supple-

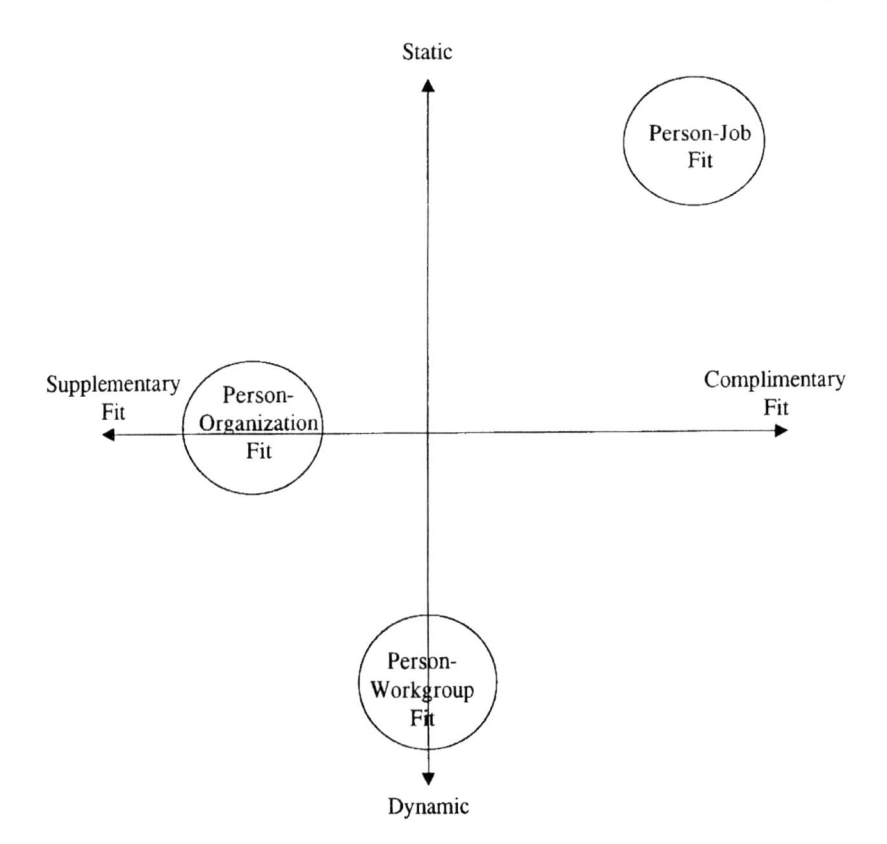

Figure 1. A multidimensional view of the different types of fit.

mentary fit refers to the possession of characteristics similar to other individuals in an environment. Muchinsky and Monahan (1987) suggested that this type of fit is the basis for vocational choice. People prefer to work in environments with others who share similar qualities. Complementary fit concerns deficiencies in the environment that are compensated by individual strengths. They argued that complementary fit based on person-job fit has traditionally driven employment selection practices. In essence, both supplementary and complementary fit should be important in selection processes. The three types of person-environment fit we are considering (i.e, person-job fit, person-organization fit, and person-workgroup fit) vary with respect to complementary and supplementary fit as well as a static and dynamic dimension (see Figure 1).

As previously discussed, person-job fit is the traditional conceptualization of fit and the model that has dominated research and practice in employment selection. Hitt and Barr (1989) viewed this as the congruence of applicant's knowledge, skills, and abilities (KSAs) with the task requirements of the job. Job analysis is used to identify the task requirements and KSAs for a job, and then a variety of selection procedures (e.g., job applications, selection tests, & interviews) are used to gather information on applicants' relevant KSAs. From this information, employers make inferences about applicants' job proficiency. As mentioned previously, this is a complementary fit (Muchinsky & Monahan, 1987).

Person-organization fit refers to the congruence of applicants' needs, goals, and values with organizational norms, values, and reward systems. It has recently received some attention in the selection literature (Schneider, 1987), and exploratory procedures (Bowen et al., 1991) as well as new statistical techniques (Edwards, 1994) are being considered to assess person-organization fit. The exploratory procedures entail identifying relevant aspects of the organizational culture and identifying applicant characteristics that are most compatible with these aspects of organizational culture. Current efforts also attempt to identify compatible value systems (Meglino et al., 1989; O'Reilly, Chatman, & Caldwell, 1991), or match applicants' needs with organizational reward systems (Scarpello & Campbell, 1983). From this information, structured evaluations can be made about applicants' fit with the organization. The identification of commonly shared characteristics of employees in an organization would best be represented by supplementary fit as defined by Muchinsky and Monahan (1987).

Person-workgroup fit refers to the match between the new hire and the immediate workgroup (i.e., coworkers and supervisor). This type of matching entails both supplementary fit and complementary fit. Based on traditional management theory (Barnard, 1938; March & Simon, 1958), cooperation between managers and subordinates is a necessary condition for organizational effectiveness. Cooperation is also important among coworkers or team members. This can best be achieved through supplementary fit. Complementary fit is also important for person-workgroup fit. One of the advantages of groups over individuals is that the human resource assets are diverse. The performance weaknesses of one individual may be offset by the performance strengths of a second individual. To enhance the overall group performance, there is a need to have workgroup members with complementary proficiencies that are beneficial for team performance.

A potential problem is that the supplemental person-workgroup fit could be based on legally protected group characteristics such as gender, age, and ethnicity. It is important to note however, that the complementary aspects of person-workgroup fit imply heterogeneity. A workgroup may decide that they need a new perspective and therefore the fit process involves looking for an applicant who will add heterogeneity to the group.

The three types of fit are similar in that each is based on matching applicant attributes with defined attributes required in the vacant position. However, the metaphor of fit may be somewhat different for different fit types. Person-job fit can be likened to fitting a peg into a hole or fitting a piece into a jigsaw puzzle. In either metaphor, the process involves defining the shape of the vacancy (hole or jigsaw space) and then seeking an appropriate candidate (peg or jigsaw piece) to fill that vacancy. Job analysis is typically the tool used to define the shape of the vacancy.

With person-workgroup fit, the fit process is less static and more dynamic. The group may initially identify the role that the new hire will fill, but the precise nature of this role may only emerge through interactions with appropriate candi-

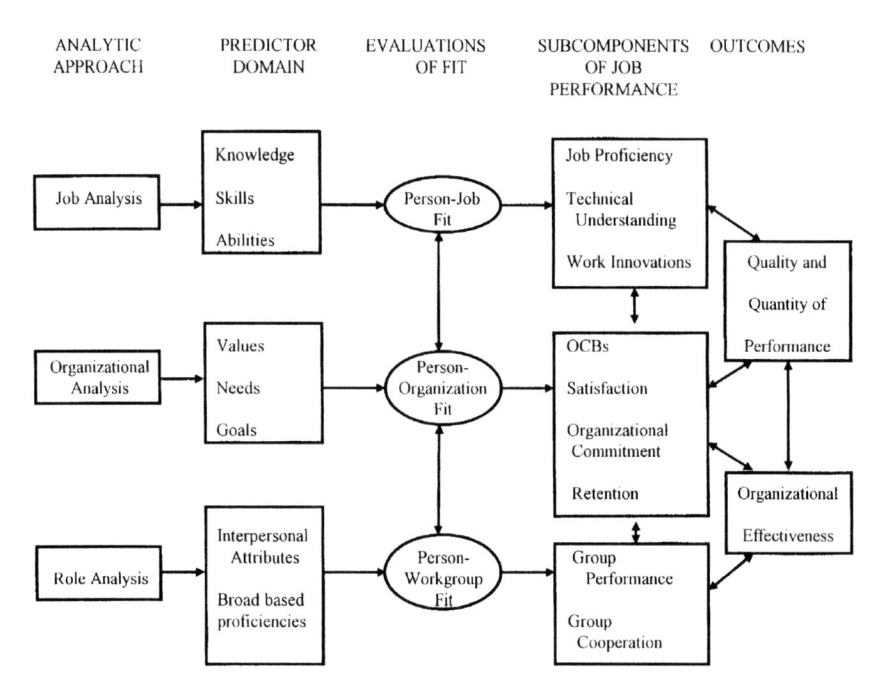

Figure 2. A proposed model of fit in the selection process.

dates. The appropriate metaphor for person-workgroup fit may be more akin to making music. A musical group may look for a musician with certain qualities, but the precise nature of the fit can only be determined through an interactive process in which the nature of the group may actually change. Existing members' roles may get renegotiated to adjust to the strengths and weaknesses of the applicant who will provide the greatest value added role to workgroup. The final roles for the new group member and existing group members may continue to evolve after the member has joined the group. Person-organization fit likely falls somewhere between person-job fit and person-workgroup fit in terms of the appropriate fit metaphor. Supplementary fit may fail to have the precision of complementary fit. But, it is less fluid than person-workgroup fit.

Figure 2 presents our proposed facet model of fit in personnel selection. Each of the proposed types of fit is associated with a different analytic approach, predictor domain, and criterion or outcome domain. With regard to the criterion domain, this model suggests that employee selection needs to be linked to multilevel analysis. Rousseau (1985) argued that organizational issues need to be explored on an individual, group, and organizational level. We suggest that job performance is a multifaceted and that aspects of job performance are relevant at the individual, group, and organizational levels. Consequently, the different types of fit for employee selection correspond to individual, group, and organizational level aspects of performance.

Figure 2 suggests that different types of inferred applicant characteristics and predictor measures are associated with the different types of fit. To assess person-job fit, recruiters evaluate applicants' knowledge, skills, and abilities to perform the job. The primary predictor measures associated with this would be job proficiency and technical understanding of the job. The ability to initiate work innovations and to achieve greater job efficiency would also influence job proficiency.

To assess person-organization fit, recruiters evaluate applicants' needs, goals, and values. The greater the correspondence between applicant needs and organizational reward systems, the greater the motivation to perform for the organization. The greater the correspondence between applicant goals and values with organizational expectations and culture, the greater the satisfaction and organizational commitment. Therefore, the criterion associated with person-organization fit should include organizational citizenship behaviors, organizational satisfaction, organizational commitment, and retention.

Person-organization fit is typically conceptualized as congruence between individual and organizational goals and values. However, it is important to realize that this fit can exist at a number of different levels within the organization. Person-organization fit can refer to congruence between the values and goals of an individual and those of the entire organization, the individual's department, or division. Past research has even defined person-organization fit as the congruence between subordinate and supervisor goals (Vancouver & Schmitt, 1991), although we believe this is more appropriately labeled person-workgroup fit. Kristof (1996) argued for a distinction between person-organization fit and person-workgroup fit since they reflect different levels of the organization, and in some cases reflect different norms and values. In our conceptualization, we are not arguing that different levels of person-organization fit are indiscriminable, rather we suggest that the predictor domain and criterion domain are largely the same for different levels of person-organization fit. Given that person-organization fit can exist at different levels within the organization, issues of levels of analysis must be considered when conceptualizing and operationalizing this type of fit. We discuss a number of these issues in later sections of this paper.

To assess person-workgroup fit, recruiters need to identify the workgroup needs and prioritize those which are most critical to group effectiveness. The foundation for an assessment of person-workgroup fit would be a role analysis. Role analysis is broader than job analysis because the activities of some work groups tend to be fluid with different team members contributing to group success in different ways. In order to do role analysis, the workgroup needs to identify both complimentary and supplementary types of fit that would improve the effectiveness of a workgroup. Attributes that may define the supplementary fit include values, personality, and interpersonal skills. Attributes that may define role complementary fit require the addition of a value added human resource asset to the group that is not prevalent in other team members. This may entail interpersonal networks, some unique technical skills, or some basic skills such as creativity, oral communication skills, or strategic thinking skills. Some of these predictor criteria such as communication skills and conflict resolution may be common in many teams as a requirement to establish person workgroup fit (Stevens & Campion, 1994). However, groups also have unique needs and distribution of human resource assets. Team effectiveness is partially based on having diversity in human resource assets. Thus, teams may need to identify the unique human resource assets that could be brought to bear to enhance group performance that are not readily available from current group members.

Given the potential for fit to emerge through a negotiation process, group members may have some latitude over work roles and opportunities to pursue work roles that are most interesting to them. This should be associated with intrinsic motivation and cooperation. Thus, the criterion domain for person-workgroup fit should include workgroup cooperation and the performance of all group members.

The objective outcomes of positive behavior in the criterion domain should include individual performance (quality and quantity), workgroup performance, and organizational effectiveness. Note that we are not suggesting that only person-job fit leads to individual performance or that only person-organization fit lead to organizational effectiveness. Fit at one level can influence performance at a variety of levels.

While efforts are made to clearly maintain a distinct nomological network with different levels of analysis, some isomorphisms across levels of analysis may exist. This poses some problems in establishing discriminant validity for the different types of fit. The isomorphisms across levels of analysis are most likely to occur with the predictors of fit. For example, personality characteristics have been used to establish person-job fit (Barrick & Mount, 1991; Behling, 1998) and person-organization fit (Bowen et al., 1991). They may also appear to be relevant for person-work group fit (McLane, 1991). Bretz, Rynes, and Gerhart (1993) reported that grade point average was used by some recruiters to determine person-organization fit. Normally one would suspect that this would be more likely to be related to person-job fit.

Additionally, there may be some isomorphisms with the sub-components of job performance. For example, organizational citizenship behaviors and cooperation with team members are likely to closely related constructs focusing on different levels of analysis. Furthermore, fit at one level can influence performance at a variety of levels. Having an effective group fit may enhance individual performance and organization commitment. Having job proficiencies may impact group performance.

In spite of these problems, one of the issues of levels in organizational research is to identify the relevant isomorphisms and cross level impacts. By doing so, one can then more fully understand the richness of behavior in organizations.

Because there are theoretical distinctions concerning fit across the levels of analysis, there may be some utility in considering the unique contributions that each type of fit has to job performance. Thus, the following discussion assumes that the types of fit and corresponding nomological networks are largely independent constructs. Expanding the dimensions relevant in predicting job performance may enhance the ability to accurately validate selection criteria and improve the effectiveness of employee selection processes.

In summary, we have proposed a facet model of selection that expands both the predictor and criterion domains by considering three types of fit. To provide support for the model, the next two sections address methods for determining predictor domain variables for the different types of fit, and research support for some of the consequences of fit assessment.

METHODS FOR DETERMINING PREDICTOR DOMAIN VARIABLES FOR FIT ASSESSMENTS

As discussed earlier in this paper, historically selection research has approached the predictor domain from the perspective of person-job fit. With our expanded model of fit, there is a need to expand the domain of predictors. There is considerable employee selection research examining the relationship between applicant characteristics and job performance. Although we briefly mention some of this research, most of the literature we review examines the specific links between the applicant characteristics and employers' fit assessments. There is also a need to redefine the process of identifying predictors. Job analysis is the process that emerged from person-job fit, but this process requires considerable reconceptualization when considering person-organization and person-workgroup fit.

Person-Job Fit

Our model suggests that person-job fit assessments are based on evaluations of applicants' knowledge, skills, and abilities. Job analysis is used to identify the important components of the job and the human attributes that are necessary to perform those components. Research examining person-job fit from a decision-making perspective, looks at individual attributes such as test scores, education background associated with major, degree, and grades, years of work experience, and relevant work experience as indicators of knowledge, skills, and abilities. To some extent personality may influence assessments of person-job fit. Unfortunately, research has not commonly asked recruiters to assess person-job fit, but instead has typically asked recruiters to assess job suitability, perceived job competence, or simply willingness to hire. All of these should be correlated with assessments of person-job fit with the first two being most closely related to person-job fit.

Test Scores

Perhaps one of the most common ways to assess applicant abilities and consequently person-job fit is through testing. General ability tests, physical ability tests, and work sample tests have been found to be effective predictors of performance (e.g., Hunter & Hunter, 1984; Schmitt, Gooding, Noe, & Kirsch, 1984). However, only two studies have linked test scores with evaluations of job suitability. Heneman (1977) had undergraduate students use test scores to evaluate hypothetical applicants for insurance agent positions. Test scores were positively associated with evaluations of job suitability and willingness to hire. With recruiters evaluating applicants for accounting positions, Carlson (1971) found that test scores explained 30% of the variance in assessments of perceived job competence. This occurred even when recruiters were told that the tests had low validity. This latter finding suggested that a link may exist between test scores and evaluations of person-job fit, even in the absence of an overall relationship between test scores and job performance.

Background Experience

In addition to test scores, background experiences and qualifications are also commonly used to assess person-job fit. Biodata research has linked these background indicators with valid predictors of job performance (e.g., Rothstein, Schmidt, Erwin, Owens, & Sparks, 1990; Schmitt et al., 1984). Additionally, considerable research has examined the impact of these indicators on evaluations of suitability and willingness to hire.

Probably the most widely researched individual attribute for evaluating the job knowledge component of person-job fit is grade point average. Six studies examined the relationship of grades with selection recommendations for hypothetical college graduates. These studies included both college students and recruiters as evaluators of paper applicants. In all studies, grades were positively associated with willingness to hire applicants (Dipboye, Arvey, & Terpstra, 1977; Dipboye, Fromkin, & Wiback, 1975; Gardner, Kozlowski, & Hults, 1991; Hakel, Dobmeyer, & Dunette, 1970; Oliphant & Alexander, 1982; Wingrove, Glendinning, & Herriot, 1984). Other studies using resume information with actual selection decisions also supported the use of grades as an important selection criterion for campus interviews (Campion, 1978; Werbel & Looney, 1994). However, this research failed to directly address the relationship between the use of grades with assessments of job competency or person-job fit.

Other important educational experience criteria related to person-job fit are major and degree. Both would be associated with person-job fit based on job knowledge. Renwick and Tosi (1978) varied undergraduate majors and degree and asked graduate administration students to assess the suitability and willingness to hire hypothetical graduating students for positions in training and development. Both degree and major influenced evaluations. Similar results were found with samples of personnel officers (Singer & Bruhns, 1991). However, two studies indicated that the position being filled may limit the impact of degree on evaluations of job applicants (Barr & Hitt, 1986; Hitt & Barr, 1989).

Work experience criteria may also be important for assessing the job skills aspects of person-job fit. Barr and Hitt (1986) varied years of work experience (10 years or 15 years) for the positions of regional sales manager and vice-president of sales. For manager evaluators, experience was correlated with willingness to hire vice-president of sales. For student evaluators, experience was correlated with willingness to hire for both job positions. A study replicating this design reported similar results (Hitt & Barr, 1989). A number of other studies found relevant work experience, especially when combined with academic achievement, explained considerable variance in evaluations of applicants' qualifications to perform a job (Dipboye et al., 1977; Hakel, Dobmeyer, & Dunnette, 1970; Rasmussen, 1984; Singer & Bruhns, 1991; Wingrove et al., 1984).

Personality

A final predictor of person-job fit that has been examined is personality. Existing research suggests that personality assessments may be related to the ability component of person-job fit. Certain personality characteristics may be more effective for some job types than others (Holland, 1985). In a meta-analysis of the relationship between personality and job performance, Barrick and Mount (1991) found that the validity of most personality dimensions vary as a function of occupational group. More recent research indicated that autonomy is one job dimension that moderates the relationship between personality and performance (Barrick & Mount, 1993). While the relationship between personality measures and job performance has been demonstrated in a number of studies, considerably less research has examined the impact of personality on evaluations of job suitability.

Rothstein and Jackson (1980) found that students evaluated applicants with job stereotype congruent characteristics more favorably for job suitability and willingness to hire than those with stereotypic incongruent characteristics. Similarly, Paunonen, Jackson, and Oberman, (1987) found that personality traits based on the Strong Vocational Interest Blank for a newspaper reporter and a payroll clerk had some impact in evaluating perceived competence of paper applicants. However, it only had a secondary impact after controlling for information about job knowledge and motivation.

Person-Organization Fit

Our model suggests that person-organization fit is based on the degree that applicants' values, needs, and goals are compatible with aspects of organizational structure and processes. As indicated earlier, these structures and processes can exist for larger intact units within the organization. This type of fit has only recently received research attention.

Bowen et al. (1991) prescribed ways for making valid and reliable assessments of person-organization fit. First, one needs to define the organization culture and values. They suggested that climate surveys could be used to do this. Q-sort methods have also been developed to assess organizational value systems, based on a representative sample of organizational members (Chatman, 1989). Selecting individuals on value congruence could then be as simple as having job candidates perform the same Q-sort task and compare candidate responses with the organizational profile. As we suggested previously, the level of organizational fit could be a functional department. In this situation, the climate survey could reflect the culture and values of the relevant level. After assessing the culture and values, one needs to infer the relevant individual differences that are compatible with culture and values of the appropriate organizational level. Finally, one needs to assess the relevant individual differences with job applicants.

The organizational analysis method outlined by Bowen et al. (1991) is basically an application of job analysis to the organizational level. It may also be worthwhile to consider the extent to which diversity and heterogeneity is valued when conducting an organizational analysis. In the process of developing reliable and valid measures for person-organization fit, there may be a tendency to narrowly define value congruence and try to identify the person most optimally congruent with an organization. Instead, what may be needed is to identify applicants with values that are truly incongruent with the dominant organizational values. One needs to assume that the socialization process will have some impact on shaping employee values. The selection process may need to screen applicants with personal values and needs that would be resistant to organizational socialization efforts.

Another limitation of patterning the organizational analysis after job analysis is that it results in a "snapshot" or static image of the organization's values and culture. Given the degree of change in organizations that has resulted from mergers, acquisitions, downsizing, and re-engineering, it may be important to capture the dynamics of an organization in the organizational analysis. Consideration of this change may suggest unique human attributes related to adaptability and learning facility. As with the issue of heterogeneity, organizational analysis techniques need to be developed to capture these dimensions of person-organization fit.

Research concerning the use of individual attributes to assess person-organization fit is limited. Rynes and Gerhart (1990) evaluated the use of person-organization fit in the selection process. Using graduating college students in a college placement center, they found that recruiters within the same company had more highly correlated assessments of applicant-organization fit than recruiters across organizations. This suggested that the concept of person-organization fit had some construct validity.

Rynes and Gerhart (1990) also reported that person-organization fit is based largely on applicants' interpersonal qualities and goal orientation. Furthermore, the objective aspects of person-job fit such as grades and work experience added little explained variance to organizational fit assessments. In trying to clarify the role of other potential criteria used to assess person-organization fit, Adkins, Russell, and Werbel (1994) examined recruiters' assessments of fit based on value congruence between applicant values and recruiters' perceptions of the dominant organizational values. Using a sample from campus interviews, they found that value congruence was unrelated to perceptions of person-organization fit and to obtaining a second interview following the campus interview.

Person-Workgroup Fit

Person-workgroup fit has not received as much research attention as person-job fit or person-organization fit, although it is clearly distinguished from these other types of person-environment fit (Kristof, 1996). Fit between an individual and a supervisor has been discussed in terms of leader-member exchange (Dienesch & Liden, 1986). Recent research has also discussed the importance of individual-team fit with respect to developing high functioning teams (Klimoski & Jones, 1995). However, person-workgroup fit also encompasses compatibility among members of a workgroup, even when the workgroup is not organized into an interdependent team.

The analysis approach for identifying predictors of person-workgroup fit may require a radical departure from the typical job analysis framework. We propose that *role analysis* should be used to select applicants based on person-workgroup fit. The essence of role analysis is to develop commonly agreed upon role expectations for a new hire. To achieve this objective, role analysis processes would focus on the method of examining workgroup goals, the necessary human resource assets that are needed to meet those shared goals, an accounting of current human resource assets, and an assessment of the means used to achieve those goals. Ideally, role analysis would be used to identify human resource production or service bottlenecks and then identify the qualities that would be needed to minimize production or service bottlenecks.

The sources for bottlenecks may stem from either a problem in the task orientation of the group or in the support functions of a group. Both are important in-group processes. The former may provide a basis for determining a complementary fit. Thus, the group may attempt to identify applicants' interpersonal networks, technical skills, or general proficiencies that are deficient with current group members. The focus on support functions may provide a basis for supplementary fit, which would be based on predictors associated with cooperative group behavior. To some extent the workgroup needs to develop collaborative relationships in order to work together effectively. These relationships would be based on effective interpersonal communications and cooperative behaviors.

In contrast to job analysis and organizational analysis, role analysis is mutually negotiated. To some extent the new hire has input into the development of role expectations. The new hire has a chance to explain how s/he could make a significant contribution to group processes and delineate an expected role within the group. The group may modify the original role expectations to accommodate the assets of the different job candidates. Thus, in contrast to other types of fit, person-workgroup fit would be mutually negotiated during (and following) the hiring process.

At this time, research is needed to assess the dimensions that would be relevant to determine person-workgroup fit. Efforts at building cross-functional teams provide an initial start for the needs to consider an array of human resource assets in order to be productive (Jacob, 1995).

Research with predictors of person-workgroup fit is limited. There is no research that directly focuses on complimentary person-workgroup fit, although some research has examined similarity between applicant and interviewer attitudes. This should be associated with supplementary fit. For example, Graves and Powell (1988) found that the perceived similarity between an applicant and the interviewer in terms of work attitudes, approaches for dealing with people, and beliefs about how people should be treated were associated with more positive interviewer evaluations of the applicant.

Indirectly, there is some evidence that communication patterns could be an important aspect supplementary fit with work dyads. In particular some communication patterns may be incompatible to create supplementary fit for work teams. Waldron (1991) investigated the impact of subordinate communication patterns on perceptions of LMX. Based on factor analysis of communication patterns, four patterns emerged. Personal communication pattern is the ability to talk freely about non-work issues. Direct communication pattern is the ability to talk freely about work issues. Contractual communication pattern shows respect for organizational rule and authority. Regulative communication pattern occurs when the employee is guarded about revealing different types of information to a superior. Waldron reported that a quality LMX was associated with personal communication, direct communication, and contractual communication in current employees.

Fairhurst (1993) identified 6 male and 6 female employees. She collected member's LMX scores and had the employee tape a 30 minute conversation with his/her leader. From transcripts, she observed that personal communication such as teasing and expressing polite disagreements were associated with a quality LMX. Competitive conflict, as represented by strong disagreements and interruptions in speech, was inversely associated with a quality LMX.

To the extent that these communication patterns would extend beyond vertical dyads is unclear. Nonetheless, it would appear that interpersonal communication patterns could be an important element for determining person-workgroup fit.

It is also possible that shared perceptions of workplace climate could influence supplementary workgroup fit. Graen and Schiemann (1978) investigated the impact of supervisory and subordinate agreement on perceptions of workplace climate with intact dyads. They reported that agreement in perceptions was significantly and positively associated with a quality LMX. Again the extent that these dyad results could extend to workgroup processes is unclear.

Additionally, existing research suggests that interpersonal behaviors could influence assessments of person-workgroup fit and are perhaps one of the most important elements in the selection process. For example, Kinnicki and Lockwood (1985) collected information from recruiters concerning their most important criteria in the selection process. The recruiters reported that interpersonal impressions were important selection criteria for job interviews. Unlike other types of evaluations, evaluations of interpersonal skills were consistently used by recruiters as a basis to make recommendations to hire hypothetical job applicants (Graves & Karren, 1992).

Potential Problems

Person-workgroup fit presents at least three problems that are less salient with the other types of fit. First, if person-workgroup fit is narrowly operationalized as supplemental fit, hiring on person-workgroup fit may lead to homogeneity within the workgroup. There is evidence that racial similarity between interviewers and applicants can positively influence interviewer recommendation, even with structured and situational interviews (Lin, Dobbins, & Farh, 1992). Any other "similar-to-me" biases that exist in recruiting and interviewing processes could also result in homogeneity rather than actual person-workgroup fit.

A second and related problem with the similar-to-me bias is the potential for employment discrimination. If the similar-to-me evaluation is based on personal characteristics such as race, sex, religious values, or ethnicity then supplementary fit is illegal and inappropriate.

Third, research indicates that applicant verbal and non-verbal cues can influence interviewer evaluations (e.g., Gifford, Ng, & Wilkinson, 1985; Tullar, 1989). In addition, ingratiating behavior and other impression management tactics have been associated with positive interview impressions (Ferris & Judge, 1991; Stevens & Kristof, 1995). The result is that in some cases the supplementary aspects of person-workgroup fit may end up being operationalized in terms of perceived likability rather that actual fit.

Perhaps the best way to avoid these problems when assessing person-workgroup fit is to carefully develop the predictor domain through role analysis. Role analysis needs to entail both supplementary and complimentary aspects of fit. Thus, a strict similar-to-me selection process would be limited. Furthermore, person-job fit can be based on impressions and "gut instincts" when the predictor domain is not structured. However, through job analysis, valid predictors of person-job fit have been identified. Similarly, if workgroup roles are clearly defined through role analysis, then structured predictors could be developed to validly assess applicants' capacity to fill a role.

Summary

The preceding review summarized methods for determining predictors of the three different fit assessments: person-job fit, person-organization fit, and person-workgroup fit. The applicant characteristics that influence person-job fit are related to knowledge, skills, and abilities. Test scores, education, and work experience have been associated with job suitability and willingness to hire. Person-organization fit appears to be influenced by applicants' interpersonal qualities and goal orientation, but research has not supported the relationship between individual-organizational value congruence and person-organization fit. Unfortunately, there is a paucity of research on person-workgroup fit. Based on ideas developed about employee selection in a team work environment (Stevens & Campion, 1994), we have proposed that predictors of person-workgroup fit may include interpersonal communication patterns, broad-based proficiencies, and cooperative behaviors. Based on the proposed model and summarized research, we develop three research propositions:

Proposition 1.   Person-job fit is associated with indicators of knowledge, skills, and abilities.

Proposition 2.   Person-organization fit is associated with indicators of values, needs, and goals.

Proposition 3. Person-workgroup fit is associated with indicators of interpersonal attributes and broad-based proficiencies.

We would like to raise four issues related to these propositions and the existing research. First, research is limited in terms of the types of applicant characteristics that have been investigated. With person-organization fit, two studies are inadequate to draw many conclusions. There is an absence of research that directly assesses person-workgroup fit. With person-job fit, there needs to be more research concerning interaction effects between different types of relevant criteria and job type. While this is beginning to occur with personality testing and job performance criteria (e.g., Barrick & Mount, 1993), other predictors have not been examined.

Second, while it appears that predictors of person-job fit have been delineated through research efforts, the relationship between applicant characteristics and fit assessments is often unclear and ambiguous. The research failed to uniformly assess the different assessments of fit directly and concurrently. Typically, related measurements (e.g., likability) or willingness to hire were used. Future research should consider directly assessing different types of fit.

Third, person-organization fit and person-workgroup fit are constructs that are beginning to receive more attention, but are still generating relatively little research from a selection perspective. The issue of selecting people for teams was discussed from a fit perspective (Klimoski & Jones, 1995). Seers (1989) proposed the development of team member exchange quality as an issue for group development. Similarly, Ferris, Youngblood, and Yates (1985) suggested that establishing person-workgroup fit is an important aspect of the employee socialization process. In a recent review of person-organization fit research, Kristof (1996) highlighted the fact that much variation exists in the conceptualization and operationalization of this type of fit. For example, fit can be assessed directly ("to what extent does this candidate fit the organization") or indirectly (by assessing values, needs, and goals of the individual examining the correspondence with organizational characteristics).

Fourth, given that we define person-organization fit to exist on multiple levels, it is important to consider suggestions from research on levels of analysis when conceptualizing and operationalizing fit (e.g., Rousseau, 1985). It is important to conceptualize an appropriate level for the particular organization. In some organizations, fit at the organizational level may be most important, whereas with highly specialized business units, fit at the business unit level may be most important, and still with other organizations, fit at multiple levels may be crucial. Additionally, when the level of conceptualization is identified, the level of operationalization should match that level of conceptualization. However, at the organizational level of analysis, it is not always clear how organizational fit can be operationalized. Should organizational values, needs, and goals be assessed from the chief executive or president, from an aggregate of perceptions from individuals at different levels in the hierarchy, or from the perceptions of recruiter, as has been the case in past research (e.g., Adkins et al., 1994). Clearly, the assessment of person-organization fit requires additional measurement considerations that are more clearly defined when assessing the other evaluations of fit (e.g., person-job fit).

CONSEQUENCES OF FIT ASSESSMENTS

There has been limited research evaluating the consequences of fit assessments. Some research has collected recruiters' fit assessments and then related these assessments to selection decisions. There has also been some research assessing the criterion validity of different fit assessments. As with the predictors of fit assessments, we consider consequences of each of the three types of fit assessment.

Person-Job Fit

Kinnicki, Lockwood, Hom, and Griffeth (1990) evaluated person-job fit with data collected from actual interviews for the position of a hospital nurse. They reported that *objective* measures of person-job fit from the job application, including years of schooling and years of work experience, were marginally related to selection recommendations and unrelated to post-entry job performance or job attitudes. However, *subjective* assessments of job skills from the interview were highly related to selection recommendations, job performance, and job satisfaction. This finding is contrary to the sizable body of research that has demonstrated the predictive validity of objective background experience measures such as biodata (Schmitt et al., 1984), and the relative lack of validity of the unstructured interview (McDaniel, Whetzel, Schmidt, & Maurer, 1994).

Person-Organization Fit

Adkins et al. (1994) reported that recruiters' subjective assessments of person-organization fit influenced the likelihood of obtaining follow up interviews. This suggested that person-organization fit may be important in the selection process. However, this research failed to indicate how person-organization fit was associated with job performance.

In terms of predictive validity of person-organization fit, Chatman (1991) found that the amount of time recruiters spent with new hires influenced perception of person-organization fit. However, the acceptance ratio of new hires was unrelated to person-organization fit. Nonetheless, employee assessment of person-organization fit was associated with job satisfaction and job retention. Because person-organization fit was unrelated to applicant selectivity, this study suggests that it was unclear if recruiters used person-organization fit in the selection process or if they were unable to effectively screen applicants based on person-organization fit. If they could effectively assess and use it, the outcomes of job satisfaction and job retention may increase.

Using a sample of public employees, Boxx, Odum, and Dunn (1991) investigated the impact of goal similarity on work attitudes. They asked employees to identify their personal goals in doing work and asked employees to identify the work related goals of their work unit. A match in goal structure was positively associated with work attitudes.

Beyond the selection domain, research has demonstrated a relationship between incumbent person-organization fit and a variety of work attitudes and behaviors. In a comprehensive review of this literature, Kristof (1996) found that person-organization fit was related to job satisfaction, commitment, intentions to turn-over, prosocial behaviors, and work performance. It is important to note that some researchers have also pointed out the "dark side of fit" in terms of organizational consequences; a high degree of person-organization fit may result in myopic perspectives, an inability to change, and a lack of organizational innovation (Schneider, 1987; Schneider, Goldstein, & Smith, 1995).

Bretz and Judge (1994) reported that the more that employees personal beliefs about work matched normative behaviors of diverse organizations, the greater the career success and organizational tenure in that organization. They suggested that this occurred because person-organization fit leads to higher levels of motivation and thus better job performance.

Person-Workgroup Fit

There is only one article directly evaluating the impact of person-workgroup fit on performance outcomes (Ferris et al., 1985). The article reported that person-workgroup fit moderated the relationships between performance and turnover and performance and absenteeism. Those better performers with better fit were more likely to stay and report to work than weaker performers with a person work-group fit.

There is also evidence that supplementary fit, as represented by effective hierarchical relationships, is important for developing a cooperative climate in workgroups. Settoon, Bennett, and Liden (1996) investigated the impact of a quality LMX on job performance and organizational citizenship behavior in a large hospital. They found that a quality LMX was associated with measurements of job performance and extra-role behaviors supportive of work-unit cooperation.

Wayne, Shore, and Liden (1997) using a sample of employees in a large corporation found that LMX based on employee assessment was related to organizational citizenship and favor doing. LMX was unrelated to work attitudes such as affective commitment and intentions to quit.

This research is obviously limited because it only focuses on one of the central dyads in a work group and because it assesses fit among incumbents rather than applicants. Nonetheless, it does provide some evidence that interpersonal fit may have an impact on cooperative group behaviors and warrants additional research with all group members.

Summary

Research addressing the consequences of fit assessments has demonstrated that (a) subjective evaluations of person-job fit are related to selection recommendations, job performance, and job satisfaction; (b) evaluations of person-organization fit are related to obtaining follow up interviews (but not selection decisions), job satisfaction, and job retention; and (c) evaluations of person-workgroup fit may be related to cooperative work group behaviors. Although these conclusions suggest that the different fit assessments are useful in evaluating job applicants, considerably more research is needed as many of these conclusions are based on single studies.

The proposed model suggests a number of additional outcomes that should be considered for each type of fit assessment. Since person-job fit is based on job relevant knowledge, skills, and abilities, evaluations should be closely tied to job proficiency. Thus, person-job fit evaluations should be related to quality and speed of performance. Since person-organization fit is based on an applicant's goals, values, and personal needs, evaluations should predict job performance criteria, but are also likely to predict citizenship behaviors, job satisfaction, organizational commitment, absenteeism, and retention, all of which contribute to organizational effectiveness. Person-workgroup fit is based on a match of interpersonal qualities and broad based job proficiencies, such that evaluations should be associated with work unit dynamics. This could influence interpersonal communication, work unit cooperation, organizational citizenship behaviors, and group performance. The theory and research on consequences of fit suggest the following propositions:

Proposition 4. Evaluations of person-job fit are related to job proficiency and work innovations.

Proposition 5. Evaluations of person-organization fit are related to job attitudes, extra-role behaviors, and attachment behaviors.

Proposition 6. Evaluations of person-workgroup fit are related to interpersonal communication, work unit cooperation, and group performance.

Clearly, more research is needed to document the predictive relationships between fit assessments and the organizational outcome suggested in the preceding propositions. In addition to the need for research on criterion-related validity, it is also worth considering the implications of the different types of fit assessment for collecting other forms of validity evidence.

ASSESSING VALIDITY OF FIT ASSESSMENTS

It has long been acknowledged that there are three different strategies for collecting validity evidence: criterion-related, construct, and content strategies. More recently, authors have argued that all validation is collecting construct validity evidence or theory development and testing (Schmitt & Landy, 1992). However, criterion-related validity has remained the dominant approach for demonstrating the adequacy of predictor measures. With the focus on person-job fit and job analysis, criterion-related validity is a natural and logical validation strategy. Content validity is also a logical strategy as it involves matching predictor content with job content. On the other hand, with person-organization fit and person-workgroup fit, content validity does not appear particularly appropriate or relevant. Criterion-related validity can be used to demonstrate the relationships between these types of fit and the consequences outlined in Propositions 5 and 6. However, perhaps with person-organization and person-workgroup fit, construct validity will become the most useful methods for collecting validity evidence.

Construct validity has not been widely used with typical predictors of person-job fit, such as ability tests. This may be because we do not have very good theories regarding the constructs that underlie jobs. Recent research has made strides in terms of defining job performance constructs (see Schmitt & Landy, 1992), but this still does not address the constructs associated with job tasks, which represent the building blocks of most job analyses. With organizational analysis and particularly role analysis, we are much closer to identifiable constructs. Through organizational analysis, we can identify values that are central to organizational fit. If we assess these same values in the predictor domain, we have established a predictor with construct validity. Similarly, the role analysis may identify key interpersonal attributes or role behaviors that are central to person-workgroup fit. Direct assessment of these attributes or behaviors establishes construct validity in much the same way that direct assessment of key job tasks establishes content validity in the person-job fit domain. In short, we are arguing that construct validity should play a much more central role in predictor validation when considering person-organization and person-workgroup fit.

Organizational Implications

One of the important organizational implications of the proposed model concerns the use of the different types of fit for different aspects of organizational effectiveness. Campbell and Pritchard (1976) suggested that individual performance is a function of a person's skills, motivation, and situational constraints and facilitators. Correspondingly, the importance of the different types of fit are likely to vary across different types of work environments and situations that stress the relative importance of skills, motivation, or situational constraints and facilitators. Relevant features of work environments include the technical requirements of the job, the distinctiveness of the organization's culture, the length of the career ladder, the presence of team-based work organization, and individual role variability within the organization.

Technical Requirements of the Job

Jobs vary by their degree of technical expertise. Some jobs are highly technical and require a certain amount of specific skill development to perform effectively in the job. These skilled jobs could include engineering positions, computer programming, nursing, and to a lesser extent some skilled blue collar positions. Other types of jobs tend to be more general and only require technical expertise that can be learned relatively quickly on the job. These jobs could include semi-skilled laborers, some sales positions, or some types of entry level management positions. When job performance is clearly tied to technical skills, skill and knowledge assessment become the most important type of fit influencing job performance.

Proposition 7. The greater the technical job requirements, the greater the importance of person-job fit for employee selection.

To some extent, research may support this proposition. Werbel and Looney (1994) reported that college recruiters seeking technical engineering applicants were more likely to use GPA as a selection criterion than college recruiters seeking retailing applicants. In this situation, GPA could serve as a proxy for skill development.

Organizational Culture

Researchers have often considered the importance of organizational culture on employee relationships (Chatman, 1989; Etzioni, 1975; Sonnenfeld & Peiperl, 1988). While these authors have different images of the components of organizational culture, they all demonstrate that employees within an organization may share a set of values, beliefs, and behaviors about appropriate work behaviors. Some organizations appear to have organizational cultures with a highly distinctive set of values, beliefs, and behaviors (Barley, Meyer, & Gash, 1988). These organizations with distinctive cultures will need to select employees who are compatible with that culture.

Proposition 8. The more distinctive the organizational culture, the greater the importance of person-organization fit for employee selection.

Van Maanen (1975) suggested that person-organization fit should be an integral part of the employee socialization process. Organizations that mandate the acceptance of unique values, beliefs, and behaviors need to address these issues prior to job entry. Furthermore, Hofstede (1984) indicated that this may be increasingly important issue as the number of multinational corporations increases.

Career Ladder Length

Career ladders have an impact on the use of fit assessments in two ways. First the use of career ladders implies that a new hire will be rotated through a series of jobs within the organization. The job rotation is seen as a type of on- the-job training that prepares one for positions higher on the career ladder. This would minimize the need to emphasize person-job fit.

Second, career ladders are used as a motivational tool. Employees who are motivated to comply with organizational expectations will be rewarded with responsibility and pay increases through a series of job promotions. This approach implies that employees need to be motivated to comply with organizational expectations to receive promotions. Individuals who demonstrate person-organization fit in terms of organizational needs and goals would likely climb the career ladders more quickly and attain greater organizational rewards than those with dissimilar needs and goals. Thus, for an organization with these long career ladders, person-organization fit should be emphasized.

> **Proposition 9.** The longer the career ladder associated with an entry-level position, the greater the importance of person-organization fit for employee selection.

Team Based Organizations

Jobs within a work unit are likely to vary based on the amount of shared work activities. Some jobs, such as sales, often require little shared work activities. Other jobs, such as project engineering, often require extensive shared work activities. When job activities are shared, it is important for employees to cooperate with each other. Some individuals may find it difficult to cooperate with a diverse mix of other individuals. For example, communication and cooperation appears to increase as the homogeneity of the work unit increases (Zenger & Lawrence, 1989). Therefore, person-workgroup fit could be an important aspect that would lead to enhanced work unit performance that is critical with a team-based work unit.

> **Proposition 10.** The greater the use of team-based systems within a work unit, the greater the importance of person-workgroup fit for employee selection.

Work Flexibility

There is increasing evidence that organizations are utilizing employees in varied capacities and demanding flexibility from both the supervisor and the subordinate. Bowen et al. (1991) suggested that the typical business environment is being defined by changing markets and technologies and that these changes create very transitory requirements for specific jobs. Given this presence of a dynamic work environment, it may be more important to select applicants for their ability to work effectively with others in the organization as opposed to a set of job skills that may become obsolete after a short period of time. Hiring on the bases of person-organization and person-workgroup fit may be one means of creating stability in a changing environment.

> **Proposition 11.** The greater work flexibility within the organization, the greater the importance of person-organization fit and person-workgroup fit for employee selection.

LIMITATIONS OF FIT ASSESSMENTS

The discussion, thus far, focuses on support for and implications of the different types of fit. Given the nature of these fit assessments, there are also some limitations to consider with their use.

One limitation of the proposed model concerns the utility of the three types of fit assessment. Each fit assessment can be problematic in some situations, which potentially limits the utility of these assessments. For example, person-job fit would be susceptible to changes in the job requirements that come from changes in technology, job redesign, or downsizing. In each circumstance, a change in the task requirements may create a lack of fit. While employers may wish to engage in workforce reductions and keep those that most appropriately fit the new task requirements, it would ideally be more effective to initiate new job search processes to maximize person-job fit. This may or may not be viewed as an appropriate use of human resources.

Additionally, person-job fit may limit organizational willingness to change labor practices. Person-job fit creates specialized labor focusing on specific tasks. If the tasks change or reassignments are made, person-job fit would diminish and job performance would decrease. Employees may resist needed changes due to expected performance decrements. This would be inappropriate in an increasingly competitive global market.

Person-organization fit is also subject to change. Increasingly, CEOs are being hired externally, and are more likely than internal hires to create radically new organizational structures. Similarly, new leadership practices that come from corporate mergers are also likely to create radically new organizational structures. In either circumstance, employees who formerly had person-organization fit may have an inappropriate fit. This may decrease work attitudes and pro-social behaviors.

Person-workgroup fit would be susceptible to turnover of personnel. As key workgroup members leave, it may be difficult to maintain existing workgroup fit. Thus, work unit conflict may increase and interpersonal communication effectiveness may decrease if there is a change in workgroup personnel. Hiring on the basis of person-workgroup fit may be particularly problematic in industries with high worker mobility.

While each of the types of fit may have limited long-term utility, organizations can always respond to these changes in a variety of ways. They may utilize retraining, relocation, progressive discipline, and dismissal to reestablish appropriate types of fit and insure effective utilization of human resources. In spite of the likelihood of significant change in the work environment, employers need to consider having a range of different types of fit to ensure effective utilization of human resources. If organizational change alters one type of fit, there are opportunities to utilize human resources effectively if other types of fit are established prior to job entry.

As mentioned previously, a second limitation is that person-organization fit and person-workgroup fit may be open to charges of employment discrimination. Organizational and workgroup fit are likely to be associated with homogeneous organizations and work units with regard to goals and values. These qualities may unintendedly covary with protected employee categories, which could lead to disparate impact. While such charges have also been made with educational experience aspects of person-job fit, these charges can be defended by establishing the business necessity of the practices that led to disparate impact (Merrit-Haston & Wexley, 1983). Similar business necessity arguments could possibly be made for other types of fit. Research should be directed toward developing ways to assess person-organization fit and person-workgroup fit using systematic and legally defensible methods. Our discussions of organizational analysis and role analysis provide a starting place for this research. However, an additional step in this process will be educating and convincing the legal system that organizational analysis and role analysis are acceptable alternatives to job analysis. Currently, job analysis provides an important component in a business necessity defense.

A final limitation of the proposed model pertains most directly to person-workgroup fit and relates to the validity of many of the predictors that could be used to assess this fit. Although we are suggesting that person-workgroup fit may have validity in predicting work unit cooperation, citizenship behaviors, and attachment behaviors, to date no research has demonstrated this validity. The problem with workgroup fit is especially salient when selection decision makers are not members of the immediate work group. When recruiters or other member of the human resource department conduct much of the screening, evaluations of person-workgroup fit potential based on applicant-recruiter congruence may not be related fit with the actual workgroup. Additionally, it is important to develop assessment procedures beyond interpersonal preferences.

CONCLUSION

We have proposed a person-environment fit model of the selection process. Specifically, we proposed that there is a need to consider (1) person-job fit, (2) person-organization fit, and (3) person-workgroup fit in the selection process. Limited research has demonstrated that recruiters appear to use different types of fit. However, future research efforts are needed to identify the situations when each of these fit assessments has the optimum utility in predicting work outcomes and organizational effectiveness. Additionally, since each of the types of fit are open to cognitive biases in the selection process, research efforts are needed to determine structured ways of assessing fit in order to reduce these cognitive biases. Presently, more efforts have been made at reducing cognitive biases associated with person-job fit than the other types of fit. Until cognitive biases can be reduced with person-organization fit and person-workgroup fit, the utility of these types of fit may be limited. If the same amount of research that has been conducted over the past fifty years to establish procedures to determine person-job fit is conducted to investigate person-organization and person-workgroup fit, significant strides could be made to develop valid and unbiased selection procedures for all types of fit.

The proposed facet model is unique in the selection literature. Research commonly emphasizes processes that can be used to create person-job fit (Arvey & Faley, 1988; Guion, 1991), or describes the cognitive decision processes used to screen and select job applicants (Dipboye, 1992). In comparison to the person-job fit approach, person-organization and person-workgroup fit approaches share a common theoretical foundation with the person-environment fit literature (Pervin, 1989). These approaches assume that a given individual can manage the demands of some environments more effectively than the demands of other environments. Thus, there is a need to select applicants based on their capabilities in managing a work environment effectively.

An important issue that warrants exploration is the utilization of the three types of fit in the selection process. Based on the propositions, we are suggesting that the relative importance of the three types of fit is likely to be contingent upon the work context. That is, employers need to identify the type or types of fit that are most essential for the work environment for which a new hire is being sought. As Gilson (1924) reported earlier, employers undergoing major cultural transformations may need to emphasize person-organization fit over person-job fit. Thus, the dominant type of fit being sought may drive the selection process. When a group of applicants has similar qualities on the dominant type of fit, then the unexamined types of fit may be used to discern the most appropriate hire given the total work context. While all types of fit have some degree of importance in the selection process, it is likely that employers will view the different types of fit in a sequential or multiple hurdles manner (Judge & Ferris, 1992).

The proposed model differs from the person-job fit literature in that it considers a broader predictor domain for establishing a person-work environment fit and a broader criterion domain to assess applicant capabilities to manage a work environment effectively. The expanded focus of this model makes the selection process considerably more complex than traditional selection processes. Research efforts should systematically define the nature of the work environment and then identify relevant personal attributes to achieve a fit with this environment. Conducting traditional job analyses alone is inadequate to assess the work environment. Standard selection procedures such as job applications and interviews may need to be expanded to acquire additional information about personal attributes. This complexity suggested by the proposed model should ultimately result in better predictors that identify effective employees and improve selection utility.

ACKNOWLEDGMENT

The authors would like to thank Bob Liden for his comments on an earlier draft of this manuscript.

REFERENCES

Adkins, C., Russell, C., & Werbel, J. (1994). *On the construct validity of organizational fit.* Paper presented to Society of Industrial and Organizational Psychologists, San Francisco.
American School of Correspondence. (1919). *Employment management and Safety Engineering.* Chicago: American School of Correspondence.
Arvey, R.D., & Faley, R.H. (1988). *Fairness in selecting employees.* Reading, MA: Addison-Wesley.
Barley, S.R., Meyer, G.W., & Gash, D.C. (1988). Cultures of culture: Academics, practitioners and the pragmatics of control. *Administrative Science Quarterly, 33,* 24-60.
Barnard, C. (1938). *The functions of the executive.* Cambridge, MA: Harvard University Press.
Barr, S.H., & Hitt, M.A. (1986). A comparison of selection decision models in manager vs student samples. *Personnel Psychology, 39,* 599-617.
Barrick, M.R., & Mount, M.K. (1991). The big five personality dimensions and job performance: A metaanalysis. *Personnel Psychology, 44,* 126.
Barrick, M.R., & Mount, M.K. (1993). Autonomy as a moderator of the relationships between the big five personality dimensions and job performance. *Journal of Applied Psychology, 78,* 111-118.
Behling, O. (1998). Employee selection: Will intelligence and conscientiousness do the job? *Academy of Management Executive, 12,* 77-86.
Bird, A. (1994). Careers as repositories of knowledge: A new perspective on boundaryless careers. *Journal of Organizational Behavior, 15,* 325-344.
Borman, W.C., & Motowildo, S.J. (1993). Expanding the criterion domain to include elements of contextual performance. In N. Schmitt & W.C. Borman (Eds.), *Personnel selection in organizations* (pp. 71-98). San Francisco: Jossey-Bass.
Bowen, D.E., Ledford, G.E., & Nathan, B.R. (1991). Hiring for the organization: Not the job. *The Executive, 5,* 35-51.

Boxx, W.R., Odum, R.Y., & Dunn, M.G. (1991). Organizational values and value congruency and their impact on satisfaction, commitment, and cohesion. *Public Personnel Management, 20*, 195-205.
Bratton, J. (1993). Cellular manufacturing: Some human resource implications. *The International Journal of Human Factors in Manufacturing, 3*, 381-399.
Bretz, R.D., & Judge, T.A. (1994). Person-organization fit and the theory of work adjustment: Implication for satisfaction, tenure, and career success. *Journal of Vocational Behavior, 44*, 32-54.
Bretz, R.D., Rynes, S.L., & Gerhart, B. (1993). Recruiter perceptions of applicant fit: Implications for individual career preparation and job search behavior. *Journal of Vocational Behavior, 43*, 310-327.
Bridges, W. (1994, September 19). The end of the job. *Fortune*, pp. 62-74.
Buckley, M.R., & Russell, C.C. (1997). Meta-analytic estimates of interview criterion-related validity: A qualitative assessment. In R.W. Eder and M.M. Harris (Eds.), *The employment interview: Theory, research, and practice*. Beverly Hills, CA: Sage.
Campbell, J.P., & Pritchard, R.D. (1976). Motivation theory in industrial and organizational psychology. In M.D. Dunnette (Ed.), *Handbook of industrial and organizational psychology* (pp. 63-130). Chicago: Rand McNally.
Campion, M.A. (1978). Identification of variables most influential in determining interviewers' evaluation of applicants in a college placement center. *Psychological Reports, 42*, 947-952.
Cardy, R.L., & Dobbins, G.H. (1996). Human resource management in a total quality environment: Shifting from a traditional to a TQHRM approach. *Journal of Quality Management, 1*, 5-20.
Carlson, R.E. (1971). Effect of interview information in altering valid impressions. *Journal of Applied Psychology, 55*, 66-72.
Carson, K.P., & Stewart, G.L. (1996). Job analysis and the sociotechnical approach to quality: A critical examination. *Journal of Quality Management, 1*, 49-66.
Cascio, W.F. (1995). Whither industrial and organizational psychology in a changing world of work. *American Psychologist, 50*, 928-939.
Chatman, J.A. (1989). Improving interactional organization research: A model of person-organization fit. *Academy of Management Review, 14*, 333-349.
Chatman, J.A. (1991). Match people and organizations: Selection and socialization in public accounting firms. *Administrative Science Quarterly, 36*, 459-484.
Deal, T.E., & Kennedy, A.A. (1982). *Corporate cultures: The rites and rituals of corporate life*. Reading, MA.: Addison-Wesley.
Dienesch, R.M., & Liden, R.C. (1986). Leader-member exchange model of leadership: A critique and further development. *Academy of Management Review, 11*, 618-634.
Dipboye, R.L. (1992). *Selection interviews: Process perspectives*. Cincinnati, OH.: South-Western Publishing.
Dipboye, R.L., Arvey, R.D.,& Terpstra, D. (1977). Sex and physical attractiveness of raters and applicants as determinants of resume evaluation. *Journal of Applied Psychology, 69*, 288-294.
Dipboye, R.L., Fromkin, H.L., & Wiback, K. (1975). Relative importance of applicant sex, attractiveness, and scholastic standing in evaluation of job applicant resume. *Journal of Applied Psychology, 60*, 39-43.
Edwards, J.R. (1991). Person-job fit: A conceptual integration, literature, and methodological critique. *International Review of Industrial and Organizational Psychology, 6*, 283-357.
Edwards, J.R. (1994). The study of congruence in organizational behavior research: Critique and a proposed alternative. *Organizational Behavior and Human Decision Processes, 58*, 51-100.
Etzioni, A. (1975). *A comparative analysis of complex organizations*. New York: Free Press.
Fairhurst, G. T. (1993). The leader-member exchange patterns of women leaders in industry: A discourse analysis. *Communication Monographs, 60*, 321-351.
Ferris, G.R., & Judge, T.A. (1991). Personnel/human resource management: A political influence perspective. *Journal of Managment, 17*, 447-448.

Ferris, G.R., Youngblood, S.A., & Yates, V.L. (1985). Personality, training performance, and withdrawal: A test of the person–group fit hypothesis for organizational newcomers. *Journal of Vocational Behavior, 27*, 377-388.

Ford, R., & McLaughlin, F. (1986). Nepotism: Boon or bane. *Personnel Administrator, 31*, 79-89.

Gardner, P.D., Kozlowski, S.W.J., & Hults, B.M. (1991). Will the *real* prescreening please stand up. *Journal of Career Planning and Employment, 51*, 57-62.

Ghorpade, J., & Atchinson, T.J. (1980). The concept of job analysis: A review and some suggestions. *Public Personnel Management, 9*, 134-144.

Gifford, R.G., Ng, C.F., & Wilkinson, M. (1985). Nonverbal cues in the employment interview: Links between applicant qualities and interviewer judgements. *Journal of Applied Psychology, 70*, 729-736.

Gilson, M.B. (1924). Scientific management and personnel work. *Bulletin of the Taylor Society, 9*, 39-50.

Graen, G.B., & Schiemann, W. (1978). Leader-member agreement: A vertical dyad linkage approach. *Jorunal of Applied Psychology, 63*, 206-212.

Graves, L.M., & Karren, R.J. (1992). Interviewer decision processes and effectiveness: An experimental policy capturing investigation, *Personnel Psychology, 45*, 313-340.

Graves, L.M., & Powell, G.N. (1988). An investigation of sex discrimination in recruiters' evaluations of actual applicants. *Journal of Applied Psychology, 73*, 20-29.

Griggs vs. *Duke Power Company*. (1971). 3 FEP 175.

Guion, R.M. (1991). Personnel assessment, selection, and placement. In M.D. Dunette & L.M. Hough (Eds.), *Handbook of industrial and organizational psychology* (2nd ed., pp. 327-397). Palo Alto, CA: Consulting Psychologists Press.

Hakel, M.D., Dobmeyer, T.W., & Dunette, M.D. (1970). Relative importance of three content dimensions in overall suitability rating of job applicants' resumes. *Journal of Applied Psychology, 54*, 65-71.

Harris, M.M. (1989). Reconsidering the employment interview: A review of recent literature and suggestions for future research. *Personnel Psychology, 42*, 691726.

Heneman, H.G. (1977). Impact of test information and applicant sex on applicant evaluations in a selection simulation. *Journal of Applied Psychology, 62*, 524-526.

Hitt, M.A., & Barr, S.H. (1989). Managerial selection decision models: Examination of configural cue processing. *Journal of Applied Psychology, 74*, 53-61.

Hofstede, G. (1984). *Culture's consequences: International differences in work-related values*. Beverly Hills, CA: Sage.

Holland, J.L. (1985). *Making vocational choices: A theory of vocational personalities and work environments*. Englewood Cliffs, NJ: Prentice-Hall.

Hunter, J.E., & Hunter, R.F. (1984). Validity and utility of alternative predictors of job performance. *Psychological Bulletin, 96*, 7298.

Iaffaldano, M.T., & Muchinsky, P.M. (1985). Job satisfaction and job performance: A metaanalysis. *Psychological Bulletin, 97*, 251-273.

Jacob, R. (1995, April 3). The struggle to create an organization for the 21st century. *Fortune*, pp. 90-100.

Judge, T.A., & Ferris, G.R. (1992). The elusive criterion of fit in human resources staffing decisions. *Human Resource Planning, 15*, 47-64.

Klimoski, R.J., & Jones, R.G. (1995). Staffing for effective group decision making: Key issues in matching people and teams. In R. Guzzo & E. Salas (Eds.), *Team effectiveness and decision making in Organizations* (pp. 291-332). San Francisco: Jossey-Bass.

Kinnicki, A.J., & Lockwood, C.A. (1985). The interview process: An examination of factors recruiters use in evaluating job applicants. *Journal of Vocational Behavior, 26*, 117-125.

Kinnicki, A.J., Lockwood, C.A., Hom, P.W., & Griffeth, R.W. (1990). Interviewer predictions of applicant qualifications and interviewer validity: Aggregate and individual analyses. *Journal of Applied Psychology, 75*, 477-486.

Kristof, A. (1996). Person-organization fit: An integrative review of its conceptualization, measurement, and implications. *Personnel Psychology, 49*, 1-50.

Lin, T.R., Dobbins, G.H., & Farh, J.L. (1992). A field study of race and age similarity effects on interview ratings in conventional and situational interviews. *Journal of Applied Psychology, 77*, 363-371.

March, J.G., & Simon, H.A. (1958). *Organizations*. New York: Wiley.

McClane, W.E. (1991). The interaction of leader and member characteristics in the leader-member exchange model of leadership. *Small Group Research, 22*, 283-300.

McDaniel, M.A., Whetzel, D.L., Schmidt, F.L., & Maurer, S.D. (1994). The validity of employment interviews: A comprehensive review and metaanalysis. *Journal of Applied Psychology, 79*, 599-616.

Meglino, B.M., Ravlin, E.C., & Adkins, C.L. (1989). A work values approach to corporate culture: A field test of the value congruence process and its relationship to individual outcomes. *Journal of Applied Psychology, 74* 424-432.

Meritt-Haston, R., & Wexley, K.N. (1983). Educational requirements: Legality and validity. *Personnel Psychology, 36*, 743-754.

Montgomery, C.E. (1996, January). Organization fit is a key to success. *HRM Magazine*, pp. 94-96.

Muchinsky, P.M., & Monahan, C.J. (1987). What is person-environment congruence? Supplementary versus complementary models of fit. *Journal of Vocational Behavior, 31*, 268-277.

Oliphant, V.N., & Alexander, E.R. (1982). Reactions to resumes as a function of resume determinateness, applicant characteristics, and sex of raters. *Personnel Psychology, 35*, 829-842.

O'Reilly, C.A., Chatman, J.A., & Caldwell, D.M. (1991). People and organizational culture: A Q-sort approach to assessing person-organization fit. *Academy of Management Journal, 34*, 487-516.

Paunonen, S.V., Jackson, D.N., & Oberman, S.M. (1987). Personnel selection decisions: Effects of applicant personality and the letter of reference. *Organizational Behavior and Human Performance, 41*, 96-114.

Pervin, L.A. (1989). Persons, situations, interactions: The history of a controversy and a discussion of theoretical models. *Academy of Management Review, 14*, 350-360.

Rasmussen, K.G. (1984). Nonverbal behavior, verbal behavior, resume credentials, and selection interview outcomes. *Journal of Applied Psychology, 69*, 551-556.

Ree, M.J., & Earles, J.A. (1992). Intelligence is the best predictor of job performance. *Psychological Science, 1*, 86-89.

Renwick, P.A., & Tosi, H. (1978). The effects of sex, marital status, and educational background on selection decisions. *Academy of Management Journal, 21*, 93-103.

Roethisberger, F.J., & Dickson, W.J. (1933). *Management and the worker*. Cambridge, MA: Harvard University Press.

Rothstein, M., & Jackson, D.M. (1980). Decision making in the employment interview: An experimental approach. *Journal of Applied Psychology, 65*, 271-283.

Rothstein, H.R., Schmidt, F.L., Erwin, F.W., Owens, W.A., & Sparks, C.P. (1990). Biographical data in employment selection: Can validities be made generalizable. *Journal of Applied Psychology, 75*, 175-184.

Rousseau, D.M. (1985). Issues of level in organizational research: Multi-level and cross level perspectives. In L.L. Cummings & B.M. Staw (Eds.), *Research in organizational behavior* (Vol. 7, pp. 1-38). Greenwich, CT: JAI Press.

Rynes, S., & Gerhart, B. (1990). Interviewer assessments of applicant "fit": An exploratory investigation. *Personnel Psychology, 43*, 13-35.

Scarpello, V., & Campbell, J. P. (1983). Job satisfaction and the fit between individual needs and rewards. *Journal of Occupational Psychology, 56*, 315-328.

Schmitt, N., Gooding, R.Z., Noe, R.A., & Kirsch, M. (1984). Metanalyses of validity studies published between 1964 and 1982 and the investigation of study characteristics. *Personnel Psychology, 37*, 407422.

Schmitt, N., & Landy, F.J. (1992). The concept of validity. In N. Schmitt & W.C. Borman (Eds.), *Personnel selection in organizations* (pp. 275-309). San Francisco: Jossey-Bass.

Schneider, B. (1987). The people make the place. *Personnel Psychology. 40*, 437-454.

Schneider, B., Goldstein, H.W., & Smith D.B. (1995). The ASA framework: An update. *Personnel Psychology, 48*, 747-773.

Scott, S. (1993). *The influence of climate perceptions on innovation behavior.* Unpublished Ph.D. dissertation, Department of Management, University of Cincinnati.

Seers, A. (1989). Team member exchange quality: A new construct for role making research. *Organizational Behavior and Human Decision Processes, 43*, 118-135.

Settoon, R.P., Bennett, N., & Liden, R.C. (1996). Social exchange in organizations: Perceived organizational support, leader-member exchange, and employee reciprocity. *Journal of Applied Psychology, 81*, 219-227.

Singer, M.S., & Bruhns, C. (1991). Relative effect of applicant work experience and academic qualification on selection interview decisions: A study of between-sample generalizability. *Journal of Applied Psychology, 76*, 550-568.

Sonnenfeld, J.A., & Peiperl, M.A. (1988). Staffing as a strategic response: A typology of career systems. *Academy of Management Review, 13*, 588-600.

Stevens, C.K., & Kristof, A.L. (1995). Making the right impression: A field study of applicant impression management during job interviews. *Journal of Applied Psychology, 80*, 587-606.

Stevens, M.J., & Campion, M.A. (1994). The knowledge, skill, and ability requirements of teamwork: Implications for human resource management. *Journal of Management, 20*, 503-530.

Taylor, W. (1911). *The principles of scientific management.* New York: Harper.

Tullar, W.L. (1989). Relational control in the employment interview. *Journal of Applied Psychology, 74*, 971-978.

Uniform Guidelines. (1978, August 25). *Federal Register*, pp. 38295-38309.

US Department Of Labor. (1991). *What work requires of schools.* Washington, DC.

US Department of Labor. (1992). *Learning for a living: A blueprint for high performance.* Washington, DC.

Van Maanen, J. (1975). Police socialization: A longitudinal examination of job attitudes in an urban police department. *Administrative Science Quarterly, 20*, 207-228.

Vancouver, J.B., & Schmitt, N.W. (1991). An exploratory examination of person-organization fit: Organizational goal congruence. *Personnel Psychology, 44*, 333—352.

Waldron, V.R. (1991). Achieving communication goals in superior-subordinate relationships: The multi-functionality of upward maintenance tactics. *Communication Monographs, 58*, 289-306.

Wayne, S.J., Shore, L.M., & Liden, R.C. (1997). Perceived organizational support and leader-member exchange: A social exchange perspective. *Academy of Management Journal, 40*, 82-111.

Werbel, J.D., & Looney, S. (1994). The use of selection criteria for campus interviews. *International Journal of Selection and Assessment, 2*, 28-36.

Wingrove, J., Glendinning, R., & Herriot, P. (1984). Graduate pre-selection. A research note. *Journal of Occupational Psychology, 57*, 169-171.

Zenger, T.R., & Lawrence, B.S. (1989). Organizational demography: The differential effects of age and tenure distribution on technical communication. *Academy of Management Journal, 32*, 353-376.

LIFE EXPERIENCES AND PERFORMANCE PREDICTION:
TOWARD A THEORY OF BIODATA

Michelle A. Dean, Craig J. Russell, and
Paul M. Muchinsky

ABSTRACT

Fleishman (1988) described biodata selection technology as among the most promising avenues for generation of new knowledge in personnel selection. However, researchers generally hold biodata selection systems in low regard due to their perceived atheoretical nature. Further, surveys indicate biodata is used in less than 5% of personnel selection decisions. We argue that biodata systems are no more atheoretical than other popular selection technologies. We review aspects of biodata instruments that make them unique among selection devices and biodata theory as embodied in the ecology model (Mumford, Stokes, & Owens, 1990) before offering two extensions. First, we propose to extend the ecology model by focusing on negative life events, reviewing diverse literatures addressing affective and cognitive reactions to these events. Second, an individual difference variable labeled "moxie" is proposed as a key mediator and/or moderator of latent negative life event-job performance relationships. Specific directions for needed research are presented.

INTRODUCTION

Performance prediction is one of the fundamental tasks of management. Specifically, managers align raw materials, capital, and human resources with market opportunities in ways that yield targeted performance outcomes. Performance predictions made in the context of personnel selection systems thus become a primary contribution of human resource managers. Similarly, development of theories or models that provide insight into why and when applicants are able to perform on the job becomes a primary objective of human resource management research.

The literature on performance prediction suggests three selection technologies tend to achieve the highest predictive power or criterion validity across situations: cognitive ability tests, work sample tests (e.g., assessment centers), and biographical information inventories (hereafter referred to as biodata). Ample meta-analytic evidence exists to support use of these technologies in a wide variety of personnel selection settings (Reilly & Warech, 1990; Russell & Dean, 1994b; Schmitt, Gooding, Noe, & Kirsch, 1984). However, only an extremely small portion of the research literature has been devoted to explaining why these technologies demonstrate criterion validity.

Hunter (1986) speculated that paper and pencil tests of general cognitive ability (g) demonstrate criterion validity due to controlled cognitive processing requirements found on the job, consistent with his finding that g criterion validities increase with job complexity. To be sure, an immense literature yields insight into the nature and development of cognitive ability (e.g., Kanfer, 1990; Kanfer & Ackerman, 1989). The same cannot be said for insight into latent causal processes between an individual's cognitive ability and subsequent work performance. It remains somewhat humbling to inform a lay audience that "we just recently have been able to conclude that smart people do better on the job, though we aren't really sure why."

Work sample and simulation criterion validities have led investigators to generate competing explanations for latent causal processes. For example, Klimoski and Strickland (1976) identified numerous competing explanations for assessment center criterion validity. Unfortunately, Klimoski and Brickner (1987) were able to generate an even larger set of competing explanations in the virtual absence of any systematic tests of the issues during the preceding 10 years (see Russell & Domm, 1995, for a test of two competing explanations). The absence of compelling theory becomes most troublesome when the focus of work sample information shifts from performance prediction to training needs assessment; that is, if we do not know what we are measuring we cannot infer what interventions (e.g., training or otherwise) might be appropriate. Note that a discussion of a common explanation of assessment center criterion validity, that is, the "consistency principle" (Wernimont & Campbell, 1968), and its shortcomings appears somewhat later in the paper.

It should not be surprising then, that the third selection technology (i.e., biographical information inventories) can also be described as lacking strong theory. What might be surprising is the verve with which labels such as "dustbowl empiricism" have been selectively applied to this technology when other technologies provide equally viable targets (Childs & Klimoski, 1986; Dunnette, 1962; Nickels, 1994; Owens, 1976). Common use of pejorative labels in the biodata literature (e.g., "atheoretical" and "dustbowl empiricism") may explain its low usage rates among practitioners; of 348 firms surveyed, Hammer and Kleinman (1988) found only 6.8% had ever used biodata in employment decisions and only 0.4% currently used biodata. Notably, 40% of Hammer and Kleinman's respondents indicated invasion of applicant privacy was a primary concern in biodata use. Similarly, a Bureau of National Affairs survey of human resource specialists reported only 4% used biodata. The human resource specialists indicated perceived invasiveness of applicant privacy as a major reason to avoid biodata selection systems (as reported in Mael, Connerley, & Morath, 1996). However, while most consider an individual's performance on standardized tests to be "personal" information,[1] at the item level few would consider the question "What does 2 + 2 sum to?" as invasive as "How many magazines did your parents subscribe to while you were in high school?"[2]

We suspect a number of unique characteristics contribute to how often authors reference biodata's "atheoretical" nature. For example, investigators have only recently started to employ traditional psychometric construct validity techniques in biodata item development (Mumford, Costanza, Connelly, & Johnson, 1996; Russell, 1994). An early emphasis on sorting biodata items into taxonomies did not evolve into a traditional psychometric assessment of construct validity. Further, empirical keys are common in biodata applications, generally devoid of theoretical rationale, and virtually unheard of in other selection technology arenas.

Regardless, Fleishman (1988) argued that the greatest opportunities for advancing our understanding of human performance in organizations lie in examining relationships between life experiences and subsequent job performance. Fleishman was not alone in his observations, they have been echoed over the years by many others (Dunnette, 1962, 1966; Guion, 1965; Owens, 1976). Biodata has likely been viewed as so promising due to the (as yet unrealized) potential of a biodata theory to guide life experience interventions (i.e., the purposeful, theory-guided exposure of individuals to certain quantities and qualities of life experiences).

The primary purpose of this paper is to review existing biodata theory development and propose a partial extension. Our approach builds on prior efforts as a first step toward specification of life experience interventions hypothesized to causally influence subsequent performance. Toward these ends, the remainder of this paper is structured in three sections. First, we review characteristics of biodata unique to this performance prediction technology, including efforts at developing biodata item taxonomies and empirical keying procedures. Fleishman's (1988) vision of biodata's "promise" will only be realized with an increase in research activity, which will require an understanding of unique biodata characteristics by a broader research audience. Second, we describe development of existing biodata theory. Third, we present an extension to this theory which specifies an additional life history construct domain and a relatively unexamined personality characteristic. A secondary purpose in reviewing biodata characteristics and theory is to provide a single source of such information for those interested in contributing to the body of research.

UNIQUE ASPECTS OF BIODATA PERFORMANCE PREDICTION TECHNOLOGY

Biodata Item Types

A number of authors have taken it upon themselves to develop taxonomies of biodata item "types" or characteristics. Perhaps most well known is Asher's (1972) taxonomy, initially offered as a means of "improving" biodata items. "Improvement" was defined in terms of eliminating potential sources of systematic error rather than increasing scale convergence with some latent construct domain, an important but unfortunate distinction that appears throughout the biodata literature. While other performance prediction procedures focus on latent constructs underlying the predictor measures (e.g., cognitive ability and Big-5 personality measures), biodata item taxonomies tend to focus on minimizing sources of prediction error.

Biodata item characteristics discussed by Asher (1972) included:

- Verifiable vs. unverifiable
- Subjective vs. objective
- Historical vs. futuristic
- Actual behavior vs. hypothetical behavior
- Memory based vs. conjecture based
- Specific vs. general
- Response vs. response tendency
- External vs. internal event

} Hard vs. soft (joint verifiable & factual vs. unverifiable & interpretive)

While Asher's (1972) taxonomy was originally presented as a *description* of biodata item characteristics, subsequent authors have used these and other characteristics in a *prescriptive* manner (e.g., Gandy, Outerbridge, Sharf, & Dye, 1989).

A number of research efforts have examined taxonomic characteristics. For example, Owens, Glennon, and Albright (1966) found that consistency in subjects' responses to biodata items was related to item brevity, the presence of an escape option, and phrasing in a neutral or pleasant connotation. Mael (1991) recently echoed Asher's (1972) earlier suggestions that some biodata item attributes aid in obtaining accurate responses. Mael suggested biodata items should ask for respondent's first-hand knowledge, avoiding asking individuals about how others would evaluate the respondent. For example, asking, "How did your parents evaluate your academic achievement?" would be second-hand information in which the respondent is asked to speculate. Verifiable items have been found to reduce the likelihood that applicants will provide bogus answers in hopes of achieving higher scores (Atwater, 1980; Cascio, 1975; Mosel & Cozan, 1952). However, Hough, Eaton, Dunnette, Kamp, and McCloy (1990) found that simply warning respondents that answers can be verified may act as a faking deterrent regardless of item content.

Mael (1991) also considered legal and moral issues surrounding the use of biodata. Items may vary in terms of controllability, accessibility, and visible job relevance. Controllability refers to the degree to which a person chose to perform or not to perform an action. This label parallels Owens and Schoenfeldt's (1979) distinction between prior behaviors (i.e., behaviors that were chosen by the applicant) and input variables (i.e., things that happened to the applicant). Mael suggested all life events, whether consciously chosen or not, have the ability to shape a person's future behavior, and should be included on a biodata instrument.

Opposition to including noncontrollable items such as parental behavior and socioeconomic status typically speaks to the "fairness" of these items, due to applicants lack of control over their early environment. This leads to a potential legal question regarding whether items should tap skills and experiences that are equally accessible to all applicants (Stricker, 1987; 1988). Stricker (1988) suggested items asking about experience as a football team captain would be unfair because individuals of a particular gender, size, or size of school may not have the opportunity to engage in this role. In contrast, Mael (1991) suggested the concept of equal access is irrelevant to prediction or test fairness as operationalized by the *Uniform Guidelines on Employee Selection Procedures* (Equal Employment Opportunity Commission, 1978). Rather, what is relevant is that the person who had access to the role was changed in some meaningful way by the role while others who were not in-role received no benefit nor harm (i.e., individuals who were not football captains are not penalized for non-exposure).

Finally, biodata items vary in visible job relevance. Many view all life experiences as potentially developmental. However, Gandy et al. (1989) used only items with a point-to-point relationship between item content and job content on their public sector biodata instrument in order to avoid the accessibility issue and maximize job relevance.

Research has examined the influence of item attributes on biodata results. Barge and Hough (1988) analyzed biodata items in terms of item heterogeneity (i.e., degree to which items measure more than one construct), behavioral discreteness (i.e., degree to which items address "a single, perhaps verifiable behavior rather than a more abstract or summary characteristic" pp. 3-4), and behavioral consistency (i.e., degree of congruency between the content domain of the biodata item and the content domain of the target job, or the degree to which an item is a sign v. sample of behavior, Wernimont & Campbell, 1968). Barge (1988) analyzed 103 items taken from Owens and Schoenfeldts' (1979) Biographical Questionnaire (BQ) and found more homogenous items (i.e., in terms of consistency and discreteness) demonstrated higher criterion validity. In a different vein, Shaffer, Saunders, and Owens (1986) found "soft," or subjective, non-verifiable items, are nearly as predictive and reliable as "hard," or verifiable, factual items. Average test-retest reliability five years after initial administration was higher for objective than subjective items. The Barge and Shaffer et al. efforts are noteworthy because they are among the first to evaluate the influence of biodata item characteristics on criterion validity (Stokes & Reddy, 1992).

However, Asher's (1972), Mael's (1991), and others' item taxonomies, like all taxonomies, can be evaluated in terms of (1) ability to reduce variance in measurement of the latent construct domain, and (2) prediction of some criterion (e.g., job performance) of interest (Fleishman & Quaintance, 1984; McCall & Bobko, 1990). Biodata item taxonomies tend to fall short in terms of the former due to emphasis on the latter; indeed, most biodata research has focused on enhancement of criterion validity with a distinct lack of attention to latent construct domains. Decreasing the presence of random and systematic measurement error in subjects' responses will certainly contribute to biodata criterion validity and ease of use. Further, taxonomic efforts pinpoint the types of items typically found on a biodata instrument and differentiate biodata items from other closely related, but conceptually distinct, measures such as personality scales. However, there is no theoretical basis for "correctness" of any of these attributes. The attributes need to be understood in terms of their contribution to measurement of some latent construct domain. While biodata taxonomies may prove useful for generating some theory or model of inter-related construct domains, item taxonomies cannot be considered theory itself (Bacharach, 1989). Unfortunately, investigations conducted in the 26 years since Asher's (1972) study suggested taxonomic efforts have not lead to an understanding of latent predictor constructs and causal processes. It would seem that such insight will not be forthcoming without a shift in discussion of "biodata item types" to a discussion of latent biodata construct domains.

Empirical Keying Procedures

Empirical keys constitute a second unique aspect of biodata performance prediction systems. Empirical keying procedures assign "optimal" weights to items or response options for purpose of performance prediction. Subject responses are typically multiplied by the respective weight associated with each item/response option and summed into a "biodata score." Weights are derived in ways that reflect (1) some empirical relationship between the item/response option and the criterion of interest (externally referenced) or (2) empirical relationships among the item/response options (internally referenced). Empirical keys in many ways might be considered a "defining characteristic" of typical biodata systems because no other behavioral science applications in organizational settings use similar scoring procedures. For example, no published research breaks down Likert scale job satisfaction items into dummy coded (0 = did not select, 1 = did select) response options before determining the optimal weighting required to predict job performance. Indeed, the statistical sophistication associated with empirical keying procedures may serve as a "barrier to entry" for many investigators and practitioners. Empirical keying is one of the primary reasons for criterion validities summarized in the meta-analytic estimates reported in Table 1 (Mitchell & Klimoski, 1982).

A wide variety of empirical keying methods exist. Unfortunately no single study has compared all methods in a single data set. Further, even if such a study existed, its findings would not generalize beyond the biodata items and latent criterion construct domains (whatever they might be) captured in that study. A brief review of the most frequently used externally referenced methods (i.e., where weights are a function of biodata-criterion relationships) is presented below, including the vertical percent difference, horizontal percent difference, mean criterion, phi coefficient, and regression methods. One internally referenced method (i.e., where weights are a function of relationships among responses to biodata items) is also discussed, as are a few emerging methods.

The vertical percent difference method (Stead & Shartle, 1940) historically appears to be most frequently used in empirical biodata key construction (Mumford & Owens, 1987). This method calculates differences in response percentages for high- and low-criterion groups in a three-step process applied to a "key development" portion of the original sample. First, high- and low-criterion groups are established using individuals who are most or least successful on the criterion of interest. Second, the percentage within each of the extreme criterion groups who chose each item response option is computed. Third, a difference in percentages for each alternative is calculated by subtracting the percentage for the low-criterion group from the percentage for the high criterion group. Strong's (1926) tables (see Stead & Shartle, 1940, p. 255) can then be used to transform percentage point differences to net weights, though alternate weighting procedures are common (e.g., simple unit weights when some arbitrarily large percentage point difference is present). Once derived in the key development sample, subject responses in the remaining portion of the original sample, called the cross validation sample, are multiplied by the respective weights, summed, and correlated with the target criterion.

Table 1. Meta-analyses of Biodata Criterion-Related Validities

Study	Criterion	K	N_i	$\bar{r}$	σ_r^2	σ_e^2	σ_ρ^2	ρ
Hunter and Hunter (1984)								
	Supervisor ratings	12	4429	-	-	-	.10	.37
	Promotion	17	9024	-	-	-	.10	.26
	Training success	11	6139	-	-	-	.11	.30
	Tenure	2	2018	-	-	-	.00	.27
Re-analysis of Dunnette (1972)	-	115	-	-	-	-	-	.38
Re-analysis of Reilly and Chao (1982)	-	44	-	-	-	-	-	.34
Re-analysis of Vineberg & Joyner (1982)	Supervisor ratings Global	12	-	-	-	-	-	.20[a]
	Supervisor ratings Suitability	4	-	-	-	-	-	.29[a]
	Supervisor ratings All ratings	16	-	-	-	-	-	.24[a]
Reilly and Chao (1982)								
	Tenure	13	5721	-	-	-	-	.32
	Training	3	569	-	-	-	-	.39
	Ratings	15	4000	-	-	-	-	.36
	Productivity	6	661	-	-	-	-	.46
	Salary	7	680	-	-	-	-	.34

Schmitt, Gooding, Noe, and Kirsch (1984)

Across all criteria	99	58107	.243	.0183	.0015	.0168	-
Performance ratings	29	3998	.317	.0357	.0059	.0298	-
Turnover	28	28862	.209	.0144	.0009	.0136	-
Achievement/grades	9	1744	.226	.0784	.0047	.0738	-
Productivity	19	13655	.203	.0036	.0013	.0023	-
Status change	6	8008	.332	.0014	.0006	.0009	-
Wages	7	1544	.525	.0157	.0024	.0133	-

Russell and Dean (1994b)
(update of Schmitt et al., 1984, all criterion validities reported in *JAP* and *Personnel Psychology* from 1965-1992)

Across all criteria	107	59172	.291	.0151	.0015	.0136	-

Note: Meta-analytic estimates of ρ_{xy} where corrected *only* for sampling error.

Perhaps the most flexible application of the vertical percent difference method is captured by a custom keying procedure developed originally by Abrahams (1965) and subsequently crafted into a computer program labeled "KEYCON" at the U.S. Navy Personnel Research and Development Center (Abrahams, 1998). The KEYCON program permits the investigator to systematically vary the following parameters:

1. The number of subjects randomly selected for the key development versus cross validation samples.
2. The cut points for determining "high" and "low" performing criterion groups in the key development sample.
3. The percent difference in response frequency between high and low performing criterion groups required for a response option to enter the key.
4. The weight associated with each response option entering the key (e.g., use of unit weights for all percentage differences larger some specified level, use of percentage difference as the weight, etc.).

Thus, an investigator can explore the tradeoff between increasing key development sample size (increasing expected reliability of the key) while decreasing cross validation sample size (increasing expected sampling error in estimating criterion validity), and vice versa. Programs such as KEYCON greatly facilitate investigator judgment calls in construction of vertical percent difference empirical scoring keys.

The horizontal percent difference method (Guion, 1965; Stead & Shartle, 1940) also arrives at weights for each response option by identifying subjects in a key development sample who were high on the performance criterion. Response option weights are derived by dividing the number of subjects selecting a particular response option from the high criterion group by the total number of subjects. Weights for response options selected by each subject in the cross validation sample are summed to form a biodata score. A criterion validity estimate is derived by correlating the key-based biodata score with the criterion in the cross validation sample.

The mean criterion technique (Devlin, Abrahams, & Edwards, 1992) does not require use of high- and low-criterion groups, rather weighing each response option by the average criterion score of respondents in the key-development sample who chose that response option. The key is then applied to a cross validation sample. The correlation between the biodata score and criterion in the cross-validation sample estimates criterion validity in the population.

Several correlation/regression methods of empirical keying are also available, including the phi coefficient method, regression technique, rare response method (Neidt & Malloy, 1954), and multiple regression method (Malone, 1977). All typically develop empirical weights in a key development sample before estimating criterion validities in a cross validation sample.

The phi coefficient method (Lecznar, 1951; Lecznar & Dailey, 1950) uses correlations between each item response option (response versus no response) and a binomial criterion as response option weights. A variation of this method uses the point biserial correlation coefficient when the criterion is continuous. Regression techniques are similar in that standard regression analysis coefficients weight items or response options on the basis of their ability to add to the prediction obtained from items already in the regression equation (Malone, 1977). Interpretation of regression weights derived for each item is subject to typical restrictions associated with common least square linear regression procedures. However, when response options are used as independent variables, the parametric assumption of independence among the predictors is violated (e.g., subjects selecting response option "1" on a question by definition did not select response option "2"), and interpretation of regression weights becomes impossible.

Russell and Dean (1994a) used a variation of the phi coefficient method in which point biserial coefficients between response options and the criterion were derived. However, they noted when distributional characteristics of two variables do not conform with parametric assumptions, distributional characteristics of phi, point biserial, or Pearson product moment correlations between the two variables are unknown and no probablistic inferences can be drawn. As a result, two response options with equal point biserial correlations with a criterion might have different standard errors (Dean, Russell, & Broach, 1998). For example, $s_{r_{pb}}$ is expected to be the same for two response options with equal r_{pb} if assumptions of bivariate normality hold (e.g., Fischer, 1970 found the best estimate of

$$\sigma_\rho = s_r = \frac{\sqrt{(1-r^2)}}{N-2}.$$ However, r_{pb} between response options and performance criteria cannot exhibit bivariate normality because one of the two variables (i.e., the response option) is discrete and usually highly skewed. In this instance, true values of $\sigma_{r_{pb}}$ might vary across response options yielding the same estimate of ρ_{pb}. Russell and Dean (1994a) argued the estimated criterion-related validity of a response option with a smaller $s_{r_{pb}}$ is less subject to random sampling error and should receive more weight in the key (e.g., by dividing r_{pb} by a bootstrap estimate of $s_{r_{pb}}$).

Consequently, Russell and Dean (1994a) developed an empirical key based on both estimates of effect size (r_{pb}) and accuracy of those estimates ($\sigma_{r_{pb}}$), using the bootstrap technique to estimate the population value of $\sigma_{r_{pb}}$. They found marginal increases in criterion validity using response options weighted by r_{pb} and $\sigma_{r_{pb}}$. Further, examination of bootstrapped frequency distributions of r_{pb} indicated that a single population value of ρ_{pb} was unlikely for a number of response options. Subsequent analyses by Brown (1995) indicated meaningful differences in ρ_{pb}.

between applicants from urban versus rural areas. Addition of this configural relationship (i.e., moderation of ρ_{pb} by urban versus rural applicant point of origin) incrementally increased criterion validity.

Telenson, Alexander, and Barrett (1983) proposed the rare response method in which item responses are weighted based on how few individuals select a given response option. This method differs from others discussed in this section in that it does not weigh response options by their criterion relationships. Instead, this method derives weights using all candidates who completed the biodata inventory regardless of whether subsequent criterion measures were available. Infrequently endorsed responses are assigned larger weights than frequently chosen responses because infrequent responses were presumed to convey more information regarding individual standing on a particular characteristic (Telenson et al., 1983). Webb (1960) and Malloy (1955) both found rare response procedures added more to prediction than a criterion-based empirical key, though Webb (1960) suggested the predictive power of deviant response keys may not hold up under cross-validation. This procedure is rarely used in biodata practice, though Smith and McDaniel (1998) recently used it to infer which subjects were giving nonsense answers.

Recently, Brown and McManus (1994) investigated an application of classification-regression tree procedures in key development. The statistical technique has its roots in Automatic Interaction Detection (AID) procedures developed by Morgan and Sonquist (1963) and is described in detail by Breiman, Friedman, Olshen, and Stone (1984). The readers are referred to Breiman et al. (1984) for a thorough discussion of this technique, though for brevity we provide a simplified description that permits comparison to other keying methods. Specifically, regression tree applications are sensitive to the existence of linear, nonlinear, and configural (interactive) relationships between biodata items/response options and the criterion. All other empirical keying procedures at best only capture linear and nonlinear predictor-criterion relationships. Brown and McManus (1994) reported a comparison of this procedure to more traditional empirical keying procedures in a large sample of applicants for life insurance sales positions. Findings failed to demonstrate a significant or meaningful difference in estimates of population criterion validity. Again, recall that any such comparisons cannot be generalized beyond the specific biodata items and criterion construct domain used in that comparison. It remains to be seen whether procedures using configural predictor combinations incrementally increase prediction or understanding at a level that justifies their complexity and cost.

Empirical Key Comparisons

While inferences are necessarily limited to the biodata items, population, and criterion domains examined, a handful of studies have directly compared alternate keying procedures. Malone (1977) compared two regression methods and the vertical percent difference keying method, finding significantly higher cross-validities for the regression methods in predicting absenteeism and no difference between methods when predicting tenure. Aamodt and Pierce (1987) found the vertical percent method superior to the rare response method, though this study failed to report cross-validities. However, in analyses of multiple biodata selection applications in the U.S. Navy, Devlin et al. (1992) found different levels of shrinkage across empirical procedures. Hence, no firm conclusion can be drawn from Aamodt and Pierce's results. Telenson et al. (1983) compared vertical percent, horizontal percent, and rare response methods, finding only one rare response cross-validity to be significant, and none of the horizontal percent or vertical percent methods yielding significant results. Telenson et al. (1983) remains one of the few studies reporting evidence of criterion validity using the rare response method.

Devlin et al. (1992) compared nine empirical keying procedures in terms of shrinkage and cross-validity. The mean criterion method yielded the highest validities in keying samples but also the greatest amount of shrinkage. Devlin et al. found vertical percent methods to generally yield the highest cross-validities. The lowest cross-validities were obtained from the rare response and mean criterion techniques. Devlin et al. (1992) suggested non-criterion-based procedures, such as the rare response method, are unlikely to be useful.

In sum, methods that directly estimate strength of relationships between response options and criteria consistently yield higher criterion validities in cross validation samples. Mitchell and Klimoski (1982) noted that theory-based[3] keys will necessarily yield lower expected criterion validities unless the theory or model fully specifies all causal influences. When performance prediction is the primary goal, the vertical percent difference method seems to capture latent linear and nonlinear predictor-criterion relationships that maximize criterion validity (Devlin et al., 1992). However, advanced configural keying procedures involving classification/regression trees and neural networking hold promise for incremental increases in predictive power (Brown, 1994).

Implications for theory have rarely been explored in the application of empirical keying procedures. In one example, Russell and Domm (1990) reported items loading on a "negative life event" factor tended to have middle-range response options (i.e., 2 through 4 on a 1 to 5 point Likert scale) enter a vertical percent difference-based empirical key. They noted this was consistent with findings from the goal setting literature, i.e., that moderately difficult goals tend to yield higher performance than too easy or difficult goals. Unfortunately, empirical keying remains an under-traveled bridge between the classic prediction versus explanation issues. A strong theory linking life experiences to future job performance should result from a combination of (1) grounded theory building (Glaser & Strauss, 1967) from inspection of empirical key derived weights, *and* (2) examination of theory-based predictions regarding which response options are likely to enter empirical keys are needed. We now turn to a brief review of theories and models put forth to explain biodata criterion validities before presenting our extension.

BIODATA THEORY

Whereas labels such as "dustbowl empiricism" and "atheoretical" have followed biodata since its inception, meaningful efforts *have* been mounted to develop explanatory models. Our extension is best understood after tracing the development of these efforts from their roots in the "consistency principle" through more recent nomological propositions.

The Consistency Principle Revisited

The consistency principle has been routinely called upon as an explanation of biodata, work sample, and assessment center criterion validities (Klimoski & Brickner, 1987; Russell & Domm, 1995; Wernimont & Campbell, 1968). Owens (1976) stated "one of our most basic measurement axioms holds that the best predictor of what a man will do in the future is what he has done in the past" (p. 625). It can be expressed as a simple time series forecast:

$$Y_t = f(Y_{t-1})$$

where human performance (Y) at time t is a function of performance at time t-1.

The consistency principle provides an elegant and quick answer to the question "Why do biodata selection systems work?" Unfortunately, this is where the discussion often ends even though the consistency principle does not *explain* anything. A closer look at exactly what the consistency principle implies should make this more clear.

Specifically, the state of economic and psychometric assessment indicates we cannot measure human performance (Y) without error, hence:

$$\hat{Y}_t = f(Y_{t-1} + e_1) \text{, and}$$

$$\hat{Y}_t = f[(Y_{t-1} + e_1) + e_t] \text{ .}$$

Importantly, measurement error cumulates as the time between performance measures increases; that is, Y_t predictions about Y_{t+50} are much less accurate than predictions about Y_{t+1}. Both predictions are attenuated by the sum $\sum_{t=1}^{k} e_t$, though k = 1 for the former prediction, while k = 50 for the latter. As measurement error occurs at each point in time, the further apart in time two performance measures

Life Experiences 259

occur the larger the cumulative error component and the smaller the correlation between them. This is a very common phenomena observed in the operations research and production planning literatures, yielding what has become known as a "simplex" matrix of correlations among performance measures taken at adjacent periods over time (Hulin, Henry, & Noon, 1990). In this matrix, time lag between performance measures increases as one moves off of the main diagonal and correlations systematically decrease.

Importantly, the simplex matrix suggests past performance will, as error accumulates over time, become an increasingly poorer predictor of future performance. In a recent example of this phenomena, Deadrick and colleagues (Deadrick, Bennett, & Russell, 1997; Deadrick & Madigan, 1990) reported a decrease in correlations between weekly performance measures among $N = 82$ sewing machine operators from $r = .92$ with a one week time lag to $r = .55$ and $r = .25$ with 26 and 51 week time lags, respectively. The simplex pattern has been a major issue since the early 1960s in debates over how to interpret changing crite-

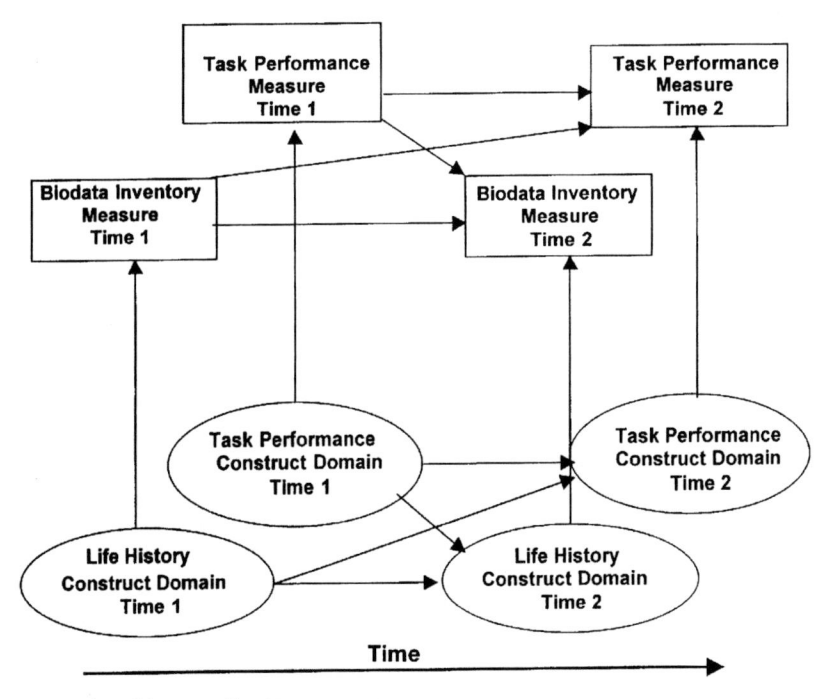

Source: Adapted from Russell and Domm (1990).

Figure 1. Model of biodata and performance construct domains and their relationship over time.

rion validities with criterion measures taken over long periods of time (Hulin, 1962).

Hence, the simplex phenomena would seem to be at odds with predictions made by the consistency principle. How can (1) performance or behavior in the distant past as captured by biodata responses accurately predict future performance or behavior *and* (2) performance measures obtained on the exact same job exhibit decreasing correlations as the time lag between measures increases? How can both the simplex pattern of successively less correlated performance measures and biodata criterion-related validity co-occur?

The answer would appear to be in the operations research distinction between forecasting and time series. Time series equations assume no causal link between the independent variable (e.g., prior performance measures) and dependent variable (e.g., future performance). Instead, time series predictions assume whatever (generally unknown) causal processes in effect at time t-1 will tend to operate the same way at time t. No insight into latent causal processes is needed to explain predictive power for equations 1 or 2 above, one only need assume causal processes change slowly over time. In the absence of strong theory, forecasting involves harvesting causal "candidate" variables for use in powerful statistical optimization algorithms in hopes of capturing an isomorphic (exact) or paramorphic (parallel but not identical to) model of latent causal influences. For example, for an Iowa pig farmer, a time series predicts change in pork belly prices from trends in past pork belly prices, while forecasting involves predicting pork belly prices from the cost of feed corn, subsoil moisture levels (which causally influence feed corn crop yields and subsequent feed supply and price), natural disasters in other large pork or pork substitute (meat) producing markets (e.g., mad cow disease), and so forth.

Those familiar with biodata selection systems should immediately see the parallel; biodata generally does *not literally* involve the prediction of future performance (or behavior) from past measures of the same. Instead, biodata is hypothesized to capture causal events or correlates of causal events that influence job candidates' future behavior.[4] The most well developed explanation of biodata criterion-related validities, that is, Mumford et al.'s (1990) ecology model, describes causally recursive sequences of life events as a learning process. Specifically, Mumford et al. (1990) described a causally related sequence of "resources" (e.g., knowledge, skills, and abilities) and "affordances" (e.g., values, expectancies, or what Kanfer & Ackerman, 1989, labeled "distal motivation") whereby individuals act in a way consistent with their environment, resources, and affordances at t_1. Individual actions at t_1 yield a new environment (including performance outcomes resulting from t_1 activities), and (possibly) a new profile of knowledge, skills, and abilities (resources) and modified preferences or desires for outcomes resulting from application of knowledge, skills, and abilities (affordances) at t_2.

The inability of a strict interpretation of the consistency principle to explain biodata criterion validity becomes clear when one considers the prediction problem faced by Russell, Mattson, Devlin, and Atwater (1990). Russell et al. faced predicting performance of Naval officers five to six years after graduation from the U.S. Naval Academy for candidates who are generally still in high school. High school seniors tend not to have prior performance records as Naval officers, so Russell et al. developed measures of past behaviors and performance in domains in which high school seniors had been exposed. As described by Binning and Barrett (1989), a strong theory of performance prediction must explain latent nomological relationships between prior life events and future job performance. The latent nomological network clearly must relate more than prior performance to future performance, given no prior performance as Naval officers existed. Russell and Domm (1990) presented the model portrayed in Figure 1 as a way of conceptualizing what such a nomological net must minimally look like. We now review efforts aimed at developing such a theory and present an extension.

The Ecology Model

Efforts have been mounted to develop theories to explain biodata criterion validity that reach beyond the consistency principle (Mael, 1991; Mumford & Owens,

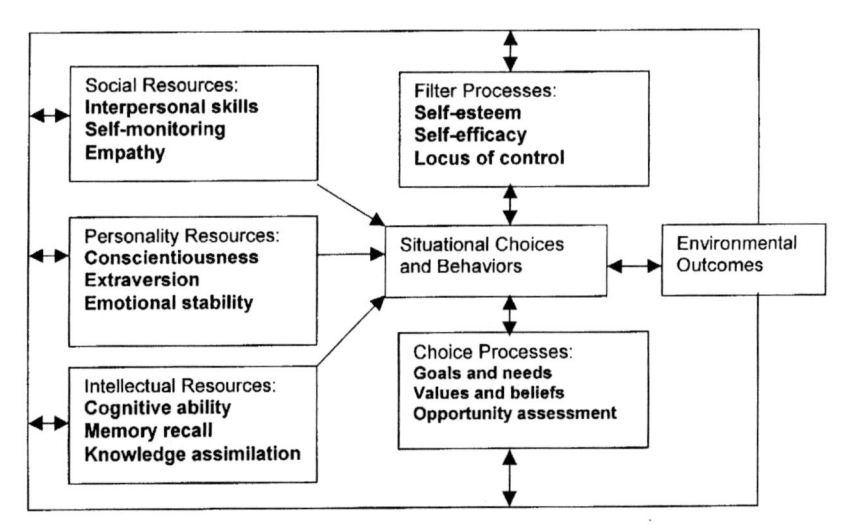

Note: Adapted from Mumford, M. D. & Stokes, G. S. (1992). Developmental determinants of individual action: Theory and practice in applying background measures. In M.D. Dunnette and L. Hough (Eds.), *Handbook of industrial and organizational psychology* (2nd ed., vol. 3, pp. 61-138). Palo Alto, CA: Consulting Psychologists Press.

Figure 2. The ecology model.

1987; Mumford & Stokes, 1992; Mumford et al., 1990; Owens 1968, 1971, 1976; Owens & Schoenfeldt, 1979). One such theoretical rationale for biodata is the ecology model (Mumford et al., 1990). The ecology model acknowledges individuals have their own unique hereditary characteristics and exposures to environmental circumstances that determine initial individual differences, focusing specifically on how individual difference characteristics shape the subsequent choices individuals make. The ecology model grew from Owens' developmental-integrative (D-I) model (Owens, 1968, 1971, 1976; Owens & Schoenfeldt, 1979), which initially proposed biodata items need to capture prior behaviors and experiences affecting personal development on individual difference characteristics (e.g., knowledge, skills, and abilities). These individual differences were hypothesized to subsequently affect a person's performance on organizational criteria of interest.

Mumford, Stokes, and Owens (Mumford & Owens, 1987; Mumford & Stokes, 1992; Mumford et al., 1990) continued to refine the ecology model framework. The ecology model not only considered individual differences as predictors of future performance, but also considered the processes that motivate and influence an individual's choices. Specifically, the model suggested people select themselves into situations based on the value of expected outcomes as well as pre-existing individual difference characteristics. Each choice requires adaptation to new situations and represents a developmental experience. The model represents an iterative process of choice, development, and adaptation. People are constantly faced with making choices and over time will tend to develop characteristic patterns of choices and behaviors.

The ecology model posits that "life events indicating successful engagement in activities requiring the application of KSAOs similar to those required on-the-job might prove to be useful predictors" (Mumford & Stokes, 1992, p. 81), as well as those events that play a role in developing knowledge, skills, abilities, and other personal characteristics (KSAOs). Note, the former capture consistency principle or time series-based performance prediction, while the latter tap latent causal processes captured in forecasting applications.

In an early attempt to understand dimensions underlying the ecology model, Nickels (1990) identified a framework of characteristics and individual differences posited to influence performance later in life. Nickels reviewed over one hundred and fifty citations of individual differences and known predictive relationships between past behavior/experience and later performance, yielding a preliminary list of 500 possible dimensions. The 500 dimensions were subjected to a series of reviews by subject matter experts to obtain a manageable and interpretable number of dimensions. Dimensions were excluded from further investigation based on a consensus decisions that a dimension...

> (a) demonstrated an obvious content overlap with another dimension (e.g., gregariousness and sociability); (b) could not feasibly be rated given the information provided by background data

items (e.g., attractiveness); (c) was inappropriate with respect to the population (e.g., paranoia in a normal population); or (d) seemed unlikely to influence the life history of individuals in adolescence and young adulthood (Nickels, 1990, pp. 28-29).

This process resulted in the dimensions being reduced in number from 500 to 44. Five general dimensional categories emerged from the reduced set of 44. Three captured general categories of individual differences posited to influence subsequent performance: personality resources, interpersonal (social) resources, and intellectual resources. Two other categories covered motivation and beliefs/attitudes, labeled choice and filter processes, respectively.

The Nickels (1990) study was one of the first attempts to operationalize the ecology model. Mumford et al. (1990), the primary ecology model architects, subsequently elaborated this framework as described in Figure 2, changing some of the labels though not the substance of Nickels dimensions (Mumford & Stokes, 1992). The ecology model suggested individual difference constructs "facilitate the attainment of desired outcomes while conditioning future situational choice by increasing the likelihood of reward in certain kinds of situations" (Mumford & Stokes, 1992, p. 81). The first three categories (i.e., personality, social, and intellectual resources) are personal characteristics posited to influence future behavior and decisions. The remaining two categories are motivational variables that might affect situational selection and resource application (i.e., choice and filter processes; Mumford & Stokes, 1992; Nickels, 1990). These five constructs are discussed below.

Personality Resources

Nickels suggested this category represented "stylistic or emotional attributes thought to impact effective environmental interactions" (1990, p. 29) such as adaptability, emotional stability, and persistence. These resources closely resemble the "Big Five" personality constructs (cf. Barrick & Mount, 1991). The five factors include: extraversion (e.g., sociable, assertive, ambitious), emotional stability (e.g., secure, anxious, well-adjusted), conscientiousness (e.g., dependable, efficient, achievement oriented), and openness to experience (e.g., cultured, curious, broad-minded; Barrick & Mount, 1991; Mount, 1997). Construct validity evidence supportive of these five personality factors has been found consistently across longitudinal studies, raters, personality inventories, and protected subgroups (Digman, 1990; Mount, 1997). Results reported by Caspi and associates (e.g., Newman, Caspi, Moffitt, & Silva, 1997) have demonstrated differences in personality temperament as early as age three predict meaningful differences in interpersonal functioning in work contexts at age 21.

Social Resources

Nickels (1990) posited that social resources influence effectiveness of interpersonal relations and therefore play a role in situation selection and subsequent behavior/performance. Some examples of constructs in this category include self-monitoring, dominance, and empathy. Mael's integration of social identity theory with the ecology model also speaks to the influence of group membership on one's own personal identity (Mael, 1991; Mael & Ashforth, 1995; Mael & Hirsch, 1993). It could be argued that the more group memberships held, the greater one's interpersonal adeptness. Large numbers of group memberships may suggest individuals are high in self monitoring and behavioral flexibility, that is, able to adjust behavior to match expectations generated from a variety of groups.

Intellectual Resources

Intellectual resources represent attributes that enable one to assimilate and retain knowledge affecting one's ability to make choices and perform efficiently and effectively. An example of an underlying construct that biodata items might capture is general cognitive ability, or "g." The ecology model explicitly includes g in intellectual resources, but also speaks to the iterative, bi-directional relationship of g to life events and performance over time. Many extant biodata items seem to tap g (Dean et al., 1998). For example, questions on Owens' (1971) Biographical Questionnaire asked individuals about academic achievement, academic attitude, and intellectualism. Biodata instruments may also tap g by asking about past experiences requiring general cognitive ability or associated with its development. In contrast, paper and pencil cognitive ability tests infer g from the number of correct answers selected by an applicant to questions tapping various knowledge content domains. Answers deemed "correct" reflect some universal agreement as to what is truth within each item content domain (e.g., "4" is the correct answer to the arithmetic knowledge question, "What is 2 plus 2?"). The "correct" answer to a g-loaded biodata item is any response option that (1) is related to the latent life history construct of interest (i.e., exhibits construct validity relative to g), and (2) contributes to explanation of subsequent performance differences.

Choice Processes

The choice processes domain represents the "differential motivational influences with respect to individual differences in performance" (Nickels, 1990, p. 43). Example components of the choice processes domain include goal orientation, personal performance standards, and desirability of the reward (e.g., "valence" in expectancy theory terminology). Research on performance prediction suggests individuals must have motivation and ability to generate performance outcomes (Campbell, Dunnette, Lawler, & Weick, 1970). Gottfredson

(1997) suggested biodata items capture motivational components of task performance better than paper and pencil mental ability tests, speculating that cognitive ability measures may best estimate what applicants "can do," but measures not specifically targeting cognitive ability (i.e., "non-cognitive" measures) such as biodata may best estimate what applicants "will do." Some investigators suggested high criterion-related validities may be partially due to biodata's ability to tap ability constructs, motivational constructs, and past examples of the interaction between the two domains (Mael, 1991; Mitchell, 1996).

Filter Processes

Nickels (1990) suggested this category represents values, beliefs, and attitudes which may influence self-perception and, consequently, decisions an individual makes. Constructs included in this category are self-esteem, self-efficacy, and locus of control.

Model Deficiencies

After reviewing biodata theory development, one might ask "What latent causal process(es) explain biodata predictive power?" The D-I and ecology models hypothesize coarse groupings of construct domains deemed relevant, on the basis of theory and research in non-biodata venues, to the development of important individual differences. This coarseness becomes clear when one attempts to generate items unique to each domain. For example, while "dominance" is clearly an individual difference characteristic associated with social interaction (and hence part of the "social resources" category), it is equally clearly a personality characteristic that might be found in the "personality resources" category. The possibility that a single life event can be causally linked to more than one outcome (i.e., what von Eye & Brandtstädter, 1998, referred to as a "fork" dependency) makes traditional convergent and discriminant assessments of biodata construct validity difficult.

Unfortunately, the models portrayed in Figures 1 and 2 also exhibit a shortcoming common to many path model conceptualizations; that is, the absence of any explanation of latent processes underlying the paths themselves (cf. Russell & Van Sell, 1986). At least one explanation of processes linking prior life experiences reflected in the ecology model involves learning or knowledge acquisition. Prior life experiences that lend themselves to learning *and* biodata items demonstrating criterion validity could be defined as situations where (1) people create meaning out of patterns of information where no meaning and/or different meanings had been previously applied, and (2) newly created meaning causally influences subsequent knowledge acquisition, behaviors, and performance. To be sure, any learning could be ecologically valid (i.e., subjects learned the correct lessons from experience) or demonstrate any one of a number of common errors. Hence, the forecasting model becomes:

$$Y_t = f(X_{1_t}, X_{2_t}, X_{3_t}, ..., X_{k_t}, X_{k+1_{t-1}}, ...) + e$$

where the various Xs are causal influences on performance Y. Note, Y occurs at time t, while life event X_k can be chronologically coincident (time t) or earlier (t-1). Integrating this model with a consistency principle-based time series forecast, performance Y_t is generated from past causal influences (captured by Y_{t-1}), any systematic changes in those causal influences (e.g., learning driven by X_t and X_{t-1} variables), and random error.

Models of how and when people assign meaning to life experiences are needed to provide guidance for biodata item development. Such a model would likely describe multiple alternative sequences of experiences that may or may not yield similar meaning across individuals. If experience sequences and attendant patterns of meaning are consistent within or across labor markets, industries, jobs, and so forth, a model of performance prediction will result linking profiles of "lessons learned" predictor constructs to subsequent job performance. In turn, systematic examination of "lesson acquisition event" domains would address latent processes within the ecology model and serve to guide both biodata item development and keying.

TOWARD A MODEL OF LIFE EXPERIENCE LEARNING

The training and development literature suggests at least three characteristics of life experiences influence learning: learning aids in the environment, time needed to reflect on experience, and failures (Goldstein, 1986). Learning aids might include sources of passive stimulation (e.g., number of magazines subscribed to by parents) and active intervention (e.g., task versus consideration-oriented leadership style of an adult role model). Quantity and quality of reflection time has been shown to influence learning in educational settings, where overstimulation decreases reflection time in preventing pattern recognition and meaning making. The literature suggests individual differences in stimulus seeking behavior are likely moderators of reflection time influence (Cacioppo, Petty, Feinstein, & Jarvis, 1996). Environmental support for learning and time needed to reflect on experience would seem to have fairly straightforward implications for theory-based biodata item development and application (see Russell, 1986, for a description of how biodata items were generated from life history essays to reflect environmental support).

Negative Life Experiences

What we find to be most interesting, however, are implications drawn from the third facet: failures, or negative life experiences. Negative life events are imbedded in theory and empirical findings in a wide array of literatures, though there appears to be few bridges between these concurrent research streams. We present a sampling of these research streams as well as literature bearing on a relevant individual difference construct. Implications for extending our understanding of processes underlying the ecology model are drawn.

Before doing so, it seems prudent to ask whether any evidence exists suggesting negative life event-oriented biodata items exhibit criterion validity. Two biodata criterion-related validity studies suggest negative life events are both recalled particularly well and demonstrate strong criterion-related validity when compiled in a biodata inventory. First, Russell et al.'s (1990) factor analysis of biodata items created for selection of midshipmen into the U.S. Naval Academy yielded a dominant initial factor containing unpleasant life events. Example items included "How often have you worked really hard to achieve something and still came up short?," "How often have you wondered whether people like you because of something you've done (e.g., sports, good grades, cheerleading, etc.) rather than for who you are?," and "How often has someone fooled you so that afterwards you felt stupid or naive?" Russell and Domm (1990) found a similar dominant factor in a separate study using a different labor pool, target job, and biodata items. Negative life event items generated from life history essays written by retail store managers included "How often have you found yourself in positions where you were forced to make frequent decisions even though you felt you needed more time or information?," "How often have you followed a friend's recommendation only to find out things did not turn out well?," and "How often have you lost sales or had to order items for a customer because the items were not in inventory?"

Results obtained from qualitative research methods also point to negative life events. Specifically, Lindsey, Homes, and McCall (1987) interviewed over 190 highly successful top-level executives to assess "key events" in executive lives. A frequently cited event across a heterogeneous sample involved having worked for a very demanding, almost abusive boss early in one's career. They also found 17.4% of all "key developmental events" reported by successful top-level managers were severe enough to be deemed "hardships" (p. 87). Cattell (1989) reported longitudinal evidence suggesting a substantial number of people exposed to severely abusive early life experiences survive and, indeed, thrive later in life.[5] Holman and Silver (1998) recently reported evidence suggesting individual differences in temporal orientation moderate individuals' abilities to cope with traumatic events.

Perhaps the most significant body of findings and theory suggesting the importance of negative life events is found in Lewin's (1951) field theory. The first step of the classic sequence of unfreezing, change, and refreezing generally involves receipt of some negative piece of evaluative information. Descriptions of the unfreezing stage in field theory-based interventions such as t-groups (Lewin, 1951) suggest this life event can be particularly unpleasant.

Finally, recent research on performance prediction has re-examined the influence of work experience on job performance (Quinones, Ford, & Teachout, 1995;

Tesluk & Jacobs, 1998). While tenure and seniority are the typical time-based surrogate measures of work experience, recent research indicates the concept is much more psychologically complex. Tesluk and Jacobs (1998) classified work experiences along a *density* continuum, with high density experiences providing greater developmental impact than low density experiences. Personal growth from work experience is thus related to the qualitative nature of the life event, providing a basis to differentiate "ten years of job experience from one year of job experience repeated ten times." We would argue that negative life events in work and non-work domains constitute some of the most developmentally "dense" experiences encountered over a lifetime. Research conducted in non-work arenas (e.g., Suedfeld & Bluck, 1993) has begun to also examine effects of stressful life events on integrative complexity. Integrative complexity is very similar to developmental density, defined as "a state-dependent, cognitive style variable derived from conceptual complexity and system theories...as an...ability to perceive and think about multiple dimensions or perspectives of a stimulus...and the ability to recognize connections between differentiated characteristics" (Pennell, 1996, p. 777).

Potential Causal Mechanisms

Hence, Lewin's (1951) model of individual change, recent research on work experience-job performance relationships, results reported from biodata criterion-related validity studies, and evidence from more qualitative, ethnographic approaches all suggest negative live events constitute key markers in adult development. While suggestive, these results do not explain why negative events are apparently recalled more frequently (i.e., yielding the dominant factor in each study), or the process by which these experiences come to be related to subsequent performance outcomes. A literature on different attributional styles and attentional resources allocated in the face of failure or disappointment provides an initial theoretical foundation. Taylor (1991) reviewed the literature on differences in attributional style as a function of failure versus success. She noted that "diverse literatures in psychology provide evidence that, other things being equal, negative events appear to elicit more physiological, affective, cognitive, and behavioral activity and prompt more cognitive analysis than neutral or positive events" (Taylor, 1991, p. 67).

Results suggest negatively valent events initially lead to mobilization of personal resources (e.g., physiological arousal, affect, attention, and differential weighing of negative cues in decision making). Peeters and Czapinski (1990) found negative events elicited more frequent and complex causal attributional activity than positive events, while others have shown negative events are considered longer (e.g., Abele, 1985) and elicit more extreme attributions (e.g., Birnbaum, 1972). Models of affiliation (Schachter, 1959) and social support (House, 1981) also suggest negative or threatening events cause people to seek companionship, support, and assistance from others. Consistent with criterion-related validities obtained for biodata items tapping negative life events (Russell & Domm, 1990; Russell et al., 1990), Brunstein and Gollwitzer (1996) reported experimental results suggesting task failure caused higher levels of subsequent task performance when the task was relevant to subjects' self definitions (i.e., self-defined goals).

Schulz and Heckhausen (1996) proposed several general principles regulating human development across the life span. One principle pertains to compensating for and coping with failure. Life presents a myriad of negative events which frustrate goal attainment and foster negative self-perceptions. Schulz and Heckhausen postulated the capacity to compensate for frustration is critical, for without it we could not persist when faced with adversity. At one level frustration can foster commitment to a goal. However, at another level failure experiences can undermine self-esteem and individuals' self-ascribed competencies. Thus, negative life events can increase our resolve for goal attainment, yet not compensated for can undermine our self-concept.

Taylor, Pham, Rivkin, and Armor (1998) recently described a program of research focusing on mental stimulation as a core construct driving (1) resolution of stressors (i.e., negative life events) and/or development of goals or visions of desirable future events, (2) anticipation and management of emotions, and (3) initiation and maintenance of problem-solving activities. Taylor et al.'s (1998) efforts target identification of effective and ineffective sources of mental stimulation in relation to the learning/self-regulation process. Initial findings suggest mental stimulation brought about by either process or outcome "mental simulation" help individuals manage affective reactions and subsequent problem solving activities (Rivkin & Taylor, in press) associated with negative life events.

As noted above, negative life events are associated with physiological arousal, affect, attention, differential weighing of negative cues in decision making, more frequent and complex causal attributional activity, are considered longer, elicit more extreme attributions, and cause people to seek companionship, support, and assistance from others. All of these responses are expected to help individuals take action to end or attenuate negative life events, and learn how to avoid similar negative events in the future. Snell (1988) described how exposure to negative events at work socialized managers to be "graduates of the school of hard knocks" (p. 27). The hard knocks consisted of situations and events in which the managers made embarrassing mistakes, or felt overextended in role responsibilities. The essence of the hard knocks was to produce distress, caused by having encountered impasses, having suffered defeats or injustices, and having come under attack.

The effect of these experiences was to teach the managers to be more aware of potential trouble, priming oneself for challenges and anticipating areas of doubt. Specific reactions acquired by the learner-managers included anticipating likely problems, preparing for unfavorable circumstances, obtaining early readings of differences of opinions, and being counseled by a trusted colleague in advance of predicaments. These learning practices can be emotionally unsettling, but they are seldom distressing. The explanatory basis of their effectiveness is that the discomfort they produce is felt currently in mild form because it is expected, rather than later, in a more severe form, when it is not expected.

There would seem to be times when people are particularly receptive to learning from negative life events. Rovee-Collier (1995) proposed the concept of *time-windows* in cognitive development. A time-window is a critical period where information about a current event is integrated with previously acquired information. However, if the same information is encountered outside of the time-window, it will not be integrated. Time-windows are not restricted to a particular age or stage of development. Nonetheless, they are open for a limited duration before closing. Discrete events that occur outside of a time-window are treated as unique and thus are not assimilated into the reservoir of collective memory.

Rovee-Collier (1995) speculated that time-windows may be the cornerstone of individual differences in cognitive domains involving integration of successive experiences. She asserted that to the degree to which personal experiences of same-age individuals differ from moment to moment, so will their time-windows, what they remember from those time-windows, and whether the new information will be integrated with existing information in the future. Negative life events that occur when time-windows are closed may be more likely to produce intensified distress upon re-exposure. The concept of time-windows may provide an explanation for the adage that "the most painful lessons in life are often those that we once learned but forgot."

Differential Recall of Negative Life Events

Curiously, Baddeley (1982), Ehrlichman and Halpern (1988), Linton (1982, 1986), and others found that people tend to remember positive life events more frequently than negative events. Further, Taylor and Brown (1988) reviewed literature indicating people actively attempt to re-interpret negative life events as positive life events. These results are inconsistent with those reported by Russell et al. (1990), Russell and Domm (1990), and Lindsey et al. (1987). However, none of the autobiographical memory research has examined *targeted* autobiographical recall. To generate initial biodata items, Russell and Domm (1990) asked incumbents to describe prior life events which were related to performance on each dimension of target job performance. Two of the life history essays used by Russell et al. (1990) targeted task performance (i.e., individual and group accomplishments), with follow-up questions that paralleled aspects of the performance appraisal criterion applied to Naval Officer performance. Finally, Lindsey et al. (1987) asked subjects to recall events viewed as central to their personal development as managers. Hence, while general autobiographical memory seems to be biased toward positive life events (Burger, 1986; Isen, 1984), quantitative and qualitative evidence (Lindsey et al., 1987; Snell, 1988) suggests negative events within the confines of career-relevant experience are vividly recalled.

One interpretation of these seemingly conflicting findings involves the public versus private nature of autobiographical recall. Specifically, the act of asking individuals for rich, narrative descriptions of prior life events that participants know will be read by others in studies of autobiographical memory may evoke self-serving attributional biases. In fact, subjects' "private" recollections, attributions, and efforts to learn and grow from prior life events may be much less subject to such biases (Csikszentmihalyi, Rathunude, & Walen, 1993). Typical biodata inventories do not request rich, detailed descriptions of prior life events, and hence may be less likely to be influenced by such biases. Lavallee and Campbell (1995) reported findings supportive of this interpretation, finding daily *goal-related* negative events elicited higher levels of self-focused attention, self-concept confusion, and subsequent self-regulatory responses. Regardless, Kluger, Reilly, and Russell (1991) found the common vertical percent difference method of biodata key development effectively eliminated self-serving biases induced by motivation to obtain a job offer from biodata *scores* (though not necessarily from biodata responses).

Individual Differences in Moxie

It seems likely that virtually every construct valid individual difference measure examined by behavioral scientists is somehow related to negative life events. We focus our discussion on one we feel is particularly relevant, "moxie." Moxie is an old term from the 1920s and 30s meaning courage or nerve. Authors in the personality literature have labeled it "ego-resiliency" (Block, 1993; Block & Block, 1980), or an openness to experience combined with ability to adapt and balance accommodation and assimilation. Ego-resilient children are "well equipped to assimilate experience—to understand and interpret events—with existing structures of thought when these structures are appropriate" (Hart, Keller, Edelstein, & Hoffman, 1998, p. 1278). Hart et al. (1998) reported evidence that ego-resilience at age seven predicts social-cognitive development at ages nine, 12, 15, and 19 independent of general cognitive ability, attention, or social participation. A similar individual difference is labeled "hardiness" in the social psychology literature and has been shown to affect relationships between negative life events and symptoms of illness (Kobasa, Maddi, & Kahn, 1982).

Why dwell on ego-resilience or moxie? First, we appear to be in an era of motivation theory focused on how behavior is sustained over long periods of time (e.g., cybernetic control theory, Bozeman & Kacmar, 1997). Motivation theory seems less concerned with one-time choices or predicting constant high-levels of effort devoted to a single task. Focus has shifted toward persistence of behavior over prolonged time periods. Note, unless an individual has been uniformly successful at all endeavors, persistence of behavior over prolonged periods of time must occur with some degree of failure or negative life events. Moxie would appear to be a critical individual difference in predicting persistence in the face of inevitable negative feedback received as one learns.

Second, if there is any constant in life it seems to be adversity. Adversity may be operationalized similarly for an entire generation (e.g., survivors of the Great Depression, the "global economic change" of the 1990s, etc.), or it may be uniquely private. Some individuals find ways to rise above the adversity; others enter prolonged periods of "victimhood;" while still others become relatively immobilized in the web of consequences. What varies is *how* the adversity manifests itself, *when* it occurs in an individual's life span, and *how* individuals respond to it.

Last, given our prior focus on negative life events, any model that does not address the presence of relevant individual difference variables and the possibility of person-situation interaction is deficient. We cannot think of an individual difference characteristic more relevant in the presence of negative life events than "resilience to negative life events," or moxie. Its potential role as a moderator and/or mediator in negative life event-job performance relationships needs to be examined.

Given these reasons for attending to moxie, where does it come from and how can we measure it? What life experiences promote/inhibit development of moxie? We suggest a number of research directions in the conclusion below.

CONCLUSION AND FUTURE RESEARCH DIRECTIONS

We have reviewed the literature on one of the most powerful performance prediction technologies currently available, showing why traditional reliance on the consistency principle must give way to more theory-rich models of biodata. We summarized characteristics of this prediction technology that make it unique (e.g., empirical keying) and troublesome (e.g., absence of strong theory). Research focusing on the nature of latent theory-based relationships in the ecology model is needed to extend our understanding of why biodata exhibits criterion-related validity. We argued that, based on learning theory predictions, emphasis on examination of negative life events is likely to shed insight into these processes. Finally, a critical individual difference, moxie, was hypothesized to moderate how negative life experiences influence the five ecology model construct domains and subsequent job performance.

We feel this evidence overwhelmingly suggests that learning how to respond effectively to negative life events constitutes key, often defining, developmental life experiences. Examination of the capacity to learn from "hard knocks" (i.e., moxie) may be a more valuable line of inquiry than examination of particular sequencing of these life events. As noted in our discussion of the consistency principle, classic biodata emphasis on "behavioral consistency" adds little to our understanding of human development. It is not necessarily the consistency of the behavior we seek to understand, but the underlying psychological constructs which ultimately produce the behavior in question. Perhaps it is useful to think of negative life events as "high density conditions" which fashion our proclivities to either rise above them through resilience and moxie or be diminished by their deleterious effects.

We believe an analysis of life events and our responses to them is, in essence, the chronicle of our individual existence. From a research perspective, such an analysis would provide the cognitive and developmental explanation for behavior we index through biodata inventories. Thus, positive and negative life experiences can be construed as providing an on-going series of arenas through which our identities become defined. Our position is best captured by the following quote:

> Every experience in life, everything which we have come in contact in life, is a chisel which has been cutting away at our life statue, molding, modifying, shaping it. We are part of all we have met. Everything we have seen, heard, felt, or thought has had its hand in molding us, shaping us. (Orison Swett Marden)

We conclude by suggesting lines of future research that might shed light on the nature of negative life events, moxie, and construct domains tapped by the ecology model.

Basic Construct Development

First, it would seem appropriate to create a taxonomy of negative life events. Such a taxonomy would differ meaningfully from biodata item taxonomies described above in that it would target a latent construct domain, specifically, aspects of negative life experiences. Some research in this area exists. The Social Readjustment Scale (Holmes & Rahe, 1967) contains a list of 43 stress-inducing life events, including death of a spouse, divorce, being fired from work, financial difficulties, and personal injuries and illnesses. Certainly other life events could be identified through open-ended questions or life history essays (cf. Russell, 1986). Lindsey et al.'s (1987) interviews with successful top-level executives suggested working for a nasty, almost abusive boss early in one's career constituted one such event. Regardless, we suspect residents of Bosnia would come up with additional adverse life experiences. Further, the term "downsizing" as an adverse life event had not been coined when the Social Readjustment Scale was developed. At the risk of sounding like cynics, we suspect a list of adverse life events that causally influence human development will be none too short.

Second, a taxonomy of human responses to adversity is needed. A list of such responses would target latent cognitive and behavioral reactions such as "freeze, flight, and fight" responses found in research on fear. While these categories are probably deficient in their sampling of potential human response to adversity, we would hope such an effort would yield identifiable profiles of response patterns.

At least one such profile is likely to be dominated by high moxie and resilience. Taylor et al.'s (1998) approaches to "mental simulation" may also prove useful.

Third, a taxonomy of affective responses to negative life experiences is needed. Emotions associated with loss of a loved one would conceivably be different than emotion associated with the loss of a job. Both may involve grief, though they may differ in terms of remorse versus anger. How remorse or anger becomes channeled into future behavior could be the prime, if not primal, basis of resiliency. Resilience born of anger (or any other emotion) may yield grossly different subsequent behavioral manifestations relative to resiliency born of the will to survive.

Toward an Integrative Model

Simply put, negative life events appear to represent the "what," cognitive and behavioral responses to negative life events represent the "how," and emotional responses represent the "why" behind key development life events. Each of these questions could be conceivably asked in the context of the five construct domains captured by the ecology model. However, we feel the true value of biodata measurement technology is in (1) assessing these three phenomena, and (2) constructing an integrative model showing how reactions to negative life events drive change and growth in the five domains underlying the ecology model. Of course, these suggested lines of research could start with existing biodata instruments, sorting items into negative, neutral, and positive life event categories. These life event categories could then be examined for differences in attributional styles and attentional resources. Other things being equal, negative life event items are expected to contribute to empirical scoring keys with greater frequency and incrementally increase biodata inventory criterion-related validities by a larger amount than neutral or positive life events. Russell et al. (1990) and Russell and Domm (1990) reported results consistent with this expectation, though replication is needed.

Given the apparent developmental centrality of negative life events, subsequent research efforts must identify the array of behavioral and affective reactions associated with negative life event categories that contribute most to performance prediction. Identification of profiles of negative life events and reactions that are differentially related to performance outcomes will yield strong implications for biodata theory development (i.e., the "what, how, and why" of latent causal processes underlying the ecology model). An extended model of such processes would permit strong tests of theory-based inferences, specifically, that interventions designed to inject an appropriate quantity/quality of adversity into an individual's life during an appropriate time-window *and* guide reactions to that adversity will yield higher performance than any other combination of adversity and adversity reactions (cf. Taylor et al., 1998). Again, the cynic in us suggests most will never need adversity injected into their lives; it will come in one form or another to everyone. We can influence reactions to adversity, though current biodata theory provides little guidance. Basic biodata research holds promise for providing such guidance, but richer and more refined theory development is needed.

NOTES

1. Note the Buckley amendment to the 1974 Privacy in Education Act makes it illegal for schools to share this information without written permission from students or their guardians.

2. This point was first made known to the authors in discussion comments made by Joel Lefkowitz at a symposium presented at the 1994 Society of Industrial and Organizational Psychology meetings.

3. We will not use the label "rational keys" to describe scoring systems derived from a priori theory-based expectations drawn from some theory or model. The label "rational" implicitly infers that alternative procedures (e.g., empirical keys) are "irrational," and nothing could be further from the truth. We are unaware of any biodata inventories whose architects did not have some reason for selecting weights found in a key.

4. Of course, some biodata applications do contain items that mimic traditional time series predictions, for example, a biodata inventory common within the life insurance industry asks experienced candidates for sales positions "How much life insurance did you sell last year?" However, these items tend not to dominate inventory content or subsequent empirical keys.

5. We would like to thank Alan Mead for directing us to this research stream.

REFERENCES

Aamodt, M.G., & Pierce, W.L., Jr. (1987). Comparison of the rare response and vertical percent methods for scoring the biographical information blank. *Educational and Psychological Measurement, 47*, 505-511.

Abele, A. (1985). Thinking about thinking: Causal, evaluative, and finalistic cognitions about social situations. *European Journal of Social Psychology, 15*, 315-322.

Abrahams, N. (1965). *The effect of key length and item validity on overall validity, cross-validation shrinkage, and test-retest reliability of interest keys.* Unpublished doctoral dissertation, University of Minnesota, Minneapolis.

Abrahams, N. (1998, June 27). Personal communication.

Asher, J.J. (1972). The biographical item: Can it be improved? *Personnel Psychology, 25*, 251-269.

Ashforth, B.E., & Mael, F. (1989). Social identity theory and the organization. *Academy of Management Review, 14*, 20-39.

Atwater, D. C. (1980). *Faking of an empirically keyed biodata questionnaire.* Paper presented at the annual meeting of the Western Psychological Association, Honolulu, HI.

Bacharach, S.B. (1989). Organizational theories: Some criteria for evaluation. *Academy of Management Review, 14*, 496-515.

Baddeley, A.D. (1982). Emotional factors in forgetting. In A. Cope (Ed.), *Your memory: A user's guide* (pp. 65-73). New York: Macmillan.

Barge, B.N. (1988). *Characteristics of biodata items and their relationship to validity.* Unpublished doctoral dissertation, University of Minnesota, Minneapolis.

Barge, B.N., & Hough, L.M. (1988). Utility of biographical data for the prediction of job performance. In L. M. Hough (Ed.), *Literature review: Utility of temperament, biodata, and interest assessment for predicting job performance* (ARI Research Note 88-020). Alexandria, VA: U. S. Army Research Institute.

Barrick, M.R., & Mount, M.K. (1991). The big five personality dimensions and job performance: A meta-analysis. *Personnel Psychology, 44*, 1-26.

Binning, J.F., & Barrett, G.V. (1989). Validity of personnel decisions: A conceptual analysis of the inferential and evidential bases. *Journal of Applied Psychology, 74*, 478-494.

Birnbaum, M.H. (1972). Morality judgments: Test of an averaging model with differential weights. *Journal of Experimental Psychology, 93*, 35-42.

Block, J. (1993). Studying personality the long way. In D.C. Funder, R.D. Parke, C. Tomlinson-Keasey, & K. Widaman (Eds.), *Studying lives through time: Personality and development* (pp. 9-41). Washington, DC: American Psychological Association.

Block, J., & Block, J.H. (1980). *The California Child Q-Set.* Palo Alto, CA: Consulting Psychologist Press.

Bozeman, D.P. & Kacmar, K.M. (1997). A cybernetic model of impression management processes in organizations. *Organizational Behavior and Human Decision Processes, 69*, 9-30.

Brieman, L. Friedman, J.H., Olshen, R.A., & Stone, C.J. (1984). *Classification and regression trees.* Belmont, CA: Wadsworth.

Brown, S.H. (1994). Validating biodata. In G.S. Stokes, M.D. Mumford, & W.A. Owens (Eds.), *Biodata handbook: Theory, research, and use of biographical information in selection and performance prediction.* (pp. 199-236). Palo Alto, CA: Consulting Psychologists Press.

Brown, S.H (1995, April). Personal communication.

Brown, S.H., & McManus, M. (1994, April). *Comparison of additive and configural biodata keying methods.* Presented at the eighth annual meeting of the Society of Industrial and Organizational Psychology, Nashville, TN.

Brunstein, J.C., & Gollwitzer, P.M. (1996). Effects of failure on subsequent performance: The importance of self-defining goals. *Journal of Personality and Social Psychology, 70*, 395-407.

Burger, J.M. (1986). Temporal effects on attributions: Actor and observer differences. *Social Cognition, 4*, 377-387

Cacioppo, J.T., Petty, R.E., Feinstein, J.A., & Jarvis, W.B.G. (1996). Disposition differences in cognitive motivation: The life and times of individuals varying in need for cognition. *Psychological Bulletin, 119*, 197-253.

Campbell, J., Dunnette, M.D., Lawler, E.E., & Weick, K. (1970). *Managerial behavior and performance effectiveness.* New York: McGraw-Hill.

Cascio, W.F. (1975). Accuracy of verifiable biographical information blank responses. *Journal of Applied Psychology, 60*, 767-769.

Cattell, H.B. (1989). *The 16PF: Personality in depth.* Champaign, IL: Institute for Personality and Ability Testing, Inc.

Childs, A., & Klimoski, R.J. (1986). Successfully predicting career success: An application of the biographical inventory. *Journal of Applied Psychology, 71*, 3-8.

Csikszentmihalyi, M. Rathunude, K., & Walen, S. (1993). *Talented teenagers: The roots of success and failure.* Cambridge: Cambridge University Press.

Deadrick, D.L., Bennett, N., & Russell, C. J. (1997). Using hierarchical linear models to examine performance trends: Implications for selection utility. *Journal of Management, 23*, 745-757.

Deadrick, D.L., & Madigan, R.M. (1990). Dynamic criteria revisited: A longitudinal study of performance stability and predictive validity. *Personnel Psychology, 43*, 717-744.

Dean, M.A., Russell, C.J., & Broach, D. (1998, April). *Sample size needed to establish criterion validity for air traffic controller selection: Comparing bootstrap vs. parametric estimation.* Presented at the annual meetings of the Academy of Management, San Diego, CA.

Devlin, S.E., Abrahams, N.M., & Edwards, J.E. (1992). Empirical keying of biographical data: Cross-validity as a function of scaling procedure and sample size. *Military Psychology, 4*, 119-136.

Digman, J.M. (1990). Personality structure: Emergence of the five-factor model. *Annual Review of Psychology, 41*, 417-440.

Dunnette, M.D. (1962). Personnel management. *Annual Review of Psychology, 13*, 285-313.

Dunnette, M.D. (1966). *Personnel selection and placement.* Belmont, CA: Wadsworth.

Dunnette, M.D. (1972). *Validity study results for jobs relevant to the petroleum refining industry.* Washington, DC: American Petroleum Institute.

Ehrlichman, H. & Halpern, J.N. (1988). Affect and memory: Effects of pleasant and unpleasant odors on retrieval of happy or unhappy memories. *Journal of Personality and Social Psychology, 55,* 769-779.

Equal Employment Opportunity Commission, Civil Service Commission, Department of Labor, and Department of Justice. (1978). Uniform guidelines on employee selection procedures. *Federal Register, 43*(166), 38295-38309.

Fischer, R.A. (1970). *Statistical methods for research workers* (14th ed.). New York: Hafner Press.

Fleishman, E.A. (1988). Some new frontiers in personnel selection research. *Personnel Psychology, 41,* 679-701.

Fleishman, E.A., & Quaintance, M.K. (1984). *Taxonomies of human performance: The description of human tasks.* Orlando, FL: Academic Press.

Gandy, J.A., Outerbridge, A.N., Sharf, J.C., & Dye, D.A. (1989). *Development and initial validation of the Individual Achievement Record (IAR).* Washington, DC: U.S. Office of Personnel Management.

Glaser, B.G., & Strauss, A.L. (1967). *The discovery of grounded theory: Strategies for qualitative research.* Chicago: Aldine.

Goldstein, I.L. (1986). *Training in organizations: Needs assessment, development, and evaluation.* Monterey, CA: Brooks/Cole.

Gottfredson, L. (1997). *Racially gerrymandering the content of police tests to satisfy U.S. Justice Department: A case study* (Working paper).

Guion, R. M. (1965). *Personnel testing.* New York: McGraw-Hill.

Hammer, E.G., & Kleiman, L.A. (1988). Getting to know you. *Personnel Administrator, 34,* 86-92.

Hart, D. Keller, M., Edelstein, W., & Hofmann, V. (1998). Childhood personality influences on social-cognitive development: A longitudinal study. *Journal of Personality and Social Psychology, 74,* 1278-1289.

Holman, E.A., & Silver, R.C. (1998). Getting "stuck" in the past: Temporal orientation and coping with trauma. *Journal of Personality and Social Psychology, 74,* 1146-1163.

Holmes, I.H., & Rahe, R.H. (1967). The social readjustment rating scale. *Journal of Psychosomatic Research, 4,* 189 - 194.

Hough, L., Eaton, N.K., Dunnette, M.D., Kamp, J.D., & McCloy, R.A. (1990). Criterion-related validities of personality constructs and the effect of response distortion on those validities. *Journal of Applied Psychology, 75,* 581-595.

House, J.A. (1981). *Work stress and social support.* Reading, MA: Addison-Wesley.

Hulin, C.L. (1962). The measurement of executive success. *Journal of Applied Psychology, 46,* 303-306.

Hulin, C.L., Henry, R.A., & Noon, S.L. (1990). Adding a dimension: Time as a factor in generalizability of predictive relationships. *Psychological Bulletin, 107,* 328-3400.

Hunter, J.E. (1986). Cognitive ability, cognitive aptitude, job knowledge, and job performance. *Journal of Vocational Behavior, 29,* 340-362.

Hunter, J.E., & Hunter, R.F. (1984). Validity and utility of alternative predictors of job performance. *Psychological Bulletin, 96,* 72-98.

Isen, A.M. (1984). Toward understanding the role of affect in cognition. In R.S. Wyer Jr., & T.K. Srull (Eds.), *Handbook of social cognition* (Vol. 3, pp. 179-236). Hillsdale, NJ: Erlbaum.

Kanfer, R. (1990). Motivation theory in industrial and organizational psychology. In M.D. Dunnette & L.M. Hough (Eds.), *Handbook of industrial and organizational psychology* (Vol. 1, pp. 75-170). Palo Alto, CA: Consulting Psychologist Press.

Kanfer, R., & Ackerman, P.L. (1989). Motivation and cognitive abilities: An integrative/aptitude-treatment interaction approach to skill acquisition. *Journal of Applied Psychology, 74,* 657-690.

Klimoski, R.J., & Brickner, M. (1987). Why do assessment centers work? The puzzle of assessment center validity. *Personnel Psychology, 40,* 243-260.

Klimoski, R.J., & Strictland, W. (1976). Assessment centers: Valid or merely prescient. *Personnel Psychology, 30,* 353-363.

Kluger, A.N., Reilly, R.R., & Russell, C.J. (1991). Faking biodata tests: Are option-keyed instruments more resistant? *Journal of Applied Psychology, 76,* 889-896.

Kobasa, S.C., Maddi, S.R., & Kahn, S. (1982). Hardiness and health: A prospective study. *Journal of Personality and Social Psychology, 42,* 168-177.

Lavallee, L.F., & Campbell, J.D. (1995). Impact of personal goals on self-regulation processes elicited by daily negative events. *Journal of Personality and Social Psychology, 69,* 341-352.

Lecznar, W.B. (1951). *Evaluation of a new technique for keying biographical inventories empirically* (Research Bulletin No. 51-2). San Antonio, TX: Lackland Air Force Base, Human Resources Research Center.

Lecznar, W.B., & Dailey, J.T. (1950). Keying biographical inventories in classification test batteries. *American Psychologist, 5,* 279.

Lewin, K. (1951). *Field theory in social science.* New York: Harper & Row.

Lindsey, E.H., Holmes, V., & McCall, M.W. (1987). *Key events in executives' lives* (Technical Report 32). Greensboro, NC: Center for Creative Leadership.

Linton, M. (1982). Transformations of memory in everyday life. In U. Neisser (Ed.), *Memory observed: Remembering in natural contexts* (pp. 77-91). San Francisco: Freeman.

Linton, M. (1986). Ways of searching and the contents of memory. In D.C. Rubin (Ed.), *Autobiographical memory* (pp. 50-67). Cambridge: Cambridge University Press.

Mael, F.A. (1991). A conceptual rationale for the domain and attributes of biodata items. *Personnel Psychology, 44,* 763-792.

Mael, F.A., & Ashforth, B.E. (1995). Loyal from day one: Biodata, organizational identification, and turnover among newcomers. *Personnel Psychology, 48,* 309-333.

Mael, F.A., Connerley, M., & Morath, R.A. (1996). None of your business: Parameters of biodata invasiveness. *Personnel Psychology, 49,* 613-650.

Mael, F.A., & Hirsch, A.C. (1993). Rainforest empiricism and quasi-rationality: Two approaches to objective biodata. *Personnel Psychology, 46,* 719-738.

Malloy, J. (1955). The prediction of college achievement with the life experience inventory. *Educational and Psychological Measurement, 15,* 170-180.

Malone, M.P. (1977). *Predictive efficiency and discriminatory impact of verifiable biographical data as a function of data analysis procedure.* Unpublished doctoral dissertation, Illinois Institute of Technology, Chicago.

McAdams, D.P., Diamond, A., St. Aubin, E.D., Mansfield, E. (1997). Stories of commitment: The psychosocial construction of generative lives. *Journal of Personality and Social Psychology, 72,* 678-694.

McCall, M., & Bobko, P. (1990). Research methods in the service of discovery. In. M.D. Dunnette and L.M. Hough (Eds.), *Handbook of industrial and organizational psychology* (2nd ed., Vol. 1, pp. 381-418). Palo Alto, CA: Consulting Psychologists Press.

Mitchell, T.W. (1996). Can do and will do criterion success: A practitioner's theory of biodata. In R.B. Stennett, A.G. Parisi, & G.S. Stokes (Eds.), *A compendium: Papers presented to the first biennial biodata conference* (pp. 2-15). Athens: Applied Psychology Student Association, University of Georgia.

Mitchell, T.W., & Klimoski, R.J. (1982). Is it rational to be empirical? A test of methods for scoring biographical data. *Journal of Applied Psychology, 67,* 411-418.

Morgan, J.N., & Sonquist, J.A. (1963). Problems in the analysis of survey data, and a proposal. *Journal of the American Statistical Association, 58,* 415-438.

Mosel, J. L., & Cozan, L.W. (1952). The accuracy of application blank work histories. *Journal of Applied Psychology, 36,* 356-369.

Mount, M.K. (1997). Big five personality tests. In L.H. Peters, C.R. Greer, & S.A. Youngblood (Eds.), *Encyclopedic dictionary of human resource management* (p. 25). Oxford: Blackwell.

Mumford, M.D., & Owens, W.A. (1987). Methodological review: Principles, procedures, and findings in the application of background data measures. *Applied Psychological Measurement, 11,* 1-31.

Mumford, M.D., & Stokes, G.S. (1992). Developmental determinants of individual action: Theory and practice in applying background measures. In M.D. Dunnette (Ed.), *Handbook of industrial and organizational psychology* (2nd ed., Vol. 3, pp. 61-138). Palo Alto, CA: Consulting Psychologists Press.

Mumford, M.D., Stokes, G.S., & Owens, W.A. (1990). *Patterns of life history: The ecology of human individuality.* Hillsdale, NJ: Lawrence Erlbaum.

Mumford, M.D., Costanza, D.P., Connelly, M.S., & Johnson, J.F. (1996). Item generation procdures and background data scales: Implications for construct and criterion-related validity. *Personnel Psychology, 49,* 361-398.

Neidt, C.O., & Malloy, J.P. (1954). A technique for keying items of an inventory to be added to an existing test battery. *Journal of Applied Psychology, 38,* 308-312.

Newman, D.L., Caspi, A., Moffitt, T.E., & Silva, P.A. (1997). Antecedents of adult interpersonal functioning: Effects of individual differences in age 3 temperament. *Developmental Psychology, 33,* 206-217.

Nickels, B.J. (1990). *The construction of background data measures: Developing procedures which optimize construct, content, and criterion-related validities.* Unpublished doctoral dissertation. Atlanta: Georgia Institute of Technology.

Nickels, B.J. (1994). The nature of biodata. In G.S. Stokes, M.D. Mumford, & W.A. Owens, (Eds.), *The biodata handbook: Theory, research, and applications* (pp. 1-16). Palo Alto, CA: Consulting Psychologists Press.

Owens, W.A. (1968). Toward one discipline of scientific psychology. *American Psychologist, 23,* 782-785.

Owens, W.A. (1971). A quasi-actuarial basis for individual assessment. *American Psychologist, 26,* 992-999.

Owens, W.A. (1976). Background data. In M.D. Dunnette (Ed.), *Handbook of industrial and organizational psychology* (pp. 609-644). Chicago: Rand McNally.

Owens, W.A., Glennon, J.R., & Albright, L.E. and the American Psychological Association Scientific Affairs Committee, Division 14 (1966). *A catalogue of life history items.* Greensboro, NC: The Richardson Foundation.

Owens, W.A., & Schoenfeldt, L.F. (1979). Toward a classification of persons. *Journal of Applied Psychology, 53,* 569-607.

Peeters, G., & Czapinski, J. (1990). Positive-negative asymmetry in evaluations: The distinction between affective and informational negativity effects. *European Review of Social Psychology, 1,* 33-60.

Pennell, G.E. (1996). Integrating complexity into the study of life events: Comment on Suedfeld and Bluck (1993). *Journal of Personality and Social Psychology, 71,* 777-780.

Quinones, M.A., Ford, J.K., & Teachout, M.S. (1995). The relationship between work experience and job performance: A conceptual and meta-analytic review. *Personnel Psychology, 48,* 887-910.

Reilly, R.R., & Chao, G.T. (1982). Validity and fairness of some alternative employee selection procedures. *Personnel Psychology, 35,* 1-62.

Reilly, R.R., & Warech, M.A. (1990). *The validity and fairness of alternative predictors of occupational performance.* Paper invited by the National Commission on Testing and Public Policy, Washington, DC.

Rivkin, I.D. & Taylor, S.E. (in press). The effects of mental simulation on coping with controllable stressful events. *Personality and Social Psychology Bulletin.*

Rovee-Collier, C. (1995). Time windows in cognitive development. *Developmental Psychology, 31,* 147-169.

Russell, C.J. (1986). *Review of the literature and development of a biodata instrument for prediction of naval officer performance at the U.S. Naval Academy.* Army Research Institute Report No. DAAG29-81-D-0100, Arlington, VA.

Russell, C.J. (1994). Generation procedures for biodata items: A point of departure. In G.S. Stokes, M.D. Mumford, & W.A. Owens (Eds.), *Biodata handbook: Theory, research, and use of biographical information in selection and performance prediction* (pp. 17-38). Palo Alto, CA: Consulting Psychologists Press.

Russell, C.J., & Dean, M.A. (1994a, May). *Exploration of a point biserial bootstrap method of empirical biodata keying.* Presented at the ninth annual meeting of the Society of Industrial and Organizational Psychology, Orlando, FL.

Russell, C.J., & Dean, M.A. (1994b, August). *The effect of history on meta-analytic results: An example from personnel selection research.* Presented at the annual meetings of the Academy of Management, Dallas, TX.

Russell, C.J., & Domm, D.R. (1990). *On the construct validity of biographical information: Evaluation of a theory based method of item generation.* Presented at the fifth annual meetings of the Society of Industrial and Organizational Psychology, Miami Beach, FL.

Russell, C.J., & Domm, D.R. (1995). Two field tests of an explanation of assessment center validity. *Journal of Occupational and Organizational Psychology, 68,* 25-47.

Russell, C.J., Mattson, J., Devlin, S.E., & Atwater, D. (1990). Predictive validity of biodata items generated from retrospective life experience essays. *Journal of Applied Psychology, 75,* 569-580.

Russell, C.J., & Van Sell, M. (1986). *Toward a model of turnover decision making.* Winner of the Edwin E. Ghiselli Award for Research Design, presented by the Society for Industrial and Organizational Psychology.

Schachter, S. (1959). *The physiology of affiliation.* Stanford, CA: Stanford University Press.

Schmitt, N., Gooding, R.Z., Noe, R.D., & Kirsch, M. (1984). Meta-analyses of validity studies published between 1964 and 1982 and the investigation of study characteristics. *Personnel Psychology, 37,* 407-422.

Schulz, R. & Heckhausen, J. (1996). A life span model of successful aging. *American Psychologist, 51,* 702-714.

Shaffer, G.S., Saunders, V., & Owens, W.A. (1986). Additional evidence for the accuracy of biographical data: Long term retest and observer ratings. *Personnel Psychology, 39,* 791-809.

Smith, K.C., & McDaniel, M.A. (1998, April). *Background experience correlates of job performance: An expanded predictor space.* Presented at the 13th annual conference of the Society for Industrial and Organizational Psychology, Dallas, TX.

Snell, R. (1988). Graduating from the school of hard knocks? *Journal of Management Development, 5,* 23-30.

Stead, W.H., & Shartle, C.L. (1940). *Occupational counseling techniques.* New York: American Book Company.

Stokes, G.S., & Reddy, S. (1992). Use of background data in organizational decisions. In C.L. Cooper & I.T. Robertson (Eds.), *International review of industrial and organizational psychology.* (pp. 285-321). London: John Wiley & Sons. Ltd.

Stricker, L.J. (1987). *Developing a biographical measure to assess leadership potential.* Paper presented at the annual meeting of the Military Testing Association, Ottawa, Ontario.

Stricker, L.J. (1988). *Assessing leadership potential at the Naval Academy with a biographical measure.* Paper presented at the annual meeting of the Military Testing Association, San Antonio, TX.

Strong, E.K. (1926). An interest test for personnel managers. *Journal of Personnel Research, 5,* 194-203.

Suedfeld, P., & Bluck, S. (1993). Changes in integrative complexity accompanying significant life events: Historical evidence. *Journal of Personality and Social Psychology, 64*, 124-130.

Suedfeld, P., Tetlock, P.E., & Streufert, S. (1992). Changes in integrative complexity accompanying significant life events: Historical analysis. *Journal of Personality and Social Psychology, 64*, 124-130.

Taylor, S.E. (1991). Asymmetrical effects of positive and negative events: The mobilization-minimization hypothesis. *Psychological Bulletin, 110*, 67-85.

Taylor, S.E., & Brown, J.D. (1988). Illusion and well-being: A social psychological adaptation. *American Psychologist, 38*, 1161-1173.

Taylor, S.E., Pham, L.B., Rivkin, I.D., & Armor, D.A. (1998) Harnessing the imagination: Mental stimulation, self-regulation, and coping. *American Psychologist, 53*, 429-439.

Telenson, P.A., Alexander, R.A., & Barrett, G.V. (1983). Scoring the biographical information blank: A comparison of three weighting techniques. *Applied Psychological Measurement, 7*, 73-80.

Tesluk, P.E., & Jacobs, R.R. (1998). Toward an integrated model of work experience. *Personnel Psychology, 51*, 321-355.

Vineberg, R., & Joyner, J.N. (1982). *Prediction of job performance: Review of military studies.* Alexandria, VA: Human Resources Research Organization.

von Eye, A., & Brandtstädter, J. (1998). The wedge, the fork, and the chain: Modeling dependency concepts using manifest categorical variables. *Psychological Methods, 3*, 169-185.

Webb, S. C. (1960). The comparative validity of two biographical inventory keys. *Journal of Applied Psychology, 44*, 177-183.

Wernimont, P.F., & Campbell, J.P. (1968). Signs, samples, and criteria. *Journal of Applied Psychology, 52*, 372-376.

UNION STRATEGIES FOR REVIVAL:
A CONCEPTUAL FRAMEWORK AND LITERATURE REVIEW

Marick F. Masters and Robert S. Atkin

ABSTRACT

Unions in the United States have faced severe challenges in recent decades. Under such circumstances, they have had no choice but to search for aggressive ways to revitalize their position. With new leadership at the AFL-CIO helm, unions have explored a number of alternative organizing, political, and bargaining strategies. We review such strategies, arguing that unions have conceptually departed from the servicing/business unionism that has gripped labor for decades. A new value-added unionism is advanced, in which unions consciously seek to add value to employees and employers in order to enhance the competitiveness of unionized firms. Implications for future research are explored.

INTRODUCTION

For the past several decades, organized labor has operated under siege. Its vital signs have faded. Fewer dues-paying members. Organizing setbacks. Reduced bargaining power. Replaced strikers. Frontal political attacks. Strauss, Gallagher, and Fiorito (1991, p. v) have described labor's debilitated condition in stark terms:

> It is generally agreed that U.S. unions are in a crisis. As a percentage of the work force, membership is falling off. Organizational drives are facing great opposition. Unions' economic and political clout have been greatly weakened. While unions are on the defensive in most Western countries, in almost none has the union decline occurred so rapidly as in the U.S.

Writhing near despair, labor unions have fought back. An insurgent group won control of the dormant American Federation of Labor-Congress of Industrial Organizations (AFL-CIO), campaigning to revitalize a decaying labor movement. At the helm, AFL-CIO president John Sweeney has continuously implored affiliated unions to change, reorder priorities, and boldly challenge the bitter status quo. Organizing new members and political action have received the most emphasis, and labor has recorded some impressive wins: the Teamsters' strike against United Parcel Service (UPS); the blockage of another fast-track-to-free-trade bill in Congress; and the defeat of a California ballot initiative (Proposition 226) to curb union political spending.

Major labor leaders have issued a clarion call for union "reinvention." They have urged unions to increase organizing, initiate strategic partnerships with business, and recruit more minorities and women. City-wide organizing drives (Union Cities), youth-activist recruitment programs (Union Summer), and political blitzes (Labor '96) have followed suit. In essence, unions have been urged to depart from past practice and transform their role as worker representatives.

While union change is undoubtedly occurring, its scope and import remain unclear. Moreover, beyond the evident need to do more to increase membership, there is little apparent agreement on what new model of unionism should emerge from this turmoil. Are the efforts at revitalization real departures from past practice? Do they collectively comprise a new form of unionism? What *should* the unions of the future be like in order to reestablish their more powerful role in the U.S. political economy?

We address these questions in this paper. The tack taken is to explore recent union revitalization strategies in the context of alternative conceptualizations of unionism, the genre of labor organizations. Unions are being urged to abandon the dominant "business" and "servicing" forms of representation, but receive only shallow guidance on viable substitutes (Hurd, 1998). Nebulous "organizing" models are offered, but they are insufficiently robust to capture the gamut of change that is either under way or deemed necessary for unions to reverse their downward path. Not since Hoxie (1917) has the union species been taxonomically evaluated.

Our thinking is that the presently available conceptualizations of unionism fail to represent the full range of necessary alternatives. In fact, current revitalizing strategies may be congealed to offer a more engrossing model of unionism than any alternatives that are now under discussion. Indeed, we advance "value-added" unionism as a hybrid of past activity and current changes to address the needs of the future.

More specifically, in this paper, we explore various alternative strategies and tactics unions have been using, to varying degrees, to increase their appeal and membership. Based on a selected review of recent academic and popular literature, we package these strategic and tactical approaches into a new form of union representation, to wit: "value-added." This form of unionism is contrasted to the more traditional business and servicing, as well as nascent organizing, models. We believe, as Hurd (1998) and others (Grabelsky & Hurd, 1994) have argued, that what is required for union success is not just an internal union restructuring or resource reallocation exercise, but a fundamental transformation of what unions do (Bronfenbrenner, Friedman, Hurd, Oswald, & Seeber, 1998). It is useful to begin, therefore, with a framework for understanding U.S. trade unionism. What is the nature of unionism? How might unions differ? What alternative forms of this distinguishable organism exist? What should unions be? We caution that the conceptual types presented below are intended to not only describe a range of alternatives available or conceivable but also to serve as heuristic devices, that is, opportunities to expand thinking about the role of unions as institutions in the political economy.

The paper unfolds as follows. The second part presents a conceptual framework for analyzing union strategies. Alternative types of unionism are discussed in the third part. The fourth part examines various examples of recent union revitalization strategies and tactics which collectively form the basis of value-added unionism. An agenda for future research is presented in the fifth part.

CONCEPTUAL FRAMEWORK

Unions are dynamic, complex institutions. Unfortunately, their treatment in much of the academic literature is atheoretical and nonconceptual. The descriptions and analyses of unions therefore often appear rather desultory and *ad hoc*. Hence, it is difficult to determine if a fundamental shift in what unions do is actually taking place. Perhaps the ongoing events in the labor movement are simply more (albeit perhaps a lot more) of the same. It is one thing for unions to allocate more money to organizing, but quite another to carve a new role as worker representatives. Recent strategies do suggest a paradigmatic shift. At the same time, the salience of a massive reallocation of union resources from one dominant function (e.g., servicing current members through collective bargaining and contract administration) to another (e.g., recruiting new members) is important to recognize.

Strategic Framework

Unions pursue a variety of activities to achieve multiple objectives. At any particular time, their strategic and tactical foci may be both varied and restricted. Varied in the sense that differences exist among unions; some are more politically oriented, for example, than others (Delaney & Masters, 1991; Greenstone, 1977). Restricted in that unions may deliberately choose to avoid certain pursuits, such as a militant radical tack to resolving disputes with employers. While unions make various "choices" along these lines (see Fiorito, Gramm, & Hendricks, 1991; Hendricks, Gramm, & Fiorito, 1993; Jarley, Delaney, & Fiorito, 1997; Lawler, 1990), their flexibility is at least partly determined by the environment in which they operate and their ideology and resource bases. Both environmental and organizational constraints come into play, especially in the short-term.

Figure 1 delineates the basic strategies unions may pursue within an industrial relations system (Masters, 1997). Building upon models posited by Dunlop (1958), Derber (1970), and Kochan, Katz, and McKersie (1986), the framework identifies four types of strategies unions may use to influence decision making at three levels. As mentioned, the underlying premise is that unions make certain strategic choices. One is the type of strategy *per se* to pursue: organizing, bargaining, legal enactment (political action), or mutual insurance (Webb & Webb, 1897). Another and related choice is how much effort to put into alternative strategies, (i.e., the allocation of union resources). The viability and effectiveness of any mix of strategies and concomitant tactics will depend somewhat on the environmental

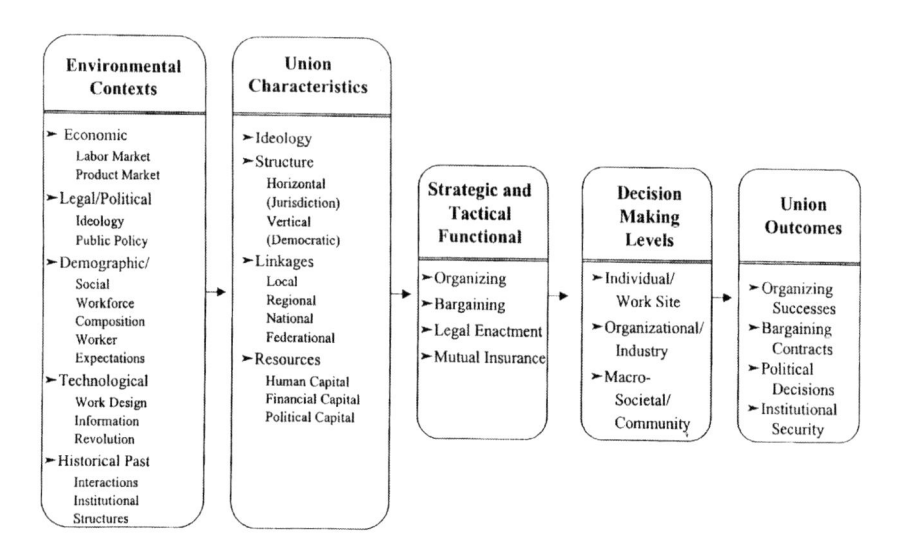

Figure 1. Strategic framework of analysis.

contexts in which unions operate and their organizational characteristics, such as ideology, structure, and the sheer availability of human, financial, and political capital at their disposal.

These types of strategic choices, moreover, are practically relevant. They are real phenomena with human consequences, not indecipherable mysteries. As Jarley, Fiorito, and Delaney (1998, p. 7) note: "unions can improve their success in organizing, the negotiation of first contracts, and political action by adopting specific structures or tactics. In combination, this implies that unions have some control over their destiny." Thus, what unions choose to do will affect decisions at the individual, organizational, and community levels, which, in turn, impact salient union outcomes pertinent to labor's institutional strength and the quality of lives of those who unions directly or indirectly represent.

Union Strategies

The strategies and tactics unions emphasize depend on their expected costs and benefits. Environmental contexts and union characteristics obviously affect this calculation. For example, unions have been widely criticized for neglecting the organizing strategy because of a inward-focused ideology geared to servicing current members (Block, 1980; Bronfenbrenner et al., 1998; Chaison & Rose, 1991; Hurd, 1998; Rose & Chaison, 1996; Voos, 1987). Although operationally different, each strategic function may be used at all decision-making levels in order to influence every set of relevant outcomes. Thus, an organizing strategy may be deployed to affect individual and organizational decisions that influence the success of recruiting drives, contract negotiations, and governmental policies.

Organizing is a basic strategy unions may use to advance their institutional position, but it also has bargaining and political implications. Although it may be operationalized in a multitude of ways that can appear disjointed, organizing is nonetheless a union function aimed strategically at increasing union membership and density (the percentage of the relevant workforce represented). Organizing is pursued not just for reasons of institutional security, however, for it also enhances the negotiating and political powers of unions (Delaney & Schwochau, 1993; Freeman & Medoff, 1984; Kochan et al., 1986).

As a functional strategy, collective bargaining focuses principally on the attainment of workplace goals through the negotiation and administration of contracts at the work site or organization/industry level, depending on the structure of bargaining (i.e., the range of employees and employers participating directly or indirectly in the negotiations). This strategy is often regarded as legalistic for its emphasis on securing and implementing favorable contractual language. In addition, it has a tangible, pragmatic rather than idealistic bent.

Legal enactment, or political action, is a third strategic approach to decision-making. Unions engage in electoral and lobbying activities (ranging from giving political action committee, PAC, money to candidates to testifying before congressional committees on pending legislation) in order to influence governmental decisions made by elected officials, administrators, regulators, and judges, each of whom, to some degree, is subject to pressures from different interest groups. Legal enactment may promote basic workplace terms and conditions across employers, thereby raising the appeal of unions (Fiorito, 1987; Kau & Rubin, 1981). In addition, unions seek laws that secure basic institutional protections, rights, and powers. At a fundamental level, legal enactment is the principal means by which employees (through unions) and employers establish their basic powers, rights, and responsibilities vis-a-vis one another. The extension of certain protections and prerogatives would arguably help unions achieve greater bargaining recognition, more members, and additional economic power. Mutual insurance is a "self-help" strategy (Fiorito et al., 1991, p. 107). Accordingly, unions find ways to benefit their members regardless of what employers or government may do. A union, for example, might finance income-maintenance benefits for survivors of members who have deceased. This type of direct union-provided assistance might not only protect members' interests but also serve to entice new recruits. Contemporary variants of this strategic approach may be found in union-financed housing construction and union-sponsored credit card privileges.

Costs and Benefits

Which set of strategies and tactics unions emphasize, as stated before, depends upon their relative costs and benefits, both of which are affected by environmental and organizational conditions. Assessing the relative costs and benefits of these options is easier said than done, and the resulting calculations undoubtedly vary among unions and over time. This calculus is complicated by the intangibility of some costs and benefits, the uncertainty or unknown probability of achieving desired results, and foregone opportunities entailed by the commitment of resources to specific pursuits. Another evaluative difficulty is the time horizon over which an analysis should apply. Unions may logically approach the question of organizing from a generational rather than annual perspective. The mere formation of the United Steelworkers of America is a process that took 50 years.

Union Characteristics

Unions are a diverse and protean organizational species. At first blush, it may seem difficult to find commonalities among unions as different as the Major League Baseball Players Association, American Federation of State, County, and Municipal Employees, the United Auto Workers, and Air Line Pilots Association beyond their basic *raison d'etre* as worker representatives. Yet, there are a set of organizational dimensions on which it is possible to measure unions by more or less common standards. Further, unions may cluster on these dimensions in ways that define their common traits as well as differences. Like organizations in general, unions may exhibit configurations that "may be represented in typologies developed conceptually or captured in taxonomies derived empirically" (Meyer, Tsui, & Hinings, 1993, p. 1175).

As shown in Figure 1, unions are commonly described in terms of four characteristics: ideology, structure, linkages, and resources. Ideology is the set of beliefs or principles which inspire organizational and individual behaviors. Structure consists of horizontal and vertical dimensions (see Fiorito et al., 1991). The former refers to the jurisdictional choices unions make. That is, unions define their own eligibility requirements (within certain legal constraints), and they may be occupationally and industrially narrow or broad. Unions also make structural choices with respect to bureaucracy, centralization, and democratic orientation (Fiorito et al., 1991; Clark & Gray, 1991). In this regard, unions may become more or less bureaucratic centralized, or democratic in order to meet the complexities of present-day challenges. There may be natural forces that push unions in one direction or the other (Bok & Dunlop, 1970; Clark & Gray, 1991; Strauss, 1991).

Apart from ideology and structure, unions are depicted in terms of their organizational autonomy and linkages. In absolute terms, there are tens of thousands of distinct labor organizations in the United States, but most choose to associate with a dramatically smaller subset of major units. The bedrock unit of the typical labor union is the local affiliate. Layered between the parent national or international union, a la the United Auto Workers, and the locals are affiliated regional organizations. Above the national/international union is the federation (e.g., AFL-CIO). In the United States, the overwhelming majority of union members belong to a relative handful (less than 30) of labor unions who are linked to the AFL-CIO (Masters, 1997). Such interconnections imply some loss of pure identity and autonomy, but parent unions retain considerable independent authority and power irrespective of their federational ties (Bok & Dunlop, 1970; Hurd, 1998). Parenthetically, the decisions unions make with respect to coordinating ongoing projects or merging together are increasingly important linkage choices. Mergers and acquisitions are one means that unions have to respond to environmental and organizational challenges (Chaison, 1996).

Available resources are also important union characteristics. The size and composition—demographically and geographically—of the union membership are ubiquitously cited dimensions. The financial wherewithal of a union in terms of assets, wealth, and income is a resource critical to supporting strategic activity and change (Bennett, 1991; Masters & Atkin, 1997; Sheflin & Troy, 1983; Troy, 1975). Another resource is political capital, which is derived from union members (volunteers and voters) and money (paid lobbyists) but separable for at least two reasons. One is that federal law requires the segregation of PAC financing from regular union finances generated by members' dues and other assessments (Epstein, 1976). A second is that the political currency a union possesses can be rapidly multiplied by interest-group coalitions and alliances.

UNION TAXONOMIES

Scholarly interest in union characteristics appears to be growing (Delaney, Jarley, & Fiorito, 1996). Numerous studies empirically connect certain characteristics with important union outcomes (Delaney, Fiorito, & Masters, 1988; Fiorito, Jarley, & Delaney, 1995; Fiorito et al., 1991). More important, to the extent unions make choices that influence their characteristics (e.g., over ideology, structure, linkages, and resources allocations) and related functional strategic foci, they may directly affect these outcomes. As Lawler (1990, p. 2) observed, an important part "of the strategic choice paradigm is that key organizational decision makers have significant discretion in selecting organizational actions and what they choose to do matters."

Unfortunately, the general approach taken to analyzing union characteristics has not promoted understanding union behaviors and whether or not behavioral changes might reflect the sort of fundamental shift that more and more observers say is unavoidable if unions are to bounce back (e.g., Tasini, 1995). A particularistic as opposed to configurational methodology is employed (Jarley et al., 1997). In fact, Hoxie's (1917) brave attempt has long represented one of the few comprehensive taxonomies of unionism available. Partly as a result, almost as much uncertainty and spirited disagreement exists over unions and their role(s) as was the case when over 80 years ago Hoxie (1917, pp. 30-31) commented:

> Thus the student honestly seeking the truth about unionism is faced at the outset with a mass of confident but contradictory interpretations. He is told that unionism is a narrow group organization designed to benefit certain favored workmen at the expense of all others; that it is an artificial monopoly of labor, an impossible attempt to raise wages by unnatural and therefore socially inimical means; that it is the creation of selfish and unscrupulous leaders primarily for their personal gain and aggrandizement, a thing foisted upon unwilling workers and designed to disrupt the natural harmony of interests between employers and employees; that it is a mere business device for regulating wages and conditions of employment, by means of collective bargaining; that it is a great revolutionary movement, aimed ultimately to overthrow capitalism and our whole legal and moral code; that it is a universal expression of working class idealism whose purpose is to bring to all the toilers hope, dignity, enlightenment, and a reasonable standard of living; that it is, in short, selfish and altruistic, monopolistic and inclusive, artificial and natural, autocratic and democratic, violent and law-abiding, revolutionary and conservative, narrowly economic and broadly social.

Hoxie (1917) strived to impose some order on this seemingly pell-mell situation. As a testament to this effort, his taxonomy (implicitly or explicitly) remains a conventional point of departure for discussing paradigms of union behavior, notwithstanding that his conceptualization of unionism fails to capture much of the essence of unions today. Still, the notions he advanced provide the (1) contours within which union behaviors may be meaningfully assorted and (2) benchmarks against which to compare emerging alternatives. Union characteristics, plus stra-

Table 1. Hoxie's Union Taxonomy

Dimension	Business Unionism	Uplift Unionism	Revolutionary Unionism	Predatory Unionism
Ideology	Pragmatic, Trade-Conscious	Idealistic	Revolutionary, Socialistic, Anarchistic	Unprincipled, Outwardly Conservative/ Business Oriented
Militancy	Conservative, Law-Abiding	Conservative, Law-Abiding	Radical, Disobedient	Corruptible Guerilla
Function	Collective Bargaining, Contract Enforcement	Collective Bargaining, Mutual Insurance, Political Action	Collective Bargaining as Expedient, Class Political Action	Collusion, Subversive
Structure	Exclusive Jurisdiction, Bureaucratic, Autocratic	Inclusive, Representative	Inclusive, Representative	Secretive

tegic functional choices, are the building blocks of paradigmatic change in unionism as a concept of worker representation.

Hoxie's Taxonomy

Hoxie (1917) developed a four-class taxonomy of trade unionism: business; friendly or uplift; revolutionary; and predatory. Unions cluster around four organizational dimensions, namely, ideology, militancy, structure, and strategic functional emphasis (see Table 1). Business unionism has become the straw person of change advocates. However limited its contemporary descriptive utility may be, this union type is blamed for much of labor's failures. If its vestiges are not jettisoned, labor's revitalization efforts cannot possibly succeed. Organizing models are therefore offered as essential replacements.

Hoxie's (1917) *business unionism* is characterized by a decidedly trade rather than class consciousness. Pragmatism dominates idealism, with an emphasis on serving the immediate work-related and economic interests of workers engaged in a common occupation or industry. Second, unions are highly instrumental in their functional pursuits, relying on collective bargaining and subsequent contract administration or enforcement to achieve material gains for represented employees. Generally speaking, political action, especially of a ideological ilk, is eschewed. Third, business unionism is fundamentally conservative in nature.

Strikes are loathed, undertaken strictly to achieve collective bargaining goals. Finally, union jurisdictions are guarded jealously. Restricting union membership and labor market entry are deemed essential to job preservation or control. Along with an exclusive rather than inclusive orientation, business unionism is structured autocratically and bureaucratically. Extensive rank-and-file involvement in union affairs is not an essential part of the ethos of business unionism.

Friendly or *uplift unionism* is also noted for its conservative, law-abiding, non-revolutionary orientation. Unlike business unionism, however, it tends to be idealistic, aspiring to the benefit of trades or classes. Elevating the quality of workers' lives is a primary aim. Collective bargaining is a means to this end, as are the functions of political action and mutual insurance. This type of unionism strives to be democratic, or genuinely representative of everyday workers' interests.

By contrast, revolutionary unionism is radical, espousing a class consciousness. The socialist variety regards collective bargaining as a means to a broader societal end, a necessary evil as it were. Broader-based political activity is encouraged to upset the social order, the capitalistic nature of which is ideologically abhorred. In the case of the quasi-anarchistic variant, the struggle for revolution is based on a wholesale rejection of conventional methodology (e.g., collective bargaining) and replaced by violent agitation.

A fourth class of unionism identified by Hoxie (1917) is labeled *predatory unionism*, of which there are the hold-up and guerilla forms. This type of unionism is unprincipled. Thus, it may be trade or class-conscious, conservative or radical. The hold-up version exists where unions essentially collude with employers for the benefit of their leadership. Corrupt union bosses use collective bargaining for personal gain. Bribes and acts of violence are used to exploit the system. Guerilla-type unions, in contrast, reject collusion, operating instead in a ruthless manner against employers for selfish ends. Both varieties operate more or less secretively.

Servicing Unionism

Hoxie's (1917) classification of unionism is arguably problematic. Obviously, it was written well before modern forces shaped today's labor movement. Thus, while U.S.-based unions are commonly described, especially in comparison to other foreign-based labor movements, as pragmatic and ideologically mellow, they do not adhere strictly to the concept of business unionism. Unions vary considerably in terms of their political orientation, militancy, and internal democratic practice/structure. Indeed, labor unions are currently under vigorous attack from ideological conservatives for their political activities, especially as they involve the expenditure of union members' dues. In short, business unionism may be convenient shorthand for organized labor in the United States, but it nonetheless provides a somewhat limited conceptual interpretation of what labor unions have done over several decades, especially since the New Deal era.

Hurd (1998) and others (e.g., Block, 1980; Voos, 1987) have argued that, in the main, contemporary unions have made the fundamental choice to emphasize servicing current members at the implicit or explicit expense of organizing new members. This choice flowed somewhat naturally from the political-democratic nature of unions, at least to the extent that union leaders depend on rank-and-file support for election and re-election. Some have claimed that this union focus has been one of the factors contributing to labor's decline in the past few decades (Chaison & Rose, 1991; Rose & Chaison, 1996). Block (1980, pp. 101-102) observed:

> that one of the main reasons that union membership as a percentage of the private labor force and employment has declined in recent years is because unions in the "traditional sectors" of manufacturing, mining, construction, and transportation...have not placed a high priority on organizing. If unions are basically democratic organizations, then it is logical to think that the reason that these unions have not placed an emphasis on organizing is because organizing new members may not be in the interests of the membership of the union.

From this perspective, a tradeoff exists perforce between organizing, on the one hand, and representational or servicing activities, with the ironic result being that historical success in the former area produces its subsequent neglect: "as unions mature and increase the extent of their organization in their primary jurisdiction, the need of the membership for union-organizing services declines relative to their need for representation services" (Block, 1980, p. 102). Thus, as union density grew rapidly during the World War II era, unions chose to concentrate on serving the increasingly complex needs of their vastly expanded membership. The allocation of finite union resources (i.e., money, time, and personnel) followed accordingly. Given the ideologically pragmatic orientation of U.S. trade unionism, a tendency substantially reinforced by the necessities of full-scale war mobilization, unions adopted a particular kind of servicing model. Simply put, unions were about securing tangible benefits for their members through mechanisms such as collective bargaining and political action. Unions pursued those activities which would further the interests of the employees in their duly recognized bargaining units. Through their membership and dues, often extracted via union-security arrangements, these employees provided the necessary resources to support union officials, staff, and services. Unions differed not so much in their explicit or implicit devotion to this model as in the way in which they operationalized relevant pursuits. Within this framework, unions made certain choices, such as vigorously pursuing, where necessary, political action, which took them beyond the strict confines of Hoxie's "business" conceptualization. They nonetheless fit into a mold that sought the "selective incentives" of unions, that is, the benefits that could be excluded from non-members (Olson, 1965).

More specifically, the servicing model of unionism is essentially a logical or evolutionary extension of business unionism (Table 2). Unions have a very results-oriented, short-term, pragmatic ideological orientation that is inwardly focused on current bargaining unit representation. The reality of extensive government

Table 2. Expanding Hoxie's Taxonomy: Servicing Unionism

Dimension	General Description
Ideology	Pragmatic, Trade-Conscious
Militancy	Conservative, Law-Abiding, Strikes Tied to Immediate Collective Bargaining Situations
Function	Collective Bargaining and Political Action Aimed to Serve Current Members; Political Action Advanced Bargaining Position
Structure	Exclusive Jurisdiction, Centralized-to-Decentralized, but Largely Nonparticipative at Rank-and-File Level, Rising Importance of Union Staff to Serve Members

involvement in economic, social, and union affairs (through policies and programs as diverse as Social Security, Medicare, and the Labor-Management Reporting and Disclosure Act of 1959), coupled with the massive growth in public-employee unionization, necessitate political action as a generally important but also very pragmatic strategic function (Aaron, Grodin & Stern, 1979; Aaron, Najita & Stern, 1988; Bok & Dunlop, 1970; Greenstone, 1977). At the same time, unions tend to be rather top-heavy or nonparticipative in terms of setting policy, particularly at the national level of organization. In addition, jurisdictions are guarded jealously, as is clearly consistent with the servicing focus on current membership.

Hurd's (1998, p. 8) elaboration upon the servicing model is worth repeating here, particularly to set up the organizing alternative that he and others have advanced:

> The national union evolved into a service organization. The detailed rule-based nature of agreements combined with the increasingly legalistic grievance and arbitration system to enhance the importance of full-time union staff who developed expertise in these functions. This development reinforced a trend toward the emergence of a union bureaucracy which had appeared in the new CIO unions in response to rapid growth and the attendant need for administrative control that expanded upon the traditional top-down hierarchy in AFL unions. Although some observers raised questions about the elitism of national union officials and/or the metamorphosis of unions into business organizations, by and large the bureaucratic system was effective at delivering what the members wanted: steadily improving economic rewards and protection from arbitrary treatment on the job....By 1955, unions were firmly committed to an approach which has come to be known as the "servicing model..." The elected officials and field staff of national unions would focus on collective bargaining and contract enforcement. The AFL-CIO would coordinate political activity. The labor bureaucracy would concentrate on supporting these functions as efficiently as possible. Under the service model, active involvement of the members would not be necessary. Missing from this framework was any clear conception of how organizing would fit.

Indeed, union membership, as a percentage of the workforce and eventually in absolute numbers as well, continually shrank as organizing lost its perceived relevance. Whereas unions allocated a large chunk of their budgets to organizing the emerging industries of steel, auto, and rubber manufacturing in the 1930s, they had spent, on average, less than 3% percent of their budgets on this strategic function in recent decades (Bernstein, 1995, 1996; Galenson, 1960; Masters, 1997). Rose and Chaison (1996, pp. 86-87) pinpointed the dilemma U.S. unions, in contrast to their Canadian counterparts, faced with a servicing mentality amidst a climate of decline:

> Canadian unions assigned higher priority to growth because the pervasive social unionism in Canada is predicated on a commitment to organizing the unorganized and to inspiring current members to make sacrifices and take risks for organizing. *The prevalent business unionism approach in the United States places a greater emphasis on the provision of representational services to the present membership than on the nonmembers* (Emphasis added)...Confronting expanded employment in lesser-unionized sectors, Canadian unions renewed their commitment to organizing, increased organizing resources, and attempted to develop new organizing strategies and appeals for those who have had little experience with collective bargaining, such as service and part-time workers. Although these efforts do not always achieve the desired results, Canadian unions did increase organizing expenditures by an average annual rate of 20 percent from 1984 to 1987...

THE EMERGING ORGANIZING MODEL

The unabated decline in union ranks has produced considerable disaffection with the servicing model. Total union membership in the United States dropped by almost six million in the 15 years between 1979 and 1993; unions like the Steelworkers, Auto Workers, Clothing and Textile Workers, and Ladies Garment Workers lost 50% or more of their members in this short period. Today, just 10% of the private sector U.S. workforce is unionized. According to some, labor's neglect of organizing has been not only short-sighted but patently irresponsible: "For too many years, the labor movement has been called a dinosaur. This is unfair and almost slanderous—not for labor, but the poor maligned dinosaur" (Tasini, 1995, p. C3).

While labor has been arguably slow to react, it did not ignore the signs of degeneration. In the mid-1980s, the AFL-CIO (1985) issued a highly introspective and somewhat self-critical report entitled *The Changing Situation of Workers and Their Unions*. It advocated union reforms and intensified organizing efforts, including experimentation with "associate membership" recruiting tactics. Nearly 10 years later, another AFL-CIO report urged labor to partner with employers to serve their mutual interests. A *Labor Perspective on the New American Workplace—A Call for Partnership* (AFL-CIO, 1994, p. 15) called upon unions to adopt new approaches to worker representation at all levels of corporate decision making and across a broader array of workplace and business issues:

> Perhaps most important of all, the new work systems require unions to embrace an expanded agenda and to assume an expanded role as the representative of workers in the full range of management decisions in which those workers are interested. In a variety of forums, at various

levels within the enterprise and the industry in which the enterprise is located, unions must represent these workers not only with respect to their terms and conditions of employment but also with respect to the strategic decisions that shape their working lives.

To some extent, both reports reflected an internal union struggle. A growing number of union leaders questioned the servicing model, with its blind focus on improving what labor has rather than enlarging the base. This psychological recognition, however, was largely inevitable. At some point, the union base began to shrink to such dangerous levels that delivering minimal services was threatened. Growing nonunion competition at home and abroad weakened labor's bargaining power. With fewer members, unions had less influence over lawmakers at all levels of government. Politically, it could turn nowhere else but to a more or less ungrateful, unreciprocating Democratic party. A declining base, furthermore, produced less revenues to sustain union services. Many unions had to cut back. Finding the money to fund major new initiatives became increasingly difficult. Eventually, unions had to reorder priorities. The option of simply adding a vast organizing program onto the menu of existing union services faded. An organizing model of unions gained popularity.

To many union observers, friends, and foe alike, the election of the Sweeney slate in 1995 heralded a paradigmatic shift from the servicing to the organizing concept of union representation. Organizing had to precede servicing because the latter could no longer be effectively delivered with an apparently irreversibly declining membership base. Sweeney promised an infusion of money and personnel into organizing, tearing a page from his Service Employees International Union book. Under the dire circumstances, few dissenters or servicing holdouts could be found:

> The American labor movement is at a watershed. For the first time since the early years of industrial unionism sixty years ago, there is near-universal agreement among union leaders that the future of the movement depends on massive new organizing. In October 1995, John Sweeney, Richard Trumpka, and Linda Chavez-Thompson were swept into the top offices of the AFL-CIO, following a campaign that promised organizing "at an unprecedented pace and scale" (Bronfenbrenner et al., 1998, p. 1).

The paramount need to organize had become manifest. Union decline had meaning significantly beyond mere lost membership. Management, which had more than passively supported the decline, had effectively exploited it to advantage (Freeman & Medoff, 1984). Increasingly, employers had acquired the power to dictate the terms of labor relations: "The 1980s witnessed a dramatic shift in bargaining power in management's favor, with employees setting the bargaining agendas in many industries..." (Voos, 1994, pp. 2-3). In both economic and political terms unions had real fears of becoming irrelevant. Certainly, major U.S. employers had unconcealed intentions of avoiding unions, combatively confronting them where necessary, and breaking them outright when feasible (Kochan et al., 1986; Voos, 1994).

Placing the organizing of union members high on the union functional agenda had clear historical precedent (Galenson, 1960). In the 1930s, unions had indeed made the strategic choice to organize (i.e., to come forcefully to the aid of nascent organizing drives in whole industries). While eventual union successes cannot be attributed to the money that followed this choice, they had everything to do with the psychological commitments that these financial investments implied.

Putting more emphasis on organizing is a straightforward proposition. But transferring a union from a servicing to an organizing model is a much less certain task. What is an organizing union? Is it principally brought about by a budgetary allocation or reordering? Or is it a psychologically based phenomenon with budgetary implications? Where do the now traditional functions of servicing members through collective bargaining and contract administration (as well as political action) fit into the organizing model?

In practice, the organizing model has been developed to include both internal and external components (Fletcher & Hurd, 1998). Internally, unions may choose to transform their existing staff functions into organizing staff, with the objective being to mobilize the current membership (Fletcher & Hurd, 1998; Hurd, 1998; Nissen, 1998; Sciacchitano, 1998; Turner, 1998). Such involvement may, incidentally, serve as a magnet to attract new members. With such an internal shift, rank-and-file involvement is often promoted to fulfill the more conventional servicing functions performed previously by union staff. The focus is, however, on organizing within the current union, with the potential benefit of adding new members being the practical fallout of such activity.

The external dimension essentially fixes the union eye on the unorganized. That is, union staff are turned into external organizers, charged with the responsibility of adding new recruits. Union rank-and-file are used to organize new members at existing sites and at nonunion sites. Servicing functions *per se* are placed on the back burner, at least in terms of the attention granted by full-time staffers. The goal is to widen the existing union market, not mobilize the current unionized workforce, although the latter may serve the former.

In a sense, the internal focus of union organizing as described above does not represent a genuinely organizing concept of unionism (in our judgment). Organizing rank-and-file is seen more as an end, with undeniably salubrious results, than a means to an end, that is, enlisting rank-and-file members to expand the dues-paying rolls, to put it crassly (Bronfenbrenner & Juravich, 1998; Grabelsky & Hurd, 1994; Lewis & Mirand, 1998; Nissen, 1998; Turner, 1998). The externally focused version colligates more with an organizing model dedicated to expanding dramatically union membership as inspired by Sweeney and company's call for transformation or reinvention.

There is an apparent philosophical premise that encompasses both the internal and external foci into a broader organizing concept of unionism. According to

Table 3. Expanding Hoxie's Taxonomy: Organizing Unionism

Dimension	Business	Servicing	Organizing Internal	Organizing External
Ideology	Trade Conscious	Trade Conscious	Trade Conscious	Trade-or-Class-Conscious
Militancy	Conservative; Law Abiding	Conservative; Law Abiding	Conservative; Law Abiding; Occasionally Civil Disobedient	Conservative; Law Abiding; Occasionally Civil Disobedient
Function	Collective Bargaining; Contrac Administration	Collective Bargaining; Contract Administration; Political Action	Organizing to Serve Other Functions	Organizing to Serve Continually Expanding Social Unit
Structure	Exclusive Jurisdiction; Hierarchical; Limited Rank-and-File Participation	Exclusive Jurisdiction; Hierarchical; Limited Rank-and-File Participation	Exclusive Jurisdiction; Democratic; Rank-and-File Participation	Inclusive Jurisdiction; Democratic; Rank-and-File Participation

Fletcher and Hurd (1998) and Grabelsky and Hurd (1994), such a conceptualization of unionism regards unions as fundamentally social units rather than principally bargaining (or political) agents. The union exists to help people solve problems, an idea akin to the mutual insurance function early advocated by the Webbs (1897). The organizing union exists to help workers help themselves. Bargaining is a means to that end, not the primary end itself. It serves a primarily organizing purpose. Unions recreate their common purpose, that is, solving workers' problems, by continually organizing those within and outside current boundaries. A seamless web of union activities is created (see Table 3):

> Rather than solving members' problems for them—the essence of the servicing model—the union transforms individual workers into a cohesive force to collectively solve problems. Following this new model, problems are seen as issues around which prospective or current members are organized, and workers learn the essence of unionism by participating in and experiencing collective action. The organizing model envisions a union that behaves basically the same before and after certification [of a bargaining unit]. The only difference is that after certification the union has access to one additional tool: the collective bargaining agreement (Grabelsky & Hurd, 1994, p. 100).

Although conceptually bold, the organizing model is defective in two serious respects. First, it is too easily confused with simply rearranging budgets. Spending more money on organizing, while necessary, does not produce success if it is spent on the failed practices of the recent past. Also, money is no substitute for the energy, confidence, and persistence that goes with genuine human commitment to an organizing philosophy. Second, and more fundamentally, it is impractical for unions to become totally organizing units. They cannot neglect essential service functions without provoking discontent. Furthermore, a strict focus on organizing limits their strategic and tactical versatility. The approach to servicing members should both be leveraged to promote organizing and advertised to recruit rank-and-file. In other words, good service is a good recruiting tool, and the bargaining and political means of servicing can generate organizing opportunities. While a genuinely holistic organizing model might conceptually encompass these possibilities, its lack of practical specificity may seriously undermine its acceptability. Therefore, an expanded conceptualization as unionism is called for; it also comports better with the scope of revitalization strategies underway.

REVITALIZATION STRATEGIES

Faced with disappointments and hostilities on multiple fronts, unions have initiated a host of strategies and tactics to put themselves on an upward trajectory. The range of related activities is extensive and diverse, as well as varied within and between unions. Because labor's revitalization efforts are neither centrally directed nor reported, they cannot be compiled into a neat compendium. On the surface, they may appear unconnected, transient, or insignificant. But taken together, several examples of strategic responses seem to reflect a paradigmatic shift.

We argue that a new concept of unionism can be lifted from these changes. Labeled *value-added unionism*, it focuses on leveraging the strategic functions and resources of unions in a coordinated effort to increase membership by emphasizing, to employers, government, and the unorganized, how unions can add value or promote their underlying interests. On the flip side of the coin, the value of supporting unions can also be shown by raising the costs of resisting their efforts to organize.

Traditional Practice

Under the business and servicing models of unionism, labor organizations focused on a rather narrow set of representational activities. Organizing efforts were characterized by reactive, *ad hoc* responses that were often resource-starved. Bargaining and political activities were directed by union leaders to goals that met the short-term interests of members and leaders. Mutual insurance became largely irrelevant, as unions relied on bargaining and government to achieve tangible benefits. In short, there was little attempt at generating genuine grass-roots, broad-based support for the cause of unions per se. These strategies reflected an inward-thinking, behaviorally insulated orientation.

Table 4. Selected Union Revitalization Strategies

Functional Dimension	Illustration

Bargaining
(Use Bargaining Clout as Leverage to Garner Members)
	• Job-Security Agreements
	   (Auto-Workers—Ford, GM, Chrysler Contracts)
	• Neutrality Agreements
	   (Steelworkers' New Directions Bargaining)
	• Promote Worker Democracy
	   (AFL-CIO Committee on Workplace Democracy)
	• Create Competitive Work Organizations
	   (Machinists' High Performance Work Organization Program)
	• Supply Qualified Workforce
	   (Laborer's Laborers-Employers Cooperation and Education Trust)
	• Forming Strategic Partnerships with Employers (Steelworkers' New Directions Bargaining)

Organizing	• Training and Mobilizing Organizers and Rank-and-File Organizing Efforts
	   (Electrical Workers' Construction Organizing Membership Education Training Program-COMET; AFL-CIO Organizing Department; AFL-CIO Union Cities; AFL-CIO Union Summer)
	• Mobilizing Community Alliances
	   (Service Employees' Justice for Janitors Campaigns; AFL-CIO Union Cities; AFL-CIO Union Summer)
	• Mobilizing Community Alliances
	   (Service Employees' Justice for Janitors Campaigns; AFL-CIO Union Cities; United Electrical Workers [UE] Steeltech Manufacturing Campaign in Milwaukee)
	• Targeted Investments to Create Jobs
	   (AFL-CIO Housing Investment Trust)
	• Alternative Forms of Representation
	   (United Food and Commercial Workers' Corporate Campaign Against Food Lion; Federation of Physicians and Dentists' Third-Party Representation of Doctors)

Political	• Form Interest-Group Alliances
	   (American Federation of State, County and Municipal Employees' Citizen Action; Laborers and the Transportation Construction Coalition)
	• Generate Popular Support for Union-Backed Causes
	   (AFL-CIO Labor '96 Campaign)
	• Promote Grass Roots Political Activism
	   (AFL-CIO Labor '96 Campaign; Teamsters UPS Strike Public Relations Campaign; Union Campaign to Defeat Proposition 226)

Ongoing Strategies

In recent years, however, many unions have initiated strategic changes to reverse the further diminution of their power as institutions. They have adopted new approaches to bargaining, organizing, and politics, involving community-based activism; economic retaliation; interest-group alliances; union consolidation; and alternative representation forms. The objective of these efforts, separately and collectively, is to create real incentives for unionization. These incentives may raise the benefits of unionism or the prospect of higher costs for opposing unionism. In addition, unions have taken a much more comprehensive view of traditional activities, using bargaining and political activities to gain more organizing leverage (see Table 4 for selected current examples).

Bargaining Strategy

This strategy leverages labor's extant relationships with employers to facilitate unionization. While in practice it revolves principally around units where unions have achieved formal recognition based on majority status, it need not be so restricted in concept. Alternative forms of worker representation might be established from which unions could extract additional advantage. The focus here, however, is leveraging current bargaining relationships in a formal sense to achieve agreements that promote union growth.

More specifically, unions have attempted to use their bargaining power to negotiate agreements to preserve and expand union jobs by limiting such practices as outsourcing and, hiring temporary or part-time rather than full-time workers. Examples include contracts negotiated between the United Auto Workers and General Motors, Ford, and Chrysler in the last full round of negotiations in the auto manufacturing industry. Indeed, the string of strikes against General Motors has resulted from alleged corporate breaches of these job-saving efforts.

In addition. unions have tried to leverage bargaining power to expedite the process of recognizing units. Neutrality agreements are a key aim of the Steelworkers' New Directions bargaining strategy. In this vein, unions like the Communications Workers and the United Food and Commercial Workers have sought card-checking agreements, in which employers are willing to recognize unions upon the presentation of signed cards authorizing union representation rather than go through a potentially exhaustive and injurious formal union certification procedure.

Union have also attempted to use bargaining approaches to give unionized firms a competitive advantage. For example, the AFL-CIO has created the Committee on Workplace Democracy to promote worker involvement "particularly in areas of decision-making which have traditionally and legally been managements' prerogative" (AFL-CIO, February 19, 1997, statement). The objective is to help companies compete by empowering workers. Such competitive advantage, facilitated directly by unions, would help firms survive and prosper, thus growing union jobs. Other prominent examples are the International Association of Machinists' high performance work organization (HPWO) programs, in which the union trains both management and labor on HPWO practices, and the Laborers International Union's efforts to assist employers in bidding successfully for lucrative government contracts, thereby promoting union work. In the Steelworkers' New Directions bargaining strategy, the union has pledged to partner with cooperative employers to grow the business: union involvement on the corporate board of directors in an important element of this partnering process.

Organizing Strategy

Numerous unions have been extremely aggressive in expanding organizing efforts and using different approaches to attract new members. To some extent, unions have returned to the grass-roots, brass-tacks approaches which served them well in the past. Only they have attempted to do so in a much more sophisticated and comprehensive way, increasing organizing budgets, expanding rank-and-file participation, and training organizers. Parenthetically, recent union mergers, in effect, announced, or being contemplated have been partly motivated to provide a more coordinated union approach to organizing. Examples include the merger of the Clothing and Textile Workers Union and Ladies Garment Workers into the Union of Needle Trades, Industrial, and Textile Employees (UNITE), the planned merger between the Auto Workers, Machinists, and Steelworkers, and still ongoing efforts to combine the ranks of the National Education Association with the American Federation of Teachers.

In particular, unions have launched organizing campaigns on a scale unmatched in nearly 60 years. To signify this renewed commitment, the AFL-CIO has created a formal Organizing Department, the first in its history. It infused $20 million into organizing campaigns, expanded its Organizing Institute Program to train grass-roots organizers, and pledged to increase union membership at a 3% annual clip by the year 2000. The federation, as suggested earlier, "also strongly urged affiliated unions to follow its example by shifting more money in their budgets to pay for well-staffed organizing campaigns" (Kelber, 1998, p. 1). Many affiliates have evidently followed suit, according to Bronfenbrenner et al. (1998, p. 1):

> The events at the AFL-CIO are not happening in a vacuum. Simultaneously, some of the nation's other large unions including the International Brotherhood of Teamsters (IBT), the Service Employees International Union (SEIU), the Communications Workers of America (CWA), the International Brotherhood of Electrical Workers (IBEW), and the newly merged Union of Needle Trades, Industrial, and Textile Employees (UNITE), have made significant structural adjustments at local and national levels to shift resources to organizing. Other unions, such as the Laborers' International Union of North America (LIUNA), the Oil, Chemical, and Atomic Workers (OCAW), and the United Paperworkers International Union (UPIU), have filled voids by establishing national organizing departments, reflecting new-found com-

mitment to organizing from the top leadership of these unions. Many other unions, at both the national and local levels, have increased their organizing activities significantly.

Several specific union initiatives are worthy of mention in this context. They reveal a four-pronged approach to organizing which departs fundamentally from what had become the more or less traditional practice of parachuting in outside organizers in largely reactive, piecemeal organizing efforts. The efforts focus on training and mobilizing organizers per se and rank-and-file organizing volunteers (Union Summer, COMET); orchestrating supportive community-based alliances, including interunion coordination (Union Cities; Justice for Janitors); targeting union-leveraged investments to create potentially unionized jobs (AFL-CIO Housing Investment Trust); and experimenting with alternative forms of representation to build union membership and fealty (Federation of Physicians and Dentists' (FPD) third-party representation of doctors (FPD, undated brochure):

> *Third party messenger services*, providing physicians with expert representation on negotiating agreements with managed care companies and health insurers; *Disciplinary representation*, representing physicians and dentist who are threatened with disciplinary action by a governmental or professional entity; *Lobbying representation* before Congress and state legislatures; *Research* on topics of special interest to physicians and dentists (Emphasis added).

The AFL-CIO's Union Summer program is an interesting attempt to train a new generation of union activists in order to perpetuate and expand the labor movement. Started in 1996, Union Summer offered a four-week internship to more than 1,000 college students, workers, and community activists "to spend three weeks in the streets and neighborhoods organizing for work-place rights and social justice" (AFL-CIO, *Union Summer 1996: Nuts & Bolts*, 1996, p. 1). The Electrical Workers' COMET program also trains organizers from existing rank-and-file. Specifically, it "is the internal organizing tool through which IBEW members are educated about the organizing process. This phase of the program serves as the foundation for creating on internal organizing culture" (Lewis & Mirand, 1998, p. 302).

To further this organizing culture, the AFL-CIO has also launched the Union Cities program. The objective is to transform existing interunion infrastructures (regional Central Labor Councils) into organizing drivers. Unions affiliated with the Central Labor Councils are urged to adopt the federation's Changing-to-Organize program, which involves:

1. Devoting more resources to organizing;
2. Developing a strong organizing staff;
3. Devising and implementing a strategic organizing plan; and
4. Mobilizing the local's entire membership around organizing.

The objective is to create "A 'Union [i.e., fully unionized] City'...where everybody wants to live, work and raise their family" (AFL-CIO, *The Road to Union*

City, 1997). A Union Cities' campaign, like the one underway in Las Vegas, encourages "organizing" Central Labor Councils to:

> Promote organizing among community allies: Using Street Heat, Rapid Response Team, turn organizing campaigns into struggles for justice for all workers in the community. Get your affiliates together to train a common pool of rank-and-file volunteer organizers who can do house calls for affiliates' campaigns (AFL-CIO, *The Road to Union City*, 1997).

Indeed, it has been suggested that Central Labor Councils, to promote community-wide organizing,

> be transformed into "community labor councils" that include as members both unions and non-union groups interested in work issues. A community labor council might well become an umbrella organization that represents workers' interests generally. The community labor council orientation would make organization of local general unions a natural activity of the council (Gapasin & Wial, 1998, p. 66).

The Justice for Janitors campaigns spearheaded by the Service Employees International Union are essentially community-spirited drives. They rely on a variety of specific aggressive organizing tactics, including rank-and-file mobilization, coalition building, public relations, and extensive regulatory oversight of recalcitrant firms. More important, however, Justice for Janitors glues factions together by stressing the theme of justice for the unjustly treated, often from the immigrant communities in a metropolitan area (Woldinger, Erickson, Milkman, Mitchell, Valenzuela, Wong, & Zeitlin, 1998). A social movement for change via unionization is created:

> As militant union activity among newcomers in a variety of industries suggests that the days of the immigrant helots are over. Emblematic of this shift is the JFJ campaign, which successfully reorganized the building services industry, ultimately bringing more than eight thousand largely immigrant workers [in the Los Angeles area] under a union contract, in what has become a model for JFJ's national organizing efforts (Waldinger et al., 1998, p. 103).

Political Strategy

In the realm of politics, unions have played a prominent role since the New Deal era (Greenstone, 1977). Generally speaking, they have provided a host of interest-group resources to Democratic candidates and lobbied elected officials through conventional channels. Recently, however, labor's political efforts have become both more ideologically oriented and targeted to securing union work. Shifting and sometimes seemingly unholy political alliances (between labor and business) have emerged to further these efforts. In this vein, labor has sought increasingly to define the political battle in ideological terms: minimum wage hikes for the poor workers; health care for the unemployed; job security for the laid off or downsized. Labor has identified itself visibly with popular working-class causes (e.g.,

Table 5. Labor '96

Theme	Labor '96 is an effort to put working family issues in the forefront of the national debate, inform working Americans about candidates' voting records on issues and hold elected leaders accountable.
Financing	Allocated 15 cents per AFL-CIO member per moth to raise $25 million; another $10 million in other funds were added to create a $35 million budget.
Issues	Focused on living wages, secure jobs, retirement security, health care, education, job safety, and workers rights.
Grass-Roots Activists	Placed 135 coordinators in 102 congressional districts, 14 senate races, and two Gubernatorial races.
Voter Mobilization	1,000 workers from 30 international unions were deployed to register voters and get-out-the-vote.
Communication	Placed radio and television commercials in dozens of cities; mailed nearly 10 million pieces of literature; placed four million calls to union members.

Source: Adapted from AFL-CIO single-page document titled "Labor 96: Working for Working Families."

opposition to the North American Free Trade Agreement). At the same time, it has lobbied for important infrastructure-building programs attractive to politicians of all ideological persuasions (e.g., the Laborers' association with the union-business Transportation Construction Coalition lobby). These lobbying efforts, if successful, preserve or generate union jobs. The desired results of both types of efforts are working class employees politically sympathetic and economically indebted to organized labor.

Labor '96 epitomizes the strong attempt to associate unions with politically and economically meaningful causes. A massive and expensive grass-roots and media campaign, Labor '96 sought to reclaim Democratic control of the U.S. Congress and re-elect Clinton-Gore (see Table 5). It is focused on promoting living wages, secure jobs, retirement security, health care, education, and job safety, as well as worker rights.

Laborers-Employers Cooperation and Education Trust (LECET)

An important example of an effective union growth strategy is the Laborers-Employers Cooperation and Education Trust (LECET). Formed as a union of common builders in 1903, the Laborers International Union (LIUNA) has struggled "for decent wages and...a respected voice in the workplace." Formerly the International Hod Carriers, Building, and Common Laborers Union of America, the LIUNA adopted its current name in 1965 at a time when it was gaining in membership and strength due to the prosperity of the post-World War II era. Shortly thereafter, however, LIUNA began to experience some of the same problems that beset labor generally and construction unions particularly. Noncompetitive labor practices and wages, combined with costly internecine jurisdictional disputes among construction unions, promoted growth among nonunion building and construction contractors. The rate of unionization dropped from 31.6% in the construction industry in 1980 to less than 18% in 1995. LIUNA's membership fell from 475,000 in 1975 to just above 350,000 in 1995.

Through proactive, progressive-minded steps, however, the LIUNA has done a literal about face. In recent years, it has bucked the union decline and increased its membership at a time when competition has, if anything, intensified in its core industries. It added about 10,000 members in 1996, according to interviews with knowledgeable union staff. More generally, union density and membership in the construction industry have picked up recently: the rate of unionization rose from 17.7 % to 18.5 % between 1995 and 1996, as unionized construction workers climbed in number from 908,000 to 994,000.

Despite its admittedly blemished reputation, in which current and past union officials have been linked corrupt practices, the LIUNA has nonetheless zealously pursued a strategy to expand *the unionized construction industry*, becoming thereby a literally organizing union. The union has relied heavily on four existing funds: (1) The Laborers-Association of General Contractors Education and Training Fund; (2) the National Health and Safety Fund; (3) LECET; and (4) the Laborers' Political League (i.e., its national political action committee). The Education and Training Fund enhances the employment skills of laborers through joint employer-union training programs. Health and Safety promotes efforts to reduce injuries and other health risks. It also attempts to reduce employers' burdensome workers' compensation costs. The Political League is fundamentally about promoting public policies that entail government spending on infrastructure projects, such as cleaning up the environment, building roads, and public housing, where laborers will be employed.

Of particular interest is the LECET program. LECET is a local, regional, and national network of joint LIUNA-management signatories in which LIUNA seeks "'to agree to disagree' with potential management partners on issues which are contentious, and to pursue partnerships in areas of mutual interests." Fundamentally, LECET is the linchpin fund of the Laborers. Through its networking, it coordinates the activities of the LIUNA, which does the actual bargaining with employers, and the other funds (especially Education and Training and Health and Safety) at three programmatic levels.

First, LECET *markets unionized workers* to union and nonunion contractors, while aggressively pursuing union construction work. The LECET credo is "Projects = Jobs + Profits." Through its Cooperation Trust Tracking System (CTTS), LECET tracks more than 25,000 active construction projects (each worth at least $3 million) in all sectors of the construction industry: commercial building; heavy and highway; tunneling; environmental cleanup; and public housing rehabilitation. This tracking system alerts signatories to LECET as to upcoming contracts on which to bid.

As part of its marketing effort, LECET advertises extensively in key trade periodicals, including the *Engineering News-Record* (ENR). Its glossy advertisements tout various LIUNA achievements through its funds:

- $35 million on training and retraining
- 20,000 + contractors participating
- 50,000 workers trained and industry qualified each year
- 84 state of the art training facilities in the United States and Canada
- 5 mobile training units

In sum, LECET "is a cutting edge joint labor-management trust fund with a simple but potent mission—to generate business opportunities for union contractors and job opportunities for [LIUNA] members" (undated LIUNA literature).

Second, LECET fosters labor-management cooperation by leveraging fund and union activities in order to make union contractors operationally competitive and better able to submit competitive bids for construction projects. LECET helps contractors find qualified workers, reduce labor costs by raising worker productivity, minimize jurisdictional disputes among building and construction trade unions, and reduce workers' compensation costs. Through a Laborers Training Tracking System, LECET is able to identify qualified and certified workers and report on their health and drug test results. This tracking is particularly useful in the environmental cleanup and tunneling parts of the industry where there is a significant shortage of skilled labor. Such assistance was essential in enabling Radian International to bid successfully on an Environmental Protection Agency Superfund contract at a Missouri site. In this vein, LECET estimates that the nuclear power cleanup industry will generate about $20 billion worth of business over the next 20 years or so. Obviously, this business can employ many LIUNA members, which are being trained on how to work in this specialized industry.

Third, LECET energetically lobbies lawmakers at local, state, and federal levels to inform them as to the benefits of unionized work. It urges lawmakers to support important infrastructure projects that might employ LIUNA members. Along with LIUNA itself, LECET is one of the founding members of the Transportation Construction Coalition, which is a union-employer group that promotes, among other things, maximum funding for the Internodal Surface Transportation Efficiency Act (ISTEA). In addition, LECET offers signatory contractors assistance in understanding and complying with a plethora of labor, environmental, and other laws/regulations in the various construction industry sectors.

On a somewhat different matter, LECET promotes the National Maintenance Agreement (NMA). The NMA includes several provisions which are aimed at promoting competitive workplace practices in construction, thus adding value to business through unionism. Among them are: a no-strike clause; flexibility in scheduling; contractor rights to determine crew size; and provisions to enable contractors to respond to changing needs. To promote competitive work practices further, LIUNA is at the forefront advocating merging the 15 or so major building and construction trade unions into four or five. These mergers would not only give employers greater staffing flexibilities and fewer jurisdictional scuffles but would help the LIUNA reach its ambitious one-million membership goal.

VALUE-ADDED UNIONISM

In a general sense, these revitalization strategies involve either leveraging already existent union clout or generating new bases of support to promote the common goal of sustaining and expanding union membership. Beyond that, as previously suggested, they may appear rather unlinked. Indeed, no consciously procrustean force exists to impose on labor unions a representational philosophy to substitute for the servicing model. Nor is there uniform opinion on an organizing alternative.

However, it is possible to glean conceptually a paradigmatic transformation in unionism from the changes under way. That is, unions have been experimenting with ways to add value to business for operating "union." The advantages come in the form of offering positive and negative incentives. There are decidedly cooperative and confrontational elements to value-added unionism.

On the positive side, unions can offer employers several benefits: a trained and stable workforce; a progressive and enforceable disciplinary process; political alliances to influence lawmakers; a favorable public relations image; labor-management peace (through no-strike pledges); expertise in designing and implementing high-performance-work-organization practices; assistance in bidding for business; and an involved and committed workforce through workplace democracy initiatives. In the process, employees (current and prospective) gain value through enhanced training and skill development, job security, higher wages and benefits, and justice and dignity on the job. In this regard, employees acquire value in large part because of their attachment to the union regardless of their employment status. Nonunion employees become attracted to these benefits.

Conversely, the union revitalization strategies entail use of the sword as well as the carrot, if for no other reason than that unions are in dire straits, a situation many employers are willing to exploit further. Costly organizing campaigns will be launched, powerful political/community alliances will be used to damage hostile companies, and corporations and their executives will be subjected to public embarrassment when confronted by unions determined to overcome their resistance. Unions therefore add more value to cooperative employers, in concept at least.

The point is that unions can be looked upon as promoters of justice that seek, through very pragmatic and useful means, to add value to the businesses that operate "union." A requisite to this effort is to add value to the unionized workforce, which promotes both greater loyalty to the union and potential productivity to the employing business. Further, and this is a critical point, unions embrace a philosophy of promoting social justice, representing employees in union/nonunion and bargaining/nonbargaining venues to serve their social and economic needs. They help educate workers, supply employers a quality workforce, and represent workers in the civic and political communities. In essence, unions become more of a *social* institution, helping people become real problem-solvers. This value-added notion encompasses enlarging "the playing field" strategic approach discussed by Wever (1998, p. 389), which "allows unions to gain leverage by addressing unresolved social problems and market failures. Because such problems are found everywhere...this strategy holds promise for unions in a wide range of circumstances and settings."

A RESEARCH AGENDA

Philosophically, the biggest difference between the business/servicing and value-added conceptualizations of unionism is that while the former focused on promoting the immediate interests of union members, the latter emphasizes giving employees (union and nonunion alike) and employers a reason to support unions. The benefits are intended to be durable (competitively sustainable) and transferable across employment relationships. Conceptually, unions become more appealing, which is a solution to their present difficulties.

What is striking about the fields of industrial relations and human resource management, however, is the absence of conceptual thinking. Rarely in the literature does one come across conceptual discussions of unionism and how it has changed or might be changed. What should unions do is really about what should unions institutionally become to broaden their appeal and ensure their continued economic and political viability. Alternative labels (e.g., the organizing model) are tossed around, but left conceptually nebulous or confused (e.g., internal versus external focus). Undoubtedly, the literature and the broader community could benefit from addressing the "larger questions about the strategic direction of the entire labor movement...[which] go to the very basis of U.S. trade unionism" (Fletcher & Hurd, 1998, p.53). A scholarly discussion must take place over "a clear alternative to business unionism, a vision that can touch a large segment of members and be relevant to everyday life" (Fletcher & Hurd, 1998, p.53).

We propose an expansive research agenda to embellish the conceptual underpinnings of trade unionism, with a strong emphasis on interdisciplinary approaches.

First, a thorough conceptual revisitation of Hoxie's (1917) model is due. It should be based on current experiences and observations of U.S.-based and non-U.S. unions.

Second, a conceptual and empirical review of revitalization strategies and tactics cross-nationally is needed. Union decline is an international phenomenon. How have unions in different countries responded and what do their varied responses suggest about the future of unions in workplace representation? One of the unfortunate gaps in the literature is a rich empirical review of how unions, institutionally, have responded. Detailed case studies and survey methods are needed.

Third, we encourage exploration of alternative models of unionism in terms of their expected effects on membership. How do social-political models overcome the free-rider barriers to collective action? Is mutual insurance, with its membership-specific benefits, the critical factor in mobilizing nonunion workers in contexts outside of the employment relationship? How can unions afford (financially) to represent the lower-paid echelons of the workforce that might benefit most from such a social service?

Fourth, the management strategy literature on resource acquisition and allocation could be applied to unions. Of particular interest is how unions might develop new products to attract membership and change their organizational structures and operations in order to be more inclusive without merely becoming a loose, meaningless and, most probably, short-lived institutional force, like the Knights of Labor.

Fifth, research is needed on the relative costs and benefits of alternative representational and revitalization strategies. How effective is political action as a revitalization strategy? Within the context of political action, which tactics (e.g., grass-roots or media advertising) are most effective under various circumstances?

Sixth, what effects do union mergers and alliances have on the labor movement? Are their certain kinds of mergers that are likely to promote union expansion? What can be learned from the business-merger phenomena about the factors that contribute to success (e.g., future growth) or failure (e.g., collapse or stagnation)?

Finally, we propose that organizational researchers in general pay more attention to unions. Labor organizations are important workplace, economic, social, and political institutions. The practices they may promote can affect organizational performance and the stability of the economic-political contexts in which businesses operate. As non-profit organizations, unions provide insight for how to manage and motivate in financially constrained settings.

ACKNOWLEDGMENT

This research was supported by the Institute of Industrial Competitiveness (IIC) at the University of Pittsburgh Katz Graduate School of Business. We thank Professor Dan Fogel, IIC Director.

REFERENCES

Aaron, B., Grodin, J.R., & Stern J.L. (Eds.). (1979). *Public sector bargaining.* Washington, DC: The Bureau of National Affairs.

Aaron, B., Najita, J.N., & Stern, J.L. (Eds.). (1988). *Public sector bargaining.* Washington, DC: The Bureau of National Affairs.

AFL-CIO. (1985). *The changing situation of workers and their unions.* Washington, DC: AFL-CIO.
AFL-CIO. (1994). *A labor perspective on the new American workplace—A call for partnership.* Washington, DC: AFL-CIO.
AFL-CIO. (1996). *Union summer: Nuts & bolts.* http://www.afl-cio.org.
AFL-CIO. (1997). *The road to union city.* http://www.afl-cio.org.
AFL-CIO. (1997, February 19). *Workplace democracy,* AFL-CIO Executive Council Statement. http://www.afl-cio.org.
Bennett, J.T. (1991). Private sector unions: The myth of decline. *Journal of Labor Research, 12,* 1-12.
Bernstein, A. (1995, July 3). Can a new leader bring labor back to life. *Business Week,* p. 87.
Bernstein, A. (1996, June 10). Andy Stern's mission impossible: A new service workers' union head aims to triple its recruits. *Business Week,* p. 73.
Block, R.N. (1980). Union organizing and the allocation of union resources. *Industrial and Labor Relations Review, 34,* 101-113.
Bok, D.C., & Dunlop, J.T. (1970). *Labor and the American community.* New York: Simon and Schuster.
Bronfenbrenner, K. & Juravich, T. (1998). It takes more than house calls: Organizing to win with a comprehensive union-building strategy. In K. Bronfenbrenner, S. Friedman, R.W. Hurd, R.A. Oswald, & R.L. Seeber (Eds.), *Organizing to win: New research on union strategies* (pp. 19-37). Ithaca, NY: ILR Press.
Bronfenbrenner, K., Friedman, S., Hurd, R.W., Oswald, R.A., & Seeber, R.L., (Eds.) (1998). Introduction. In K. Bronfenbrenner, S. Friedman, R.W. Hurd, R.A. Oswald, & R.L. Seeber (Eds.), *Organizing to win: New research on union strategies* (pp. 1-16). Ithaca, NY: ILR Press.
Chaison, G.N., & Rose, J.B. (1991).The macrodeterminants of union growth and decline. In G. Strauss, D.G. Gallagher, & J. Fiorito (Eds.), *The state of the unions* (pp. 3-46). Madison, WI: Industrial Relations Research Association.
Clark, P.F. (1989). Organizing the organizers: professional staff unionism in the American labor movement. *Industrial and Labor Relations Review, 42,* 584-599.
Clark, P.F. (1992). Professional staff in American unions: Changes, trends, implications. *Journal of Labor Research, 13,* 381-392.
Clark, P.F., & Gray, L.G. (1991). Union administration. In G. Strauss, D.G. Gallagher, & J. Fiorito (Eds.), *The state of the unions* (pp. 175-200). Madison, WI: Industrial Relations Research Association.
Clark, P., & Masters, M.F. (1996). *Pennsylvania COPE union member survey: A final report.* Unpublished monograph.
Craft, J.A., & Extejt, M.M. (1983). New strategies in union organizing. *Journal of Labor Research, 4,* 19-32.
Cutcher-Gershenfeld, J., & McHugh, P.P. (1994). Collective bargaining in the North American auto supply industry. In P.B. Voos (Ed.), *Contemporary collective bargaining in the private sector* (pp. 225-258). Madison, WI: Industrial Relations Research Association.
Delaney, J.T., & Masters, M.F. (1991). Union and political action. In G. Strauss, D.G. Gallagher, & J. Fiorito (Eds.), *The state of the unions* (pp. 277-312). Madison, WI: Industrial Relations Research Association.
Delaney, J.T., Jarley, P., & Fiorito, J. (1996). Planning for change: Determinants of innovation in U.S. national unions. *Industrial and Labor Relations Review, 49,* 597-614.
Delaney, J., & Schwochau, S. (1993). Employee representation through the political process. In B. Kaufman & M. Kleiner (Eds.), *Employee representation: Alternatives and future directions* (pp. 265-304). Madison, WI: Industrial Relations Research Association.
Derber, M. (1970). *The American idea of industrial democracy: 1865-1965.* Urbana: University of Illinois Press.
Dunlop, J.T. (1958). *Industrial relations systems.* New York: Holt.

Epstein, E.M. (1976). Labor and federal elections: The new legal framework. *Industrial Relations, 15,* 257-274.
Fiorito, J. (1987). Political instrumentality perceptions and desires for union representation. *Journal of Labor Research, 8,* 271-290.
Fiorito, J., Gramm, C.L., & Hendricks, W. (1991). Union structural choices. In G. Strauss, D.G. Gallagher, & J. Fiorito (Eds.), *The state of the unions* (pp. 103-128). Madison, WI: Industrial Relations Research Association.
Fiorito, J., Jarley, P., & Delaney, J.T. (1995). National union effectiveness in organizing: Measures and influences. *Industrial Labor Relations Review, 48,* 613-635.
Fletcher, B. Jr., & Hurd, R.W. (1998). Beyond the organizing model: The transformation process in local unions. In K. Bronfenbrenner, S. Friedman, R.W. Hurd, R.A. Oswald, & R.L. Seeber (Eds.), *Organizing to win: New research on union strategies* (pp. 37-53). Madison, WI: Industrial Relations Research Association.
Freeman, R.B., & Medoff, J.L. (1984). *What do unions do?* New York: Basic Books.
Galenson, W. (1960). *The CIO challenge to the AFL. A history of the American labor movement: 1935-1941.* Cambridge, MA: Harvard University Press.
Gapasin, F., & Wial, H. (1998). The role of central labor councils in union organizing in the 1990s. In K. Bronfenbrenner, S. Friedman, R.W. Hurd, R.A. Oswald, & R.L. Seeber (Eds.), *Organizing to win: New research on union strategies* (pp. 54-68). Madison, WI: Industrial Relations Research Association.
Grabelsky, J. & Hurd, R. (1994). Reinventing an organizing union: Strategies for change. In P.B. Voos (Ed.), *Proceedings of the Forty-Sixth Annual Meeting of the Industrial Relations Research Association* (pp. 95-104). Madison: WI: Industrial Relations Research Association.
Greenstone, J. D. (1977). *Labor in American politics.* Chicago: University of Chicago Press.
Hendricks, W., Gramm, C.L., & Fiorito, J. (1993). Centralization of bargaining decisions in American unions. *Industrial Relations, 32,* 367-390.
Hoxie, R.F. (1917). *Trade unionism in the United States.* New York: D. Appleton and Company.
Hurd, R.W. (1998). Contesting the dinosaur image: The labor movement's search for a future. *Labor Studies Journal, 22,* 5-30.
Jarley, P., & Fiorito, J. (1990). Associate membership: Unionism or consumerism? *Industrial and Labor Relations Review, 43,* 209-224.
Jarley, P., Delaney, J.T., & Fiorito, J. (1997). *Moving beyond the craft-industrial distinction: Identifying national union configurations.* Unpublished manuscript.
Jarley, P., Fiorito, J., & Delaney, J.T. (forthcoming). Do unions control their destiny? In K. Hutchinson (Ed.), *Proceedings of the Fiftieth Annual Meeting of the Industrial Relations Research Association.* Madison, WI: Industrial Relations Research Association.
Juravich, T., & Shergold, P.R. (1988). The Impact of unions on the voting behavior of their members. *Industrial and Labor Relations Review, 41,* 374-385.
Katz, H., & MacDuffie, J.P. (1994). Collective bargaining in the U.S. auto assembly sector. In P.B. Voos (Ed.), *Contemporary collective bargaining in the private sector* (pp. 181-224). Madison, WI: Industrial Relations Research Association.
Kau, J.P., & Rubin, P.H. (1981). The Impact of labor unions on the passage of economic legislation. *Journal of Labor Research, 2,* 133-146.
Kaufman, B., & Kleiner, M.M. (Eds.) (1993). *Employee representation: Alternatives and future directions.* Madison, WI: Industrial Relations Research Association.
Kelber, H. (1998). *Labor talk: The organizing crisis.* Unpublished manuscript.
Kochan, T.A., Katz, H.C., & McKersie, R.B. (1986). *The transformation of American industrial relations.* New York: Basic Books.
Lawler, J.J. (1990). *Unionization and deunionization strategy, tactics, and outcomes.* Columbia: University of South Carolina Press.

Lewis, J., & Mirand, B. (1998). Creating an organizing culture in today's building and construction trades: A case study at IBEW Local 46. In K. Bronfenbrenner, S. Friedman, R.W. Hurd, R.A. Oswald, & R.L. Seeber (Eds.), *Organizing to win: New research on union strategies* (pp. 297-308). Ithaca, NY: ILR Press.

Masters, M.F. (1997). *Unions at the crossroads: Strategic membership, financial and political perspectives.* Westport, CT: Quorum Books.

Masters, M.F., & Delaney, J.T. (1984). Interunion variation in congressional campaign support. *Industrial Relations, 23,* 410-416.

Masters, M.F., & Delaney, J.T. (1985). The causes of union political involvement: A longitudinal analysis. *Journal of Labor Research, 6,* 341-362.

Masters, M.F., & Delaney, J.T. (1987). Union political activities: A review of the empirical literature. *Industrial and Labor Relations Review, 40,* 336-353.

Meyer, A.D., Tsui, A.S., & Hinings, C.R. (1993). Configurational approaches to organizational analysis. *Academy of Management Journal, 36,* 1175-1195.

Nissen, B. (1998). Utilizing the membership to organize the unorganized. In K. Bronfenbrenner, S. Friedman, R.W. Hurd, R.A. Oswald, & R.L. Seeber (Eds.), *Organizing to win: New research on union strategies* (pp. 135-149). Ithaca, NY: ILR Press.

Olson, M., Jr. (1965). *The logic of collective action: public goods and a theory of groups.* Cambridge, MA: Harvard University Press.

Peters, R., & Merrill T. (1998). Clergy and religious persons' roles in organizing at O'Hare Airport and St. Joseph Medical Center. In K. Bronfenbrenner, S. Friedman, R.W. Hurd, R.A. Oswald, & R.L. Seeber (Eds.), *Organizing to win: New research on union strategies* (pp. 164-178). Ithaca, NY: ILR Press.

Rogers, J. (1995). A strategy for labor. *Industrial Relations, 34,* 367-381.

Rose, J.B., & Chaison, G.N. (1995). The state of unions: United States and Canada. *Journal of Labor Research, 6,* 97-112.

Rose, J.B., & Chaison, G.N. (1996). New measures of union organizing effectiveness. *Industrial Relations, 29,* 457-468.

Sciacchitano, K. (1998). Finding the community in the union and the union in the community: The first-contract campaign at Steeltech. In K. Bronfenbrenner, S. Friedman, R.W. Hurd, R.A. Oswald, & R.L. Seeber (Eds.), *Organizing to win: New research on union strategies* (pp. 150-163). Ithaca, NY: ILR Press.

Sheflin, N., & Troy, L. (1983). Finances of American unions in the 1970's. *Journal of Labor Research, 42,* 149-158.

Strauss, G. (1991). Union democracy. In G. Strauss, D.G. Gallagher, & J. Fiorito (Eds), *The state of the unions* (pp. 201-236). Madison, WI: Industrial Relations Research Association.

Strauss, G., Gallagher, D.G., & Fiorito, J. (Eds.). (1991). *The state of the unions.* Madison, WI: Industrial Relations Research Association.

Tasini, J. (1995, February 12). Labor's last chance: What the union can do to help the American workers—and save themselves. *Washington Post,* p. C3.

Troy, L. (1975). American unions and their wealth. *Industrial Relations, 14,* 134-144.

Turner, L. (1998). Rank-and-file participation in organizing at home and abroad. In K. Bronfenbrenner, S. Friedman, R.W. Hurd, R.A. Oswald, & R.L. Seeber (Eds.), *Organizing to win: New research on union strategies* (pp. 122-134). Ithaca, NY: ILR Press.

U.S. Bureau of Labor Statistics. (1998, January 30). *Union members in 1997* (press release).

Voos, P.B. (1983). Union organizing: Costs and benefits. *Industrial and Labor Research Review, 36,* 576-591.

Voos, P.B. (1987). Union organizing expenditures: Determinants and their implications for union growth. *Journal of Labor Research, 8,* 19-30.

Voos, P.B. (1994). An economic perspective on contemporary trends in collective bargaining. In P.B. Voos (Ed.), *Contemporary collective bargaining in the private sector* (pp. 1-24). Madison, WI: Industrial Relations Research Association.

Waldinger, R., Erickson, C., Milkman, R., Mitchell, D.J.B., Valenzuela, A., Wong, K., & Zeitlin, M. (1998). Helots no more: A case study of the Justice for Janitors campaign in Los Angeles. In K. Bronfenbrenner, S. Friedman, R.W. Hurd, R.A. Oswald, & R.L. Seeber (Eds.), *Organizing to win: New research on union strategies* (pp. 102-120). Ithaca, NY: ILR Press.

Webb, S., & Webb, B. (1897). *Industrial democracy*. New York: Augustus M. Kelley (1965 reprint).

Wever, K. (1998, July). International labor revitalization: Enlarging the playing field. *Industrial Relations, 37*, 388-407.

ABOUT THE CONTRIBUTING AUTHORS

Robert S. Atkin is currently at the Katz Graduate School of Business, University of Pittsburgh. Prior to his present position, he served as Associate Dean at Carnegie-Mellon University's Graduate School of Industrial Administration for 12 years. He has worked in private industry, and consulted with various domestic and foreign companies. He holds a B.S. in Mechanical Engineering, an M.S. in Management Science, and a Ph.D. in Human Resources Management. His current research focuses on the financial aspects of labor unions, and the future of the U.S. labor movement. Among his recent research is the article "The Finances of Major U.S.-based Unions, 1979-1993," which was published in the October, 1997 issue of *Industrial Relations*.

Kevin Banning is Assistant Professor of Management at Auburn University, Montgomery. After a brief managerial career in a *Fortune* 100 company and starting a new business, he received a Ph.D. in Strategic Management from the University of Florida. He remains on the board of directors of the firm that he co-founded, and now advises small business owners and managers. His research interests include mergers and acquisitions, agency theory, executive succession, and corporate performance.

Robert A. Baron is Professor of Management and Professor of Psychology at Rensselaer Polytechnic Institute. He received a Ph.D. in Psychology from the University of Iowa. Professor Baron has held faculty appointments at Purdue University, the University of Minnesota, the University of Texas, the University of South Carolina, and Princeton University. In 1982, he was a Visiting Fellow at Oxford University. From 1979 to 1981, he served as a Program Director at the National Science Foundation in Washington, DC. He is a Fellow of the American Psychological Association and the American Psychological Society. Baron has published more than 90 articles in professional journals, and 27 chapters in edited volumes, and he is the author or co-author of 36 books. From 1993-1996, he was a member of the Board of Directors of the Albany Symphony Orchestra. He is the President of Innovative Environmental Products, Inc., and he holds three U.S. patents. His current research interests focus on: (1) social and cognitive factors in entrepreneurship; (2) workplace aggression and violence; and (3) the impact of the physical environment on productivity.

Clint Bowers is Assistant Professor of Psychology and Director of the Team Performance Laboratory at the University of Central Florida. He received a Ph.D. in Psychology at the University of South Florida in 1987. He subsequently served as staff psychologist in the U.S. Navy. Since arriving at the University of Central Florida in 1989, he has pursued his research interests in team training and performance, communication analysis, and neuropsychological bases of human performance, and he has served as an author on over 100 scholarly works. Dr. Bowers is a Fellow of the American Psychological Association. He currently serves as a reviewer for several scientific journals, including *Human Factors, Military Psychology,* and *The International Journal of Aviation Psychology.* He has recently been invited to serve as co-editor of the *Handbook of Applied Experimental and Engineering Psychology.*

Janis Cannon-Bowers is a Senior Research Psychologist in the Science and Technology Division of the Naval Air Warfare Center Training Systems Division (NAWCTSD), Orlando, Florida. She holds M.A. and Ph.D. degrees in Industrial/Organizational Psychology from the University of South Florida. As the team leader for Advanced Surface Training Research at NAWCTSD, she has been involved in a number of research projects directed toward improving training for complex environments. These have included investigation of training needs and design for multi-operator training systems, training effectiveness and transfer of training issues, tactical decision-making under stress, the impact of multi-media training formats on learning and performance, and training for knowledge-rich environments. She has authored over 50 journal articles, chapters, and technical reports. She is currently editing a book on *Making Decisions Under Stress: Implications for Individual and Team Training* (APA Books).

Michelle A. Dean is a doctoral candidate in Human Resources Management and Organizational Behavior at Louisiana State University. She is presently serving as Visiting Assistant Professor of Management at the University of North Texas. Her current research interests are in personnel selection issues, specifically focusing on the prediction of performance using biodata. Her research has been published in the *Academy of Management Journal*, and presented at the National Meetings of the Academy of Management and the Society for Industrial and Organizational Psychology.

About the Contributing Authors

Wolfgang Elsik is Associate Professor of Business Administration at the Vienna University of Economics and Business Administration in Vienna, Austria. Prior to that, he was a Visiting Scholar at the Institute of Labor and Industrial Relations at the University of Illinois at Urbana-Champaign, and he served as interim Professor of Human Resources Management and Organization at the University of Bamberg, Germany. His research interests include strategic human resources management, career plateauing, politics in organizations, and organization theory. Professor Elsik also has taught at universities in Eastern Europe. He has published two books and several articles in German research journals (e.g., *Die Betriebswirtschaft* and *Zeitschrift fuer Personalforschung*).

Maria Carmen Galang is Assistant Professor on the Faculty of Business at the University of Victoria in British Columbia. She received a Ph.D. in Labor and Industrial Relations from the University of Illinois at Urbana-Champaign, and she has had experience in establishing and managing human resources departments in the garment, construction, and nonlife insurance industries in the Philippines, as well as designing and conducting training programs on a consulting basis. She teaches courses in human resources management and cross-national management, and her research interests include power and politics in organizations, and human resources management from both a cross-cultural and macro perspective. Her publications have appeared in *Human Relations, Human Resource Planning,* and *Human Resource Management Review.*

Stephen Gilliland is Associate Professor of Management and Policy at the University of Arizona. He received a Ph.D. in Psychology from Michigan State University. His research interests include organizational justice, individual decision making, and the application of these areas to human resources policies and procedures. He has published numerous articles on these issues in the *Academy of Management Review, Journal of Applied Psychology, Personnel Psychology, Organizational Behavior and Human Decision Processes,* and the *Journal of Management*. He is currently on the editorial boards of the *Academy of Management Journal, Journal of Applied Psychology,* and *Personnel Psychology.* He was the 1997 recipient of the Ernest J. McCormick Award for Early Career Contributions from the Society for Industrial and Organizational Psychology.

Luis Gomez-Mejia is the Dean's Council of 100 Distinguished Scholar and Professor at the Arizona State University College of Business. He received a Ph.D. from the University of Minnesota. His current research interests are macro compensation issues, including executive compensation and compensation strategy.

Randall Harris is Assistant Professor of Management at California State University, Stanislaus. He received a Ph.D. in Management from the University of Florida. His research interests include compensation systems and organizational performance.

K. Michele Kacmar is Associate Professor of Management and the Director of the Center for Human Resource Management at Florida State University. She received a Ph.D. in Human Resources Management from Texas A&M University. Her general research interests include impression management and organizational politics which she often applies to the area of human resources management. She has published over 30 articles in journals such as *Journal of Applied Psychology, Organizational Behavior and Human Decision Processes, Journal of Management, Human Relations,* and the *Journal of Applied Social Psychology*. Dr. Kacmar currently serves as an Associate Editor for *Human Resource Management Journal,* and she is on the editorial board of the *Journal of Applied Psychology*. Dr. Kacmar was elected to the Executive Committee for the Human Resources Division of the Academy of Management, the Board of Governors for the Southern Management Association, and the Board of Directors of the Research Foundation for the Society for Human Resource Management.

Jeffrey Katz is Assistant Professor of Management and the Payless Shoe Source Professor of Business at Kansas State University. He received a Ph.D. in Strategic Management from the University of Florida. He teaches courses in strategic management and international management, and his research interests focus on the roles that owners and managers play in selecting strategies determining the success of global competitors.

Manuel London is Professor and Director of the Center for Human Resources Management at the State University of New York at Stony Brook. He received a B.A. in Psychology and Philosophy from Case Western Reserve University, and an M.S. and Ph.D. in Industrial and Organizational Psychology from the Ohio State University. He taught at the University of Illinois at Urbana-Champaign for three years, and then he was a researcher and human resources manager at AT&T for 12 years before assuming his present position at Stony Brook. He is the author of several books including *Career Barriers: How People Experience, Overcome, and Avoid Failure* (Lawrence Erlbaum Publishers, 1998), *Self and Interpersonal Insight: How People Learn About Themselves and Others in Organizations* (Oxford University Press, 1995), *Employees, Careers, and Job Creation* (Jossey-Bass, 1995), and *Developing Managers* (Jossey-Bass, 1985). His current research interests include career planning and development, career motivation, and multi-source feedback surveys (i.e., 360 degree feedback).

Misty L. Loughry is a Ph.D. student in Strategic Management at the University of Florida. Prior to joining the Ph.D. program, she had a ten-year career in banking. Her research interests include managerial discretion, the strategic implications of organizational behavior issues, and time management.

Marick F. Masters is Professor of Business Administration at the University of Pittsburgh's Joseph M. Katz Graduate School of Business. He received a Ph.D. in Labor and Industrial Relations from the University of Illinois at Urbana-Champaign. Dr. Masters served on the faculty of the Department of Management at Texas A&M University from 1982-1986, and he has been at the University of Pittsburgh since 1986. He has a wide range of teaching and research interests in the areas of human resources management and employee and labor relations. He has published more than 60 articles in such journals as *Academy of Management Journal, American Journal of Political Science, American Political Science Review, Harvard Business Review, Industrial Relations,* and *Industrial and Labor Relations Review.* Dr. Masters has served on the editorial boards of the *Journal of Management* and *Journal of Managerial Issues.* He has recently published a book on unions in the U.S.entitled *Unions at the Crossroads* (Quorum Books, 1997).

Paul M. Muchinsky is the Joseph M. Bryan Distinguished Professor of Business in the Joseph M. Bryan School of Business and Economics at the University of North Carolina at Greensboro. He received a Ph.D. in Industrial and Organizational Psychology from Purdue University. He has long-standing research interests in the scientific and practical value of biographical information.

Lori Rhodenizer is a research assistant in the Team Performance Laboratory at the University of Central Florida. She received a B.S. in Psychology from Washington & Lee University in 1992, and she is currently a doctoral candidate in Human Factors Psychology. Her research interests include team training, practice, and technologies to enhance training effectiveness. She is a member of the Human Factors and Ergonomics Society and the American Psychological Association.

Gail S. Russ is Assistant Professor of Management and Quantitative Methods at Illinois State University. She received a Ph.D. in Business Administration from Texas A&M University. Before receiving her Ph.D., she worked for several years in private sector businesses, primarily in the computer software industry. In addition to her appointment at Illinois State University, she was Visiting Assistant Professor for several years in the Department of Business Administration at the University of Illinois at Urbana-Champaign. Professor Russ has research interests in the areas of organizational theory and strategic management, particularly organizational politics, corporate governance, impression management, and organizational legitimacy. She has published her research in such journals as the *Journal of Applied Social Psychology, Management Communication Quarterly,* and *Human Resource Planning,* and book chapters in volumes such as *Impression*

Management in the Organization (edited by Giacalone and Rosenfeld, 1989, Lawrence Erlbaum Publishers).

Craig J. Russell is the J.C. Penney Professor of Business Leadership at the Michael F. Price College of Business, University of Oklahoma. He received a Ph.D. in Business Administration from the University of Iowa in 1982. His current research interests focus on developing and testing models of leadership and management performance. He serves on the editorial boards of the *Journal of Applied Psychology, Journal of Leadership Studies, Journal of Management,* and *Organizational Research Methods.* His research has been published in the *Academy of Management Journal, Journal of Applied Psychology, Journal of Management,* and *Personnel Psychology* among others.

Eduardo Salas is a Senior Research Psychologist and Head of the Training Technology Development Branch of the Naval Air Warfare Center Training Systems Division. He received a Ph.D. in Industrial and Organizational Psychology from Old Dominion University. Dr. Salas has co-authored over 75 journal articles and book chapters, and he has co-edited five books. He serves on the editorial boards of *Human Factors, Personnel Psychology, Military Psychology, Interamerican Journal of Psychology,* and *Training Research Journal.* His research interests include team training and performance, training effectiveness, tactical decision making under stress, team decision making, performance measurement, and learning strategies for teams. Dr. Salas is a Fellow of the American Psychological Association and a recipient of the Meritorious Civil Service Award from the Department of the Navy. He also holds courtesy appointments at the University of South Florida and the University of Central Florida.

Paula Silva is Assistant Professor at the University of New Mexico. She received a Ph.D. in Organizational Behavior from the University of Florida. Her research interests include human resources issues and executive compensation.

James W. Smither is Professor of Management at LaSalle University in Philadelphia. He received a B.A. in Psychology from LaSalle University, an M.A. in Education from Seton Hall University, an M.A. in Industrial and Organizational Psychology from Montclair State University, and a Ph.D. in Industrial and Organizational Psychology from Stevens Institute of Technology. He has published articles in journals such as *Personnel Psychology, Journal of Applied Psychology,* and *Organizational Behavior and Human Decision Processes.* He is currently Associate Editor for *Personnel Psychology.* Before coming to LaSalle, Smither worked in corporate human resources for AT&T. He continues to consult in the areas of management development and human resources.

Henry L. Tosi is the McGriff Professor of Management in the Warrington College of Business Administration at the University of Florida. He also holds an appointment as Visiting Professor of Human Resources Management and Organization at SDA-Bocconi in Milan, Italy. He received a Ph.D. from Ohio State University. His research interests include managerial discretion, organizational control mechanisms, compensation, agency theory, and meso-organizational issues.

Steve Werner is Assistant Professor of Management at the University of Houston. He received a Ph.D. from the University of Florida. His research interests include managerial compensation, compensation determinants, international aspects of compensation, and equity issues.

James D. Werbel is Associate Professor of Management and Co-Director of the Murray Bacon Center for Business Ethics at Iowa State University. He received a Ph.D. in Management from Northwestern University. He has published articles in *Administrative Science Quarterly, Journal of Applied Psychology, Journal of Organizational Behavior, Personnel Psychology,* and *Journal of Vocational Behavior.* He currently serves on the editorial board for *Human Resource Management.*

SET UP A CONTINUATION ORDER TODAY!

Did you know you can set up a continuation order on all JAI series and have each new volume sent directly to you upon publication. For details on how to set up a continuation order contact your nearest regional sales office listed below.

To view related Business, Management and Accounting series, please visit

www.ElsevierBusinessandManagement.com

30% DISCOUNT FOR AUTHORS ON ALL BOOKS!

A 30% discount is available to Elsevier book and journal contributors ON ALL BOOKS plus standalone CD-ROMS except multi-volume reference works.

To claim your discount, full payment is required with your order, which must be sent directly to the publisher at the nearest regional sales office listed below.

ELSEVIER REGIONAL SALES OFFICES

For customers in the Americas:
Customer Service Department
11830 Westline Industrial Drive
St. Louis, MO 63146
USA
For US customers:
Tel: +1 800 545 2522
Fax: +1 800 535 9935
For customers outside the US:
Tel: +1 800 460 3110
Fax: +1 314 453 7095
Email: usbkinfo@elsevier.com

For customers in Europe, Middle East and Africa:
Elsevier
Customer Services Department
Linacre House, Jordan Hill
Oxford OX2 8DP
United Kingdom
Tel: +44 (0) 1865 474140
Fax: +44 (0) 1865 474141
Email: amstbkinfo@elsevier.com

For customers in the Far East:
Elsevier
Customer Support Department
3 Killiney Road, #08-01/09
Winsland House I,
Singapore 239519
Tel: +(65) 63490200
Fax: + (65) 67331817/67331276
Email: asiainfo@elsevier.com.sg

For customers in Australasia:
Elsevier
Customer Service Department
30-52 Smidmore Street
Marrickville, New South Wales 2204
Australia
Tel: +61 (02) 9517 8999
Fax: +61 (02) 9517 2249
Email: service@elsevier.com.au